VIDEODISC
AND OPTICAL
MEMORY SYSTEMS

Dr. Jordan Isailović

Prentice-Hall, Inc., Englewood Cliffs, New Jersey 07632

Library of Congress Cataloging in Publication Data

Isailović, Jordan
    Videodisc and optical memory systems.

    Includes bibliographies and index.
    1. Optical storage devices. 2. Video discs.
I. Title.
TK7895.M4I83 1985   621.3819'5833   84–17957
ISBN 0–13–942053–3

Editorial/production supervision and interior design: Fred Dahl
Cover design: Jayne Conte
Manufacturing buyer: Gordon Osbourne

Printed in the United States of America

10 9 8 7 6 5 4 3 2 1

ISBN 0-13-942053-3   01

Prentice-Hall International, Inc., *London*
Prentice-Hall of Australia Pty. Limited, *Sydney*
Editora Prentice-Hall do Brasil, Ltda., *Rio de Janeiro*
Prentice-Hall Canada Inc., *Toronto*
Prentice-Hall of India Private Limited, *New Delhi*
Prentice-Hall of Japan, Inc., *Tokyo*
Prentice-Hall of Southeast Asia Pte. Ltd., *Singapore*
Whitehall Books Limited, *Wellington, New Zealand*

CONTENTS

PREFACE, ix

ACKNOWLEDGMENT, xiii

Chapter 1

INTRODUCTION, 1

1.1 General Introduction, 1
1.2 The Videodisc System, 2
1.3 A Brief Historical Note, 4
1.4 How Data Are Stored on a Videodisc, 6
1.5 Comparison of Videodiscs with Videotape, 10
1.6 Comparison of Videodiscs with Optical Read/Write Discs, 12
1.7 TV Signal Recording: Standard Formats Versus Extended Play, 14
1.8 Optical Versus Capacitive Discs, 17
1.9 Videodisc Standards, 18
1.10 Present and Future Uses of Videodisc Systems, 19
1.11 Organization of the Book, 22
   References, 22
   Appendix 1.1: More of the Videodisc History, 24
   Discovision Associates: Videodisc History, 24

Chapter 2

RECORDING, 29

2.1   Introduction, 29

2.2   Basic Operations, 31

2.3   Optical Videodisc, 35

   2.3.1   Mastering, 35
   2.3.1.1 Optomechanical system, 36
   2.3.1.2 Recording materials and processing, 37
   2.3.1.3 Photoresist mastering, 39
   2.3.1.4 Metal film mastering, 41
   2.3.1.5 Direct read after write (DRAW), 43
   2.3.1.6 Quality control testing, 46

   2.3.2   Galvanization: Stamper Production, 47

   2.3.3   Replication, 48
   2.3.3.1 Molding, 48
   2.3.3.2 Metalization, 50
   2.3.3.3 Overcoating, 50
   2.3.3.4 Miscellaneous, 50

   2.3.4   Cold Flow, Birefringence, and Other Problems, 50

   2.3.5   Optical Transmissive Discs, 52
   2.3.5.1 Mastering, 52
   2.3.5.2 Replication, 53
   2.3.5.3 Double-sided transmissive discs, 53

   2.3.6   Comparison Between Reflective and Transmissive Videodiscs, 54

   2.3.7   Film-Based Videodiscs, 54

2.4   Capacitive Videodiscs, 56

   2.4.1   Mastering, 57
   2.4.1.1 Electromechanical mastering, 59
   2.4.1.2 Optical mastering, 61
   2.4.1.3 Electron-beam mastering, 62

   2.4.2   Stamper Production, 64

   2.4.3   Replication, 64
   2.4.3.1 Materials and compound processing, 64
   2.4.3.2 Molding, 66
   2.4.3.3 Coatings, 68

   2.4.4   Flat Capacitive Disc, 70

2.5   Videodisc Testing, 70

2.6   Servo Systems for Mastering, 73

   2.6.1   Laser Control, 73
   2.6.2   Focus Servo, 74
   2.6.3   Spindle Servo, 77
   2.6.4   Carriage Drive, 78

   References, 79

Appendix 2.1: Standing Waves, 80
Appendix 2.2: Exposure and Development, 80
Appendix 2.3: The Diazo Process, 82

Chapter 3

OPTICAL PLAYBACK, 83

3.1   Introduction, 83

3.2   Optical Pickup: Optical Principles, 88

      3.2.1   Diffraction Grating, 88
      3.2.1.1 A pit dept of λ/4, 89
      3.2.1.2 A pit dept of λ/8, 90

3.3   Player Optics (Optical System), 93

3.4   Modulation Transfer Function and Its Compensation, 94

      3.4.1   Cosine Equalizer, 97
      3.4.1.1 Modeled response, 98
      3.4.1.2 The Influence of the instant radius on the output signal, 99

      3.4.2   Optical Feedback for the MTF Compensator, 105

      3.4.3   Comparison: Cosine Equalizer Versus Optical Feedback, 107

3.5   Time-Base Error, 107

3.6   Dropout Detector, 109

3.7   Control and Servo Systems in the Player, 109

      3.7.1   Spindle Servo, 111
      3.7.1.1 System with tachometer, 111
      3.7.1.2 System with recorded reference signal, 112

      3.7.2   Focus Servo, 113

      3.7.3   Tracking (Radial) Servo, 116

      3.7.4   Tangential (Time-Base Correction) Servo, 122

      3.7.5   Carriage Servo, 123

      3.7.6   Signal-to-Noise Ratio of the Error Signal, 124

3.8   Optical Transmissive Discs, 124

      3.8.1   U-Shaped Stabilizer, 124

      3.8.2   Stabilization by Bernouilli Effect, 125

3.9   Film-Based Videodiscs, 127

3.10 Optical Readout of the Grooved Disc, 129

      References, 130

      Appendix 3.1: Cosine Equalizer, 131
      Appendix 3.2: LC Resonant Circuit of the Radius-Compensated MTF Compensator for the Video
                    FM Signal, 133
      Appendix 3.3: The Influence of Eccentricity on the Signal Recorded on a Videodisc, 135
      Appendix 3.4: Analysis of the Time-Based Servo Error Signal, 151

Chapter 4

CAPACITIVE PICKUP PLAYBACK, 158

4.1 Introduction, 158
4.2 Principle of the Capacitive Pickup, 160
    4.2.1 Recorded Signal Elements, 160
    4.2.2 Detection of Capacitive Variations, 163
4.3 System Parameters, 165
4.4 Time-Base Error Correction, 167
4.5 Electronics in the Player, 169
4.6 Stylus, 174
4.7 Capacitive Readout from Flat Disc, 177
    References, 183

Chapter 5

FUNDAMENTALS OF OPTICS, 184

5.1 Introduction, 184
5.2 Geometrical Optics, 185
    5.2.1 Simple Lenses, 187
    5.2.2 Aberations, 191
5.3 Wave Optics, 195
    5.3.1 Waves, 195
    5.3.1.1 Complex amplitude and the quadratic-phase signal, 196
    5.3.1.2 Miscellaneous, 199
    5.3.2 Interference, 201
    5.3.3 Diffraction Theory, 203
    5.3.3.1 Fresnel diffraction, 206
    5.3.3.2 Fraunhofer diffraction, 207
    5.3.3.3 Diffraction gratings, 210
5.4 Fourier Optics, 210
    5.4.1 Basic Principles of Fourier Optics, 212
    5.4.2 The Optical Transfer Function, 214
    5.4.3 The Modulation Transfer Function, 219
    5.4.4 Resolution of Optical Systems, 224
    5.4.5 Apodization, 225
5.5 Physical Optics, 226
5.6 Reflective Videodisc Player, 229
5.7 Optical Modulator, 233
    5.7.1 Electro-optical Modulators, 234
    5.7.2 Acousto-optic Modulators, 235
    References, 238
    Appendix 5.1: Fourier Transform: Transform Pairs and Transform Properties, 239

Chapter 6

CHANNEL CHARACTERIZATION, 244

6.1   Introduction, 244
6.2   Videodisc Channel, 245
    6.2.1   Noise, 246
    6.2.2   Dropouts, 247
    6.2.2.1 Dust and scratches on the disc surface, 249
    6.2.3   Noise on a Disc Made of Photographic Film, 249
    6.2.4   Capacitive Disc Noise, 250
    6.2.5   Other Impairments, 251
6.3   Optical Modulation Channel, 252
    6.3.1   Photodetector, 252
    6.3.2   Optical Readout, 255
    6.3.3   Other Impairments in the OMC, 256
6.4   Signal Channel, 257
    6.4.1   Noise, 257
    6.4.1.1 Capacitive videodisc, 259
    6.4.2   Linear Distortion, 259
    6.4.3   Other Impairments, 260
6.5   Experimental Results: The Noise Measurements, 260
    6.5.1   Background Noise, 260
    6.5.2   Dropouts, 264
    6.5.3   Polar Figure of the Dropouts and Disc Noise, 265
    6.5.4   Discussion, 266
6.6   Catastrophic Failures and Their History Through Videodisc Process Production, 266
    References, 268
    Appendix 6.1: Mathematical Presentation of the Noise, 269

Chapter 7

OPTICAL MEMORIES, 293

7.1   Introduction, 293
7.2   Optical Disc Drive: Recording and Playback System, 297
7.3   Recording Media, 301
    7.3.1   Nonerasable Optical Recording Media, 301
    7.3.1.1 Ablative thin film, 302
    7.3.1.2 Media with postprocessing, 302
    7.3.2   Erasable optical media, 303
    7.3.2.1 Permanent storage, 303
    7.3.2.2 Limited Storage, 307

7.4   Structures for Optical Recording, 308

    7.4.1   Multilayer (Antireflection) Structures, 309
    7.4.1.1 Bilayer Antireflection Structure, 312
    7.4.1.2 Trilayer Antireflection Structure, 313

    7.4.2   Recording Modes, 315

7.5   Optical Disc Evolution: Testing and Measurements, 317

7.6   Modifications and Variations, 318

7.7   Looking Ahead, 318

    References, 320

    Appendix 7.1: Recording Codes, 322
    Appendix 7.2: ODC Laser Videodisc Recording System LVDR 610, 329

INDEX   331

PREFACE

This book is the outcome of the author's experience in research and development in the area of videodisc systems. The idea for its writing was born during numerous trips between Belgrade and Los Angeles. At the start of each trip, I wondered whether I had with me "everything I needed." Thus, the idea was to write the book that would contain the basis of videodisc systems—that is, serve as an "exterior memory." The initial intention of the author in writing this book was to encompass, in a single text, the broad area joining videodisc systems. Among other items, signal processing, modulation theory, TV systems, image processing, and optics were planned to be discussed.

The recording of the videodisc is usually performed by a laser. Playback is done on the basis of various pickup sensors. They are either optical or capacitive pickups. The disc itself is usually a metallized plastic, so that it represents an inexpensive memory medium of very large capacity. At the present state of technology, a disc provides 30–120 minutes of TV programming in color with stereo sound reproduction. For application in computer techniques, they are nondestructive memories with a capacity of over 30 million bits. It could be expected that these numbers in the near future will be surpassed. The longest movie, for example, could be located on one disc, and in the same way, the complete data about all the inhabitants of Yugoslavia on one or two discs.

The application possibilities of a videodisc are fantastic: for the reproduction of films and other TV materials, for storing large quantities of data, for reproduction of educational material, for use in library science, medicine, culinary art, in military services, and so on. In computer techniques, with the memory (ROM) of such a large capacity, coupled with the possibility of very swift access and data rates of over 10^7 bits/second, some hitherto strict principles can be changed. For example, in many applications the necessity for lengthy subprograms will be removed, because all the necessary data could be "on hand." The selection of movable targets in stationary radars could be made on the basis of complete memorization of the ambient. The list of applications is limited only by the imagination.

Optical data storage systems that utilize a highly focused laser beam to record and instantaneously playback information are very attractive in computer mass storage technology. They offer very high storage density with very high data rates, rapid random access to the data, potential achieval properties, and the projected low cost of media. These systems are also attractive for the television broadcasters and other communications professionals, whose need to share and then retrieve extremely large quantities of video information, text or moving and/or stationary pictures (for example) is growing rapidly.

This book considers theoretical as well as practical aspects of the videodisc systems and optical data storage systems. Its purpose is to help engineers, scientists, students, and technicians. An additional purpose is to make videodisc systems and optical discs more widely known so that they may be advantageously used to perform many functions. A discussion of disc versus disk is given in Chapter 7. The book could serve as an auxiliary textbook for a course in Optoelectronics as well as for various courses on digital electronics/memories.

Originally, the book was written as sixteen chapters.

The first four chapters comprise the first part of the book covering record processes, basic materials for recording, and the principles of basic playback systems.

The second part, Chapters 5 and 6, covers the basis of optics and noise in system considerations.

In the third part (Chapters 7–10), modulation fundamentals are presented.

Chapters 11–14 make up the fourth part. Topics covered are the mathematics of continuous imaging, psychophysical properties of human vision, colorimetry, and the principles of forming TV images along with existing standards. Also discussed is the basic problem of standard TV recording, as well as extended play techniques.

The last part (Chapters 15 and 16) cover audio signal recording and playback and applications of videodisc as both analog and digital channel (storage).

In the Appendix, the overview of the optical data storage (nonerasable and erasable) and recording structures are given.

Publishing constraints forced me to split the materials into two books. This is the first one.

The first chapter is a short survey of the book—the basic problems and principles. The videodisc is first compared with the magnetic media, and then a comparison of main videodisc systems is given.

In Chapter 2 the videodisc is considered from the production point of view. The survey of basic recording materials and their characteristics are quoted. The basic stages of disc production are described for both optical and capacitive discs.

Chapter 3 contains the principles and descriptions of the basic optical playback system. The functioning of the basic servo systems is also given.

Chapter 4 contains the principles and the description of the basic capacitive playback system, both for a pregrooved system and a flat disc.

The second part consists of Chapters 5 and 6. Necessary bases of optics are presented in Chapter 5, while the problem of noise in the system are considered in Chapter 6.

In Chapter 7, the optical data storage (i.e., draw, write once/read many times, optical disc, and so on)—programmable and erasable—is discussed. The intended application of the system (whether for digital data or TV imaging) is noted only where necessary for clarity and understanding.

Thus, the book presents the basic principles and technologies of videodiscs, compact discs/digital audio discs, and optical memories.

ACKNOWLEDGMENT

I am deeply indebted to my colleagues whose encouragement, discussions, inspirations, and support have contributed to the preparation of this book. These contributions have taken many forms, and an attempt to list all of their names would be impractical. I would like to single out a few to whom I am particularly indebted. They are listed more or less according to the chronology of their contributions.

From the beginning, Mr. Paul Obradović was my main inspiration to write this book. Many hours have been spent discussing different subjects in this book with Mr. Ray Dakin—one of the founders of the videodisc technology known as MCA technology. Mr. Scott Golding-Zlatić made the first version of the book readable in English.

Special thanks to Mr. John Raniseski, Director of Research and Development Laboratory at Discovision Associates who made it a very pleasant atmosphere in which to work. Mr. Ronald Clark, Patent Attorney for Discovision, was very helpful in supplying me with the information needed including several hundred patents on videodisc technology.

Considerable time was contributed by Mr. Milan Topalovic—from Belgrade TV—in helping me and discussing with me techniques on TV signal recording on videodiscs. The following colleagues have reviewed with enthusiasm, selected chapters of the manuscript, and gave to me useful comments and suggestions: Mr. John Winslow, Mr. Richard L. Wilkison, Mr. Carl Eberly, Mr. Michael Michalchik, Mr. Richard Allen, all members of Discovision Research and Development Laboratory; Dr. Tim Strand of the University of Southern California; Dr. Zrilić from Ljubljana; Prof. Mysore Lakshminarayana of the California State Polytechnic University of Pomona; and Mr. Paul Thomsen of Conrac Corporation.

Dr. William V. Smith, author or coauthor of two books in laser technology, with his tremendous experience and patience, encouraged me to complete this book. Also, Dr. Smith left to me, from his collection, a great number of papers and other material relevant to this subject.

Special thanks to those who typed and retyped all those revisions, Nada Obradović, from Electrotechnical faculty, University of Belgrade, Lenka Golding-Zlatić, and Word Processing at DVA, Shari LaBranche, who did the greatest share of it.

Chapter 1

INTRODUCTION

1.1 GENERAL INTRODUCTION

The initial aim in the development of the videodisc was to get a system that records audio-video information on a disc, replicates this disc accurately and inexpensively on plastic, and finally plays the replicas on home television screens by means of a disc player attachment. Beginning efforts were directed toward exploring the development of new types of video equipment which would permit the packaging and marketing of entertainment, educational, and industrial audiovisual programs. These initial efforts marked the beginning of a communications evolution: the first wholesale exodus away from the printed media since Gutenberg invented his crude, but innovative, press. Today, the videodisc represents one of the most advanced communications systems. It represents a revolutionary breakthrough in audiovisual communications for the 1980s. For the first time, commercial, industrial, and institutional programmers have the tools with which to tailor high-quality pictures and sound to the specific levels and interests of their audience. The result is a new dimension of persuasion and influence: a new ability to inform, educate, to sell, and to entertain.

Every so often a system appears that inspires dreams across a wide area—a technology whose potential stirs speculation in dozens of disciplines. The transistor is one example, microprocessors another, and the videodisc and optical memories appear to be another. In consumer electronics the videodisc is the most important development since the introduction of color television.

Many disciplines are involved in videodisc technology. Among them are TV systems, communications theory, signal processing, image processing, and optoelectronics. Although this field is still growing, the fundamentals are established. The set of topics presented in this book are chosen to enable the reader to develop a firm base in the fundamentals and to begin to develop an appreciation for the wide scope of applications and future directions for the field.

1.2 THE VIDEODISC SYSTEM

The enormous popularity of the phonograph record and its playback equipment, together with the rapid growth of television, led directly to investigating the possibility of recording video signals on a disc. What makes the disc so interesting is the potential combination of three properties: very low cost, very high information density, and the possibility of instant access to any portion of a long recording. In general, videodisc systems offer:

- Up to 2 hours of play on a single disc
- Color video
- Stereo sound
- Special features
- Direct, rapid access
- External computer control

Basically, the videodisc (VD) system (Fig. 1–1) is similar to the long playing (LP)–audio system. As a result of the recording process, inexpensive plastic discs in large quantity are obtained. Recorded programs are then played back through a specially designed system, a player.

The production process for a videodisc is more or less comparable with that used for conventional audio records. First, a master recording is made from a master tape (Fig. 1–2). Theoretically, any source material may be used for mastering, including 16mm or 35mm film slides, off-air TV signals, videotape, a camera, an electronically generated signal, and so on. From a practical standpoint, however, most mastering is accomplished by transferring any of the potential sources to 1-inch—C format videotape. Alternatively, 2-inch helical or 2-inch quadruplex tape can be used. In this process, material may be edited together from several sources; color correction, if any, may be accomplished; audio may be added or changed; and frame-by-frame address codes can be inserted in the programming. A master disc can be used directly as a stamper, or as a basis ("mother") for the production of many stampers. Also, in source mastering processes, the recorded pattern can be "directly read after writing" (DRAW), a very convenient property. Further, stampers are used to replicate recorded patterns in inexpensive plastic video records.

Figure 1–1   Videodisc system.

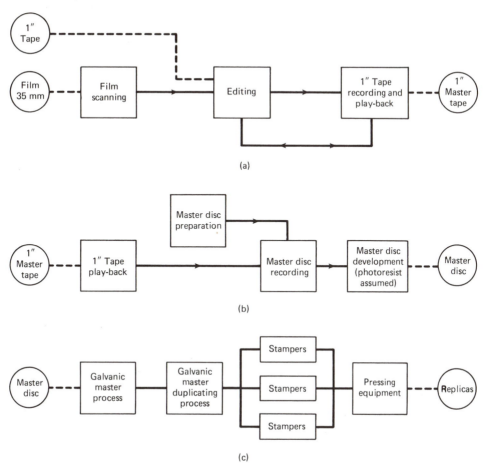

Figure 1–2   Videodisc mastering and replication.

The videodisc player is self-contained and is designed to be connected to the antenna terminals of any domestic color television receiver to provide playback of replicated videodiscs. The essential elements of the playback system are a pickup, a means for spinning the videodisc at the correct speed, a means for positioning a pickup (stylus) on the disc surface, the electronics necessary to process the signal for a TV or other receiver (or user), the controls, and the power electronics to operate the unit.

Although detailed models for the reading process depend, among other things, on the pickup, it is possible to construct a generalized model independent of these details. Such a model is shown in Fig. 1–3. An oscillator generates a signal with a very high carrier frequency. This carrier is (for example) amplitude modulated by the mechanically stored information pattern on the disc. This change on the carrier, called mechanical modulation (MM), is detected by the pit pattern to electrical signal

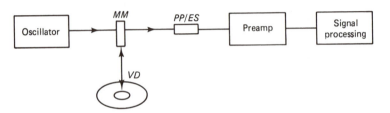

Figure 1–3   General model for the reading process.

converter (PP/ES). In the preamplifier, signal conditioning and band limiting are performed. The signal processing depends on the particular application. If a TV signal is recorded, the unit includes video and audio demodulation and radio-frequency (RF) signal formation. Dropout compensation is also included. (Because of the very small size of the information elements on the disc surface, small impurities in the disc material will cause signal interruption. A dropout detector operates an electronic switch by means of which a parallel channel, with a line-scan delay, provides the information carried on the previous line.)

The several videodisc systems have varying degrees of technological differences, but the following operational features can, in general, be found:

1. Freeze frame
2. Frame-by-frame viewing (forward or reverse)
3. Scan (forward or reverse)
4. Slow motion (forward and reverse)
5. Frame number display
6. Search (rapid access to a desired frame)
7. Autostop (stops the normal playing mode at a preselected frame, and goes into the freeze-frame mode)
8. Reject (canceling the operating mode; returning the disc to the load position)
9. Dual audio (selecting either one or both of the audio channels)
10. Programming features
  • User-programming capability
  • Self-programming capability (self-programmed discs)
  • Direct computer interface
  • External synchronization (to the composite signal and color subcarrier)

1.3 A BRIEF HISTORICAL NOTE

The concepts underlying videodisc systems have found their widest and most dramatic application in recent years, but the ideas are far from new. The basic idea behind disc recording can be recognized in the first recording and playback of voice on a wax cylinder, achieved by Thomas Edison in 1877. The idea of videodisc technology itself was explored by John Logie Baird, and others, as early as 1925. In 1927, Baird demonstrated a Phonevision system from a wax disc displayed by an optical scanner. In 1935, at Selfridge's department store in London, Baird Radiovision offered for sale 6 minutes of video display.

Optical videodisc development has roots going back to the late 1950s. At that time, work on videodisc development was generally kept secret [1], which makes it difficult to follow. In 1958, Paul Gregg started work on the idea of a videodisc based on electron-beam film recording. His work influenced Minnesota Mining & Manufacturing (3M) in the early 1960s, and MCA in 1967, to begin videodisc development efforts. In 1961, Philip Rice started the development for 3M of a videodisc system using a photographic approach. By October 1962 this work had resulted in a playable videodisc that was optically recorded and read. The concept of spinning the videodisc at 1800 rpm was developed, but the playback pickup was unable to follow a single track. In March 1962, a patent was filed on data storage and retrieval systems, and issued in December 1965 [2]. The first papers [3,4] were published in 1970 and 1971. Westinghouse Electric Co. announced the Phonovid system, a single-frame disc, in 1965.

In June 1970, a videodisc made by TELDEC, a joint venture of Telefunken and Decca of England, was announced and demonstrated to the public. The TELDEC (later TED) disc had been developed in a small laboratory in Berlin, originally set up for the purpose of improving phonograph recording techniques. The TELDEC engineering group had succeeded in recording 5 minutes of video and sound on an 8-inch thin, flexible disc by means of electromechanical cutting at a rate 25 times slower than real time. Signals were reproduced by means of a pressure pickup using a sled-shaped diamond stylus that transmitted vibrations from the recorded vertical, or "hill-and-dale," undulations to a piezoelectric element.

In September 1972, demonstration of live recording development and playback of a videodisc master was performed in a press conference held by N. V. Philips of Eidhoven, the Netherlands. This was a laser recording and playback system. In December 1972, MCA demonstrated the DiscoVision system, a laser reading a thin, pitted reflective plastic disc which had 30 minutes of play. In August 1973, Philips demonstrated the VLP system, a laser reading a thick, pitted reflective plastic disc, again with 30 minutes of play.

Technical details of the VLP system were published in 1973 [5–7]. In April 1974, the first videodisc technical session was held at the Los Angeles SMPTE conference [8–10], at which MCA, Philips, Zenith, and IO Matrics disclosed technical details of their systems. In May 1974, Thomson CSF made public technical details of their system, laser reading of a thin, transparent plastic disc, at the SID conference in San Diego [11]. In September 1974, Philips/MCA announced agreement on a software format. In October 1974, Thomson CSF demonstrated their system at VIDCA in Cannes, France.

In March 1975, RCA demonstrated the SelectaVision disc system, a capacitive stylus in a grooved plastic disc. At the same time, TELDEC introduced their product in Germany. In April 1975, RCA gave a technical paper describing their system at the SID Conference in Washington, D.C. In June 1975, the U.S. government (ARPA) began support of digital optical disc technology by backing the DRAW disc, a product of Philips Laboratories in Briarcliff, N.J. (see Section 1.6). Since then, programmable optical memories have received considerably more attention.

During 1976–1977 many other companies all over the world joined videodisc technology efforts. Development of both optical write once and audio digital optical disc systems was boosted. In November 1976, three articles were published on reflective optical videodisc technology [12–14]. In 1977, MCA and Pioneer formed Universal Pioneer Co. (UPC) to produce videodisc players for the market.

In 1978, three large collections of relevant papers were issued [15–17], UPC delivered the first industrial players, and Magnavox and MCA marketed consumer videodisc systems: MagnaVision players and DiscoVision discs. In September 1978, a paper was presented on the double-density recording method for videodisc technology [18], although experiments had been performed in 1977; more details were published later [19].

In 1979, Teldec stopped active production and marketing of TED systems. Mitsubishi showed Visco-opel, their improved Teldec-type system with 2 hours of playing time on a 9-inch disc. In May 1979, Philips demonstrated the "compact disc" (CD), a 12-cm PCM audio optical disc with 1 hour of playing time, read by a laser diode. In September 1979, General Motors purchased 11,000 Pioneer players and MCA special discs for car dealers. Also in September, DiscoVision Associates (DVA) was founded (a 50–50 joint venture between MCA and IBM).

In January 1980, Mitsubishi disclosed its use of JVC grooveless capacitive systems. A very high density (VHD) system was also announced. In February, Ardev disclosed details of a player with a 3-second still-frame audio capability—a light bulb player based on photographic foils. In March 1980, Pioneer introduced a consumer videodisc player, the VP-1000. In June 1980, Philips, Sony, and CBS announced a cooperative effort on a 16-bit, 12-inch-diameter compact digital disc with 60 minutes of audio. In the summer of 1980 digital data were recorded on a videodisc by DiscoVision Associates using multilevel coding; papers were published later [20,21].

In March 1981, RCA marketed a SelectaVision player and software. In May 1981, Sears announced an experimental videodisc sales catalog.

In January 1982 at the University of California, Irvine, the first seminar on videodisc engineering was held [22]. More historical data can be found elsewhere [23–26]. A DVA document, "Videodisc History," is given in the Appendix.

1.4 HOW DATA ARE STORED ON A VIDEODISC

The layout of a typical videodisc is shown in Fig. 1–4. The information track is a spiral of a given track pitch q (b revolutions per millimeter), which is usually constant over the entire recorded area, and a given density of information elements, a (elements per millimeter).

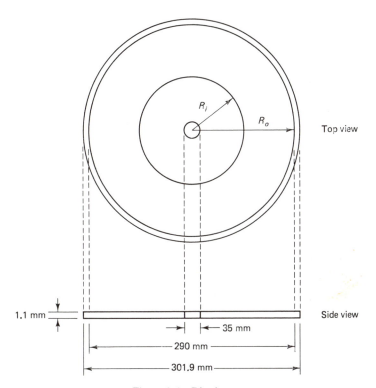

Figure 1–4   Disc layout.

There are two types of discs, called CAV and CLV. CAV stands for "constant angular velocity," which means that the speed of rotation is constant: 1500 rpm for PAL/SECAM ($n = 25$ revolutions per second) and 1800 rpm for NTSC ($n = 30$ revolutions per second) if one frame is recorded per revolution. For example, for NTSC, 900 rpm corresponds to 2 frames and 450 rpm to four frames recorded per revolution. In flat disc systems, this type of disc allows special playing modes, such as still frame and slow motion.

The second type is CLV, "constant linear velocity." Here the speed of rotation decreases inversely proportional to the readout diameter, that is a-constant. As a result, more information can be stored on the disc. On the other hand, these discs can be played only in a continuous way, that is, in the normal forward mode.

In Fig. 1–5 a cross section of an optical videodisc is shown. Transmissive [and absorbtive (not shown)] discs do not have protective coatings or (aluminum) reflective coatings. But the form of the pits is similar for transmissive and reflective optical discs (Fig. 1–6). The average track pitch is 1.6 μm (q in Fig. 1–6). This means that the period at the end of this sentence would cover more than 500 tracks. The effective pit width (γ) is 0.4 μm, and effective pit length (β) varies with radius and modulating signal. The complete TV picture requires a surface of 0.6 mm^2 on the disc. In principle, optical videodiscs with undulating tracks are possible [16, p. 2022].

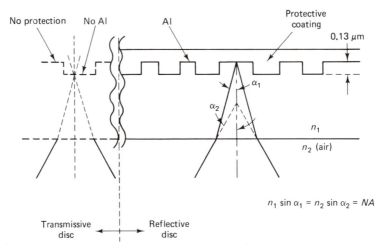

Figure 1–5 Cross section of the optical disc.

Because of the associated problems, a poor signal-to-noise ratio, for example, the method is not widely used.

A cross section of a grooved capacitive videodisc is shown in Fig. 1–7. A simplified arrangement of the information elements on the capacitive discs, whether grooved or not, is shown in Fig. 1–8. For the grooved system, the track pitch is typically $q = 2.7$ μm and is slightly larger than the pit width (γ). The pit length (β) is on the order of 0.25 μm.

A cross section of a flat capacitive videodisc is shown in Fig. 1–9. The simplified arrangement of the information elements is the same as for the grooved disc. The tracking pits, f_{p1} and f_{p2}, between the information pits are to control the pickup stylus on the specified track. The track pitch (1.35 μm) is half that of the grooved system.

When the videodisc is used as an analog channel, the information is contained in the length and displacement of the pits. For the video signal recording, pulse

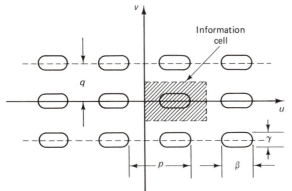

Figure 1–6 Arrangement of the information elements on the optical disc.

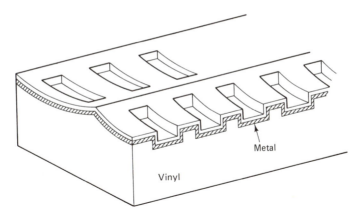

Figure 1-7 Grooved capacitive disc.

frequency modulation is used. To minimize the baseband component—the spectral component of the FM signal in the bandwidth of the original modulating (video) signal—the duty cycle is kept constant. To minimize the second harmonic of the video FM carrier, the duty cycle is chosen to be 50%. Audio information is contained in the duty-cycle modulation. Continuous FM signals—video and audio modulated— are linearly summed and then passed through a hard limiter. Before summing, amplitudes of the audio and video carriers should be properly chosen to minimize moiré patterns in the reproduced picture (because of audio crosstalk to the video channel) and to keep relatively low disturbances in the audio channel (Fig. 1-10).

It should be mentioned that because of the low-pass characteristic of the playback videodisc system, the different recorded spatial frequencies will give different amplitudes of the readout signals. This is illustrated in Fig. 1-11 for the optical videodisc system, although the same phenomena exist in the capacitive systems. Depending on the noise influence in the particular case, compensation circuitry can be included in the player electronics.

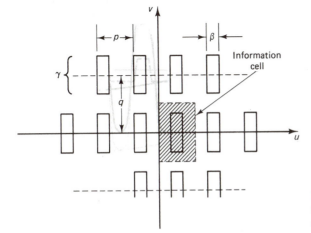

Figure 1-8 Arrangement of the information elements on the capacitive disc.

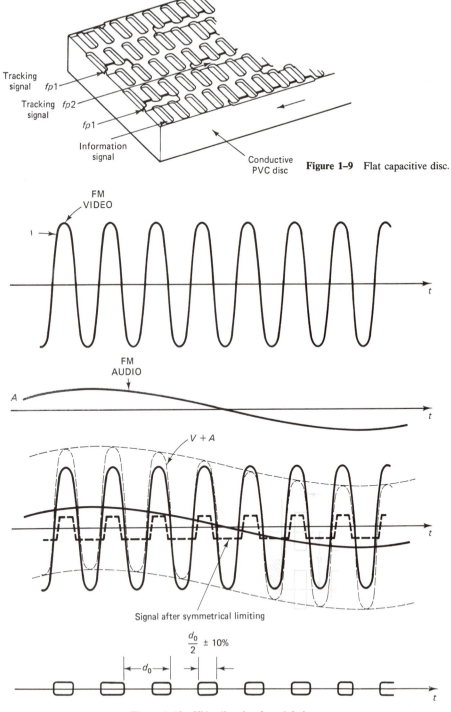

Tracking signal $fp1$

Tracking signal $fp2$

$fp1$

Information signal

Conductive PVC disc

Figure 1–9 Flat capacitive disc.

FM VIDEO

FM AUDIO

A

$V + A$

Signal after symmetrical limiting

$\dfrac{d_0}{2} \pm 10\%$

d_0

Figure 1–10 Videodisc signal modulation.

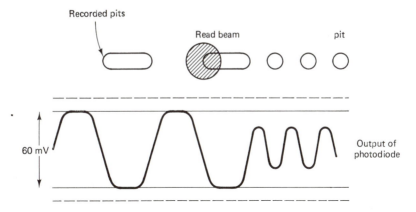

Figure 1–11   Influence of the low-pass characteristic.

1.5 COMPARISON OF VIDEODISCS WITH VIDEOTAPE

The videodisc is not the first source of recorded TV signals; we have had videotape for quite a while. Although magnetic tape appeared to have good promise as a home recording medium for off-the-air TV programs, it did not appear to be a good candidate for prerecorded video programs. If programs were to be produced as prerecorded tapes, the cost of the tape itself in a length sufficient for a feature movie seemed excessive for a mass consumer product. Further, programs could be replicated on tape only by some form of serial recording technique requiring considerable time for the transfer of each program, and therefore requiring a large number of expensive duplicating stations if programs were to be produced in quantity.

The physical information density on a videodisc is about 100 times greater than that for magnetic discs for computer applications and about 10 times greater than that for video magnetic tape. Although each of the videodisc formats, optical (reflective, transmissive, and photographic absorbative) and mechanical (capacitive, grooved, or flat discs), has varying degrees of technological difference, most of them offer (or promise) the following basic advantages:

1. Fast replication rates
2. Relatively inexpensive discs (at the manufacturing level)
3. Attractive operational features, such as still-frame and random-access retrieval
4. Relatively low cost hardware
5. Extremely high signal packing densities

By comparison, the videotape recorder can reproduce material superbly, but both the machine itself and the recorded material were in the seventies too expensive for the consumer market, and for many nonconsumer applications, too expensive as well. This is due partially to the fact that the videotape recorder provides a recording function in addition to the playback function. Complexity tends to increase costs. Although the price (and size, for example of the 8mm tape recorder) of the videotape recorder has lately been considerably reduced, it should be mentioned that low cost was one of motivations for the videodisc technology development. Another factor affecting the cost of the videotape medium is the replication aspect. Currently, it is not possible for large numbers of copies of a videotape program to be made rapidly and inexpensively, owing to technological limitations. The best that anyone has consistently been able to do is to replicate on a real-time basis. If a program runs 1 hour in length, it takes 1 hour to make a copy of that program. This cost factor is further compounded by the high cost of raw videotape stock.

In brief, it is the potential combination of properties—high speed of replication, very low cost, very high information density, and possibility of instant access to any portion of a long record—that makes the videodisc so attractive. On the other hand, videotape can be "reprogrammed" many times, whereas a videodisc is a "read-only memory." But there are enough applications for both videotape and videodiscs separately, and for many uses the two technologies can be used to support each other.

1.6 COMPARISON OF VIDEODISCS WITH OPTICAL READ/WRITE DISCS

Although the principal impetus to the development of videodisc technology has been the entertainment market, it has also been evident that the technology would be exploited to store digital information. Two approaches have been identified [27]: (1) encoding digital information on standard videodiscs, that is, obtaining the videodisc by the mastering and replication process, and (2) recording digital information on site at the computer. The latter method was developed by Philips Laboratories, North American Philips, under contract to the Department of Defense's Advanced Research Projects Agency and is designated "optical digital disc" technology to distinguish it from the digital applications of the optical videodisc. This technology utilizes a specially prepared disc, and an erasable property may be included [28], but a direct-read-after-write (DRAW) property (i.e., the data can be read immediately after being recorded) is always included. The DRAW property is not unique to one-disc-per-recording (mastering) technology; many videodisc recording processes have the DRAW property. Even so, the DRAW disc is a common name for not prerecorded discs. Some other common names are: optical digital disc, optical memory, write once-read many times (those that do not include an erasable disc), etc.

The potential for encoding digital information on the standard videodisc was recognized at an early stage of the technology's development. The economic advantages of exploiting existing videodisc production facilities for the publication of machine-readable information was obvious [29]. There are, however, many problems to be resolved before videodisc technology can be used successfully for the storage and retrieval of digital information. Paramount among these is error identification and correction (discussed in more detail in Chapter 7).

There are two approaches to digital applications of videodisc technology (Fig. 1–12):

1. Encoding digital information on the standard video signals and formats
2. Direct digital recording

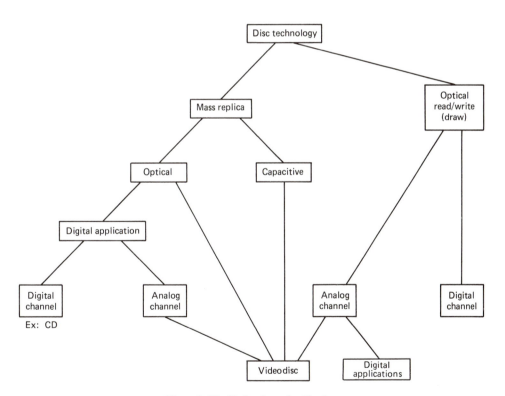

Figure 1–12 Technology classification.

Using multilevel coding, binary data were converted by DiscoVision Associates to the standard NTSC format and successfully recorded on disc and retrieved in 1980 [20,21].

Direct digital recording and replication was successfully performed by MCA DiscoVision in 1977 using a double-density coding method, the Jordan code [19,20]. A regular player was modified to recover data from the disc.

The basic difference between digital applications of videodiscs and optical digital discs is that:

- For videodiscs, a use requiring multiple (plastic) replicas is envisioned.
- For optical digital discs, specially prepared discs are utilized, and, basically, one disc per recording is obtained.

1.7 TV SIGNAL RECORDING: STANDARD FORMAT VERSUS EXTENDED PLAY

As used here, the term "standard format" means that:

- The TV signal recorded on the disc is one of the existing standards (NTSC, PAL, SECAM).
- There is one (or, in principle, an integer number of) TV picture per revolution, in order that special playing modes (freeze frame, frame-by-frame viewing, etc.) can be performed.

"Extended play" means that longer playing time is obtained with the same technology. Although there are many ways to perform this [some of which are discussed in Chapter 7 in J. Isailović's *Videodisc Systems* (to be published)], two methods are widely used: CLV recording, and compressed bandwidth. These are shown in Table 1–1. The CLV mode gives over 50% increased playing time over the CAV mode for the same signal bandwidth. The price paid is loss of the special playing modes, at least for optical videodiscs.

TABLE 1–1 TV SIGNAL RECORDING FORMATS

Disc recording	Standard TV signal	Compressed bandwidth
	Signal processing	
CAV	Standard format	Extended play (possible special modes)
CLV	Extended play (normal play)	Extended play (normal play)

When signal bandwidth is lowered, some high-frequency details are lost (Fig. 1–13) in the picture reproduced. Some high-frequency details that can be seen in the picture reproduced in a standard TV channel cannot be seen in compressed-bandwidth TV channel reproduction.

In Fig. 1–14, standard and compressed-bandwidth TV signals are shown in:

- The frequency domain before FM (NTSC standard)
- The time domain, composite video signal
- The frequency domain after FM (i.e., in the videodisc channel)

Because of the system limitation [15, p. 33] in capacitive videodisc systems [30], the compressed-bandwidth approach is widely used. Usually, the term "buried subcarrier" refers to a technique by which the chrominance subcarrier and its modulation sidebands are placed at a relatively low frequency range within the wider-band luminance channel (but the bandwidth is significantly lower than the one specified by the corresponding standard) in such a way as to cause a minimum of interference and such that the chrominance and luminance may easily be separated at the reproducer and converted to standard (NTSC, for example) format [31]. Thus, in extended-play systems, a special playing mode can be sacrificed and some high (spatial)-frequency detail can be lost.

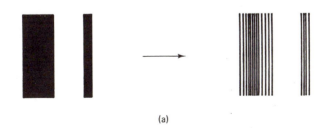

(a)

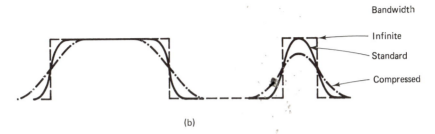

Bandwidth

Infinite

Standard

Compressed

(b)

Figure 1–13 Influence of lowering the bandwidth: (a) strip pattern; (b) reproduced signal for different bandwidths.

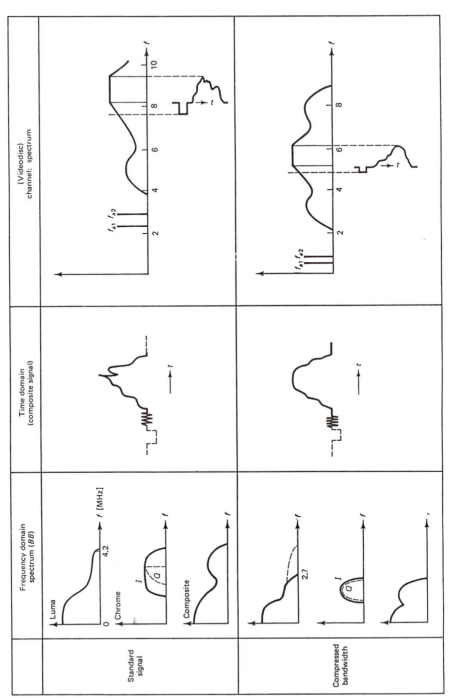

Figure 1–14 Standard and compressed-bandwidth TV signals.

1.8 OPTICAL VERSUS CAPACITIVE DISCS

Optical and capacitive videodisc systems are distinguished by the pickup used to detect patterns recorded (stored) on the videodisc surface (Fig. 1–15). Since there is no physical contact between the pickup and the disc itself in ·the optical systems, there is no wear on the disc or on the pickup. Other benefits include the ability for special playing modes such as freeze frame and slow motion, boasting a frame number random-accessing technique by which a specific frame number can be found among 54,000, for example, with the Pioneer VP-1000, within 20 seconds, and with LD-V-6000 within 3 seconds.

The grooved system uses a capacitive pickup guided by a V-shaped groove on the disc. Although large debris that would bother an optical pickup is usually brushed aside with the contact pickup, disc protection is needed (caddy), because small debris that optical pickups can keep out of focus sometimes get wedged under the stylus. Special effects are more difficult, although repeat play of a few seconds is relatively easy to do by stopping the arm cage.

The grooveless disc system, sometimes called the VHD (video high-density discs) system, incorporates a sapphire or diamond electrode that detects the capacitance variation between the disc and the stylus as the stylus moves across the surface of the disc. As with the optical approach, random-access, still, slow, and scan playback modes are available.

In general, greater flexibility and picture quality can be obtained with laser optical videodisc systems. Because there is no stylus-to-disc contact in an optical system, disc life is extended beyond that for mechanical contact systems. On the other side, the players for grooved systems are cheaper, mainly because tracking and focusing servo systems are not needed. Flat (grooveless) capacitive systems are a compromise between those two systems: cheap (capacitive) pickup with the possibility of special playing modes, but a focusing servo is required. The price per disc should not differ significantly for any of the systems. The life time is measured in the thousands of hours for laser and hundreds of hours for the capacitive pickup.

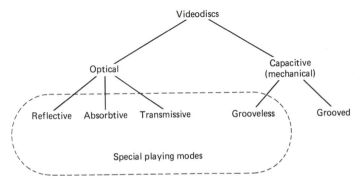

Figure 1–15 Videodisc system classification.

1.9 VIDEODISC STANDARDS

Basically, there are five types of videodiscs: three optical (reflective, transmissive, and absorbtive) and two capacitive (grooved and flat disc). In principle, two modes, CAV and CLV, can be used. Also, signal processing can be performed according to one of the existing standards (NTSC, PAL, SECAM) or in compressed-bandwidth form for each type. Thus total industrial standards are not possible. For now, the only "standards" are: (1) the outside diameter of the disc is approximately 30 cm (the VHD disc is smaller, about 25 cm); (2) the audio bandwidth is 20 kHz; and (3) pulse frequency modulation for the video, and the frequency-sharing principle (linear summation, Fig. 1–10) for adding the audio subcarriers, are used. For example, for the CAV mode, the rotational rate used is 1800, 900, or 450 rpm (30, 15, or 7.5 revolutions per second).

Next we give some examples of values for optical reflective and capacitive grooved discs.

1.9.1 Optical Reflective Systems: Standard Video (NTSC) Format

- Rotational vortex (CAV): 1800 rpm (NTSC), 1500 rpm (PAL/SECAM)
- Playing time: 30 minutes each side (CAV); 60 minutes each side (CLV or compressed bandwidth)

FM Carrier Frequencies

Blanking level	$f_b =$	8.1 MHz
White level	$f_w =$	9.3 MHz
Horizontal sync tip	$f_s =$	7.6 MHz
Maximal instantaneous	$f_{max} =$	11.3 MHz

Composite Video Preemphasis   Time constants and corner frequencies:

$$\tau_1 = 320 \text{ ns} \qquad f_1 = 497.6 \text{ kHz}$$
$$\tau_1 = 120 \text{ ns} \qquad f_2 = 1.327 \text{ MHz}$$

Audio Carriers

Channel I: $f_{a1} = 146.25 \, f_H = 2.301136 \text{ MHz}$	
Channel II: $f_{a2} = 178.75 \, f_H = 2.812499 \text{ MHz}$	
Maximal FM deviation	$\Delta f_a = 100 \text{ kHz}$
Audio bandwidth	$B_a = 20 \text{ kHz}$

Preemphasis time constant $\qquad \tau_a = 75\ \mu s$

Preemphasis corner frequency $\qquad f_{ra} = 2.12\ \text{kHz}$

Performance   (signal-to-noise ratios)

Luminance S/N $\quad = 40\ \text{dB}$
Chrominance S/N $= 30\ \text{dB}$
Audio S/N $\qquad = 60\ \text{dB}$

1.9.2 Capacitive Grooved Discs: Compressed Bandwidth (Buried Subcarrier)

- Rotational speed: 450 rpm
- Luminance bandwidth: 3 MHz
- Chrominance bandwidth: 0.5 MHz
- Color subcarrier ("buried"): $f_{sc}^* = 195 f_H/2 = 1.534091\ \text{MHz}$

FM Carrier Frequencies

$$f_b = 5\ \text{MHz}$$
$$f_w = 6.3\ \text{MHz}$$
$$f_s = 4.3\ \text{MHz}$$
$$f_{max} = 6.9\ \text{MHz}$$

Audio Carrier: $\quad f_a = f_{a1} = 91\ f_H/2 = 716\ \text{kHz}$
(if stereo used): $\qquad f_{a2} = 115\ f_H/2 = 905\ \text{kHz}$

Minimal audio FM deviation: $\quad \Delta f_a = 50\ \text{kHz}$
Audio bandwidth: $\quad 20\ \text{kHz}$
Audio S/N: $\quad 60\ \text{dB}$

1.10 PRESENT AND FUTURE USES OF VIDEODISC SYSTEMS

Considered as a communication channel, the videodisc can be used as either an analog or a digital channel. From the user's point of view, videodisc technology can be used for video, audio, or computer applications or as a combined application. In any event, mass storage is behind many of these applications as it is behind other future applications.

The main applications of the videodisc systems as a source of video material are:

- Home
- Education

- Training
- Catalogs (stores, museums)
- Libraries
- Encyclopedia
- Hotels

Originally, videodisc technology was spawned by the home entertainment industry. Applications for hotels are similar to home video. Education and training can get new dimensions and new directions as a consequence of videodisc technology influence. With or without (pre)programming, self-education and self-training are a reality. Lectures in science, language, gymnastics, or cooking can be adjusted to a given group (person) or classroom situation by choosing or recombining parts of the lecture material containing a standard repertoire. Based on the interactive customized programming, educational programs can be geared to the learning rate of individual students; sales training tailored to accommodate different levels of skills and experience; sales presentations designed to persuade at the consumer's level of interest. A combination of videodisc and microprocessor capabilities puts audiovisual programming control in the hand of viewers. Depending on the impact you want to create, you can:

- Program audiovisual materials in multiple.
- Choose a question-and-answer format giving audiences a choice of answers (the chosen answer will automatically select the corresponding program module).
- Select different program segments to meet the needs of a specific audience.
- Review some sequences while skipping others.
- Provide two narrations.

Essentially, this means that a single, prerecorded videodisc program can be resequenced and redirected to serve a variety of audiences.

Training in military, industrial, or aerospace work [32] has thus been provided with a new training tool. Catalogs of merchandise or artwork can be stored on videodiscs. Deteriorating pieces of art can be preserved on disc so that restorers always have an accurate image from which to work. In the retail store, customers can browse through a catalog with a videodisc unit and view photo images, such as action sequences and product demos, all accompanied by stereo sound. This can be used for electronic storage at home (mail order) as well as at store counters. For libraries, videodiscs might store both text and pictures from rare or expensive books. Not only would little quality be lost, but the information could be retrieved faster than with microfiche and would take up far less space. Large encyclopedias can be stored on discs and, combined with hard copy or CRT display, desired output can be obtained.

The main applications of videodisc systems as a source of audio material are:

- Video and stereo audio
- Audio only: analog and digital
- Digital only: compact disc (CD)

In the entertainment industry, audio accompanies video. This can be modified in software in such a way that audio becomes the main "message" on the disc, with video accompanying and supplementing the audio: for example, to enhance music, such as presenting classical music with a "relaxing" video content. Audio can also be combined with video in a time-sharing mode such as in a recording, and in a freeze-frame playing mode. That is, audio of considerable length can be recorded instead of a few frames; this is accompanied by one or more frames ("stop audio"), and then played until this frame(s) is played as a freeze frame on the screen. The audio can be recorded either in analog or in digital form, and the same player can be used to play both video and audio discs. Or a player can be specially designed for the audio disc only: compact disc systems, for example. Two hours of very high quality sound can be obtained from one side of a 12-cm-diameter optical disc using 16-bit PCM (pulse-code modulation).

The optical videodisc would make a fantastic computer peripheral. The technology offers enormous storage capabilities, random access, cheap reproduction, long life, and a flexibility previously unknown in data storage media. Pictures, voice, alphanumeric data, and logic could be put onto a disc with equal ease and in any desired proportion. For use as an input/output peripheral, the videodisc can be used as either an analog or a digital channel. For digital-image-processing facilities, videodisc can be used as an analog data base.

Optical videodiscs will affect on-line information retrieval services in a variety of ways, some complementary and some competitive with optical read/write discs. The videodisc, with its potential for storage and readout of machine-readable data bases, may provide the greatest competition for centralized, on-line services.

One important application of the optical videodisc will be the local provision of graphics to complement on-line information delivery, which has traditionally suffered from the absence of graphic material. The graphic material desired extends from simple line drawings, such as those used for chemical structures, to more complex line drawings, such as those used in patent submissions, and also to halftones, color images, and full audiovisual sequences. The latter will be of particular importance as a complement to full-text encyclopedic data bases and computer-based educational material. Given the availability of low-cost mass storage systems such as optical videodiscs, there will be a trend toward greater on-line availability of full text. Scientific journals and patents are examples of such use.

The videodisc could easily serve as a gigantic storehouse containing a combination of source-code programming and digital data for a host or microcomputer uses. A good example is word processing [32]. A disc could contain word-processing source information on only a fraction of its surface. The rest of the surface could hold huge vocabularies, leaving the remainder for grammar rules or biographical facts that could be drawn upon automatically to check a writer's input. As part of electronic office networks, development might go toward multi-disc-player, multiuser systems. Three-dimensional and high-resolution (high-definition) TV are also possible candidates for videodisc applications. Videodisc system applications are limited only by the human imagination.

1.11 ORGANIZATION OF THE BOOK

The task of organizing a book can be quite difficult, and often there seems to be no best way of arranging the material. For example, should principles and methods be presented independent of existing systems or through their explanation? Should common subjects, channel characterization, for example, be treated separately for the primary systems, or at once, with additional discussion of the differences? Also, supporting subjects, optics, for example, can be mixed with the primary material or presented separately and cross-referenced in the principal discussions.

The determination of the final arrangement of material was based on a number of factors, including the classroom teaching experience of the author, and suggestions of friends familiar with the material. The solution chosen for the dilemmas outlined above was to present supporting material separately; to present first principles and then discuss existing systems; to present common subjects in the same discussion but to include differences as well.

Chapters 2, 3, and 4 are devoted to videodisc systems: technology, recording, and reading. Supporting material—optics—is presented in Chapter 5, and common material—channel characterization—is presented in Chapter 6. Optical memories are discussed in Chapter 7, but common principles with optical discs already discussed in Chapters 2 and 3 are not repeated there. The basic principles of the videodisc compact disc/digital audiodisc and optical memory systems are presented together.

REFERENCES

1. P. Gregg, personal communication, Apr. 1982.

2. F. F. Dove, "Data storage and retrieval systems," U.S. Patent No. 3,226,696, Dec. 1965. See also, J. T. Mullin (3M), U.S. Patent No. 3,189,683, June 15, 1965 (filed Dec. 9, 1960); and J. E. Wolfe and R. C. Reeves Jr., U.S. Patent No. 3,247,493, Apr. 19, 1966 (filed Sept. 26, 1961). D. P. Gregg, "Electron beam recording and reproducing system," U.S. Patent No. 3,350,503 (filed March 21, 1962).

3a. P. Rice et al., "Incandescent bulb 'reads' photo-film disc," SMPTE Tech. Conf. Equip. Exhibit, 1970.

3b. P. Rice, A. Macovski, E. D. Jones, H. Frohbach, R. W. Crews and A. W. Noon: "An experimental television recording and playback system using photographic discs" *Journal of the SMPTE*, Vol. 79, Nov. 1970, No. 11.

4. D. Maydon, "Micromachining and image recording on thin films by laser beams," Bell Syst. Tech. J., Vol. 50, No. 6, July 1971, pp. 1761–1789.

5. K. Compaan and P. Kramer, "The Philips VLP system," *Philips Tech. Rev.*, Vol. 33, No. 7, 1973, pp. 178–180.

6. W. Van den Bussche, A. H. Hoogendijk, and J. H. Wessels, "Signal processing in the Philips VLP system," *Philips Tech. Rev.*, Vol. 33, No. 7, 1973, pp. 181–185.

7. P. J. M. Janssen and P. E. Day, "Control mechanisms in the Philips VLP record player," *Philips Tech. Rev.*, Vol. 33, No. 7, 1973, pp. 186–193.

8. K. D. Broadbent, "A review of the MCA DiscoVision system," 115th SMPTE Tech. Conf. Equip. Exhibit, Los Angeles, Apr. 26, 1974.

9. G. W. Hrben, "An experimental optical videodisc playback system," 115th SMPTE Tech. Conf. Equip. Exhibit, Los Angeles, Apr. 26, 1974.

10. J. A. Jerome and E. M. Koczorowski, "Film based videodisc system," 115th SMPTE Tech. Conf. Equip. Exhibit, Los Angeles, Apr. 26, 1974.

11. Series of articles in *1974 SID Int. Symp. Dig. Tech. Pap.*, Lewis Winner, New York.

12. C. Briscot, J. C. Lahureau, and C. Puech, "Optical readout of videodisc," *IEEE Trans. Consum. Electron.*, Vol. CE-22, No. 4, Nov. 1976, p. 304.

13. P. W. Bogels, "System coding parameters, mechanics, and electro-mechanics of the reflective videodisc player," *IEEE Trans. Consum. Electron.*, Vol. CE-22, No. 4, Nov. 1976, p. 309.

14. J. S. Winslow, "Mastering and replication of reflective videodiscs," *IEEE Trans. Consum. Electron.*, Vol. CE-22, No. 4, Nov. 1976, p. 318.

15. *RCA Review*, Vol. 39, No. 1, Mar. 1978: special issue on videodiscs (capacitive).

16. *Applied Optics*, Vol. 17, No. 14, July 1, 1978: seven papers on video long play systems.

17. *RCA Review*, Vol. 39, No. 3, Sept. 1978: special issue on videodisc optics.

18. J. Isailović, "A new method for encoding and decoding digital data for optical memories," ICO-12, Madrid, Sept. 1978.

19. J. Isailović, "New Code for digital data recording/transmitting," Publications of Electrical Engineering Faculty, University of Belgrade, Series: *Electronics, Telecommunications, Automation*, No. 136–141, 1980, pp. 41–51.

20. J. Isailović, "Binary data transmission through TV channels by equivalent luminance signal," *Electron. Lett.*, Vol. 17, No. C, Mar. 19, 1981, pp. 233–234.

21. J. Isailović, "Binary data transmission through TV channels by equivalent chrominance signal," SPIE Conf., Los Angeles, January 1982.

22. J. Isailović, "Videodisc engineering: a comprehensive introduction to the state-of-the-art," two-day seminar, University of California, Irvine, Jan. 29–30, 1982.

23. S. E. Poe, "Photographic video disk technology assessment report," Final Report, prepared under the Contract N00600–76–C–0505, Oct. 1976.

24. A. Kopel, "Laser applications: videodisc," in *The Laser Applications*, Vol. 4, ed. J. W. Goodman and M. Ross, Academic Press, New York, 1980, pp. 71–123.

25. E. Sigel, M. Schubin, and P. F. Merrill, *Video Discs—the Technology, the Applications and the Future*, Van Nostrand Reinhold, New York, 1980.

26. S. Jarvis, "The perilous history of the videodisk, including interactive videodisks," *Opt. Mem. Newsl.*, No. 4, July–Aug. 1982, pp. 12–13.

27. C. M. Goldstein, "Optical disc technology and information," *Science*, Vol. 215, Feb. 12, 1982, pp. 862–868.

28. H. Brody, "Materials for optical storage: a state-of-the-art survey," *Laser Focus*, Vol. 17, No. 8, Aug. 1981, pp. 47–52.

29. G. C. Kenney, "Special purpose applications of the optical videodisc system," *IEEE Trans. Consum. Electron.* Vol. CE-22, 1976, p. 327.

30. J. K. Clemens, "Capacitive pick-up and recording/playback systems Therefor," U.S. Patent No. 3,842,194, Oct. 15, 1974.

31. D. H. Pritchard, J. K. Clemens, and M. D. Ross, "The principles and quality of the buried-subcarrier encoding and decoding system," *IEEE Trans. Consum. Electron.*, Vol. CE-27, No. 3, Aug. 1981.

32. M. Edelhart, "Optical discs: the omnibus medium," *Technology*, No. 1, Nov.–Dec. 1981, pp. 42–56.

APPENDIX 1.1 MORE OF THE VIDEODISC HISTORY

Three additional historical references have recently come to my attention. Two are publicly available (as referenced). The third, an internal document from Disco-Vision Associates, is unpublished and only appears here.

1. In 1961–62 the first videodisc recorder was developed for the 3M Company at Stanford Research Institute (SRI), but publication of that work was delayed until 1970, at the request of 3M Company (P. Rice: Private Communication). It was first reported in the SMPTE Journal in November 1970 (ref. 3, in ch. 1). A more complete history of the work appears in the SMPTE Journal in March 1982: P. Rice and R. F. Dubbe—"Development of the First Optical Videodisc" pp. 277–284.

2. It is believed that the first inventor to have the idea of the optical reading videodisc was R. F. Friebus. This is discussed in the article, M. Corre and F. Pellotier: "An Optical Videodisc Invented in 1929 by Reginald F. Friebus," Memories Optiques—International Edition, Paris, France, No. 1, March 1984, pp. 16–19.

3. This account, although one-sided, contains interesting information about the world's once largest optical videodisc manufacturer.

DISCOVISION ASSOCIATES: VIDEODISC HISTORY

The revolution of applied technology, which has taken place in the second half of this century, has produced invention upon invention. Among them, the laser optical videodisc, an iridescent platter that records and stores information in the exciting format of video, along with its surrounding technology, has fascinated not only the technological community but also the consumer at home. The videodisc, disc players, and instruments that write and manufacture the discs have seen an evolutionary past; they will no doubt see a similar future.

The birth of the videodisc was a labor of the minds of several individuals. In the late 1950s an inventor, named David Paul Gregg, began his experiments with early videodisc technology in the laboratories of the Westrex Corporation. His work of putting a video signal onto a molded plastic disc in a coded, optically readable form earned him the generally accepted title of the grandfather of the modern videodisc, which uses the concept of an optically readable bi-level or digital signal format. A "typical pioneer with arrow scars in his back," Gregg never profited from his discoveries.

Gregg's ideas were further developed by two scientists, Ray Dakin and John Winslow, at MCA's research laboratories. In the late 1960s MCA had purchased a small engineering and manufacturing company which Gregg had founded. This company was principally involved in the design and manufacture of high speed audio cassette manufacturing equipment. MCA discovered that the company was also continuing Gregg's research and development of a videodisc system. MCA later sold the company but retained the Gregg patents to videodisc technology. Inventor Kent Broadbent was instrumental in creating, managing, and directing the MCA videodisc project. He hired inventors Dakin, Winslow, Slaten, and Elliott. Inventors Dakin and Winslow elaborated on Gregg's ideas and built prototype mastering machines for recording color video and sound using an FM signal; they produced the first semblance of what is known today as the videodisc. Using a casting process, a flexible disc was formed from a bismuth-coated glass master.

In December 1972 MCA DiscoVision was announced as an official MCA entity, and the company made its first public demonstration of the world's first replica disc at Universal Studios, on the same sound stage where "I Love Lucy" was filmed. The 7-minute disc played short movie clips in continuous, full color with audio; it was a flexible replica manufactured from a metal-covered master disc. MCA inventor Manfred Jarsen is credited with the development of this disc.

In June 1973 a second demonstration was held in Chicago for MCA stockholders; it was the first full-length demonstration of a regular 20-minute flexible disc replica. This caused excitement in the U.S. because the technology had been invented here and because MCA's wealth of 11,000 movie titles could easily support the consumer market. This was also the first demonstration of the stop-motion feature of the industrial players. The image quality of this disc was noticeably improved over the one produced at the demonstration held a year earlier.

The 1974 SMPTE convention in Los Angeles was the site of a third public MCA DiscoVision demonstration. An extended play flexible disc containing 40 minutes of information was shown. MCA had developed an information compression scheme that enabled 40 minutes of video and audio information to be placed on a

20-minute disc. The disc also featured random access by frame number for the first time.

Videodisc development was increasing throughout the world, and a 1974 effort to standardize the new industry by MCA DiscoVision and N.V. Philips resulted in an agreement between the two companies. N.V. Philips had an optical videodisc research program roughly paralleling MCA's. This program was allegedly inspired by Gregg's work, and it was not as far along in hardware development as MCA's program. To assist Philips in their player research and development, MCA sent a player to Philips' labs in Holland.

Over the next two years, the DiscoVision/Philips agreement resulted in a compromise between the MCA and Philips systems and set the worldwide standard for optical videodiscs—a rigid, one-sided, 30-minute disc with 54,000 tracks, 30 centimeters in diameter. Additional standards were set for encoding format, stereo carriers, bandwidth, and color information. Great growth, expansion, and insured maintenance of quality was achieved by the cooperation that took place.

Philips, MCA, and Magnavox, Philips' manufacturing subsidiary, held a fourth disc demonstration in New York City in March 1975. Philips demonstrated a rigid, one-sided disc and MCA demonstrated its flexible, one-sided disc, both on a Philips player. A clear acrylic "correction plate" was placed on the Philips player along with the flexible disc when it was played.

The disc changed during the coming year as a result of continued MCA research and development activities. In 1976 it was decided that a rigid disc would better protect the encoded information on the disc while being more reliable and easier to manufacture. An injection-molding production system was first used at this time. MCA inventor John Holmes filed several patents on mold-tooling. He invented disc tooling to produce the rigid disc in a standard injection-molding machine.

Also at this time, the bismuth-coated glass mastering process was abandoned in favor of the new "AZ Photoresist" mastering process. This new process called for application of photoresist to a glass master that was directly exposed by a visible, blue laser. As a result, a metal stamper was formed.

With the invention of MCA, employee Gary Slaten's two-sided or "2x" disc, the playing time of the videodisc was successfully doubled. Slaten realized that if two discs were pressed separately and then bonded together with the information-bearing sides together, the information on each disc could be read from the back sides of the disc, thereby "sandwiching" the encoded information and protecting the discs from temperature and humidity while doubling the length of the recording.

The industrial player was developed concurrently with the disc and manufacturing machines, led by MCA's James Elliott, the father of player technology. In response to the new rigid disc, players now employed an optical focus servo rather than the original pneumatic servo. The new optical servo "read through" the plastic substrate on the rigid disc. Most of the early players were built by hand and sold for $20,000 to $30,000 each. MCA Disco Vision manufactured 40 to 50 experimental Model 700 players; the Model 700 was the first videodisc player ever sold to customers.

The CIA, MIT, and the University of Nebraska were among the early purchasers of players.

In June 1977 an agreement was reached by Pioneer Electronic Corporation and MCA to create a joint venture named Universal Pioneer Corporation (UPC), a separate company to manufacture players. Originally, the company made only industrial players; these were manufactured in the U.S. under the MCA Disco Vision label. UPC used early hand-built players and technological know-how acquired under the DVA agreement to redesign the players for assembly line manufacture, producing the Disco Vision PR-7820 players. These industrial players featured programmable microprocessors and random access capability; the 7820 was the first industrial player ever to be mass produced. MCA's A. R. "Biff" Gale sold 10,300 of the UPC manufactured players—nearly all of UPC's first year's production—to General Motors in 1979. In 1980 UPC brought out the father of today's consumer players, the VP-1000, which featured random access capability.

Experimentation with an early form of digital data disc technology was undertaken first in 1977 with the development of the Jordan code. Additional work with audio was done later in mid-1978 in conjunction with Pioneer, who had no disc manufacturing facilities at that time. Nine months of research, using direct digital format and the optical disc, produced three different samples, the last of which performed extremely well, much like today's compact disc.

The general consumer was introduced to the laser optical videodisc system in December 1978. Magnavox introduced a player in Atlanta that MCA supported with 200 movie titles. According to *Forbes* magazine (December 1979) 3,300 players and 50,000 discs were sold the following year, the first full year the product was on the market.

As a result of an early 1979 effort to find a new encoding medium, DVA inventor Richard Wilkinson (formerly with MCA DiscoVision Laboratories) developed a "New Photoresist," a DRAW (Direct Read After Write) material which required no developing. "New Photoresist" made possible the introduction of DVA's Master V mastering machine, which reads while writing, thereby providing excellent quality control. Ironically, "New Photoresist" was in no way related to photoresist technology; the name was intended to throw industry spies off track.

MCA announced a partnership with IBM in 1979. Called DiscoVision Associates (DVA), the company's nearly 250 employees did research on videodisc application and programming as well as on improvements in writing and replicating discs. DVA also assumed marketing responsibilities for the optical disc market.

Stop motion audio research was being done by DVA's Ray Dakin (formerly with MCA Disco Vision Laboratories) and Jordan Isailović from 1980 to 1981. Several disc formats were developed which carried a 48K byte memory of digitized, time compressed audio information in video format. The audio was read by the player and then stored, to be played back at any time—a feature particularly useful during freeze frames or slow motion.

During 1980–1981, extended play was developed in two formats. Two hours per side were achieved in the CLV format and one hour per side in the CAV format. A compressed bandwidth technique was used. Isailović developed design criteria; Isailović and Dakin designed and tested a system that performed successfully.

DVA sold its production facilities to Pioneer in 1981 but continues to manage an extensive patent portfolio, holding over 1,000 patents on videodisc technology throughout the world.

Chapter 2

RECORDING

2.1 INTRODUCTION

In this chapter various recording processes are discussed. First, optical and capacitive systems are discussed in general, and then separately. Primary attention is given first to the reflective videodisc system, then to the grooved system (the capacitive electronic disc, CED). Finally, servo systems are discussed. Although the discussion is general, most of the servo systems discussed are involved in the optical system. Whenever possible, the discussion is kept nonspecific; nevertheless, a number of the methods included may have been superseded by the time the book is published.

Various steps are involved in making a master recording and producing discs from it in quantity. For each step, different techniques can be used. In general, a recording system should have the following properties:

- High yield of defect-free masters
- Capability of the basic writing process to write signals with a short spatial wavelength
- Simple recording and processing facilities
- Capability of making multiple copies from a single original master
- Immediate readout from the original master during recording (desirable)

One important feature of the videodisc is the high density of information that can be replicated on an inexpensive medium such as polymethyl methacrylate (PMMA) or polyvinyl chloride (PVC) plastic [1–3]. The discussion above leads to the following requirements for the recording:

1. Materials and technology for recording reliably with sufficient resolution to make well-defined topographical signal elements approximately 1000 Å deep and a fraction of a micrometer in width and length.

2. Recording in real time, which requires the making of signal elements at an average rate of approximately 4 to 10 MHz.

3. Recording the signal elements in a uniform spiral.

4. A process for obtaining from the original recording a flaw-free metal master qualified to be the starting point for fan-out to stampers for use in pressing thousands of finished discs.

All of the known techniques for the recording of information at extremely high storage density are compared to determine some practical means of recording the master discs. The principal contenders appear to be [2]:

- Magnetic recording
- Electron-beam recording in electron-beam-sensitive materials
- Electromechanical cutting
- Optical recording in photoresist or metal film or dyed polymers

The density of information that can be stored on a single disc of reasonable size by magnetic recording is too low, so that many discs would be required for programs of normal length. For electron-beam recording it is hard to get sufficient energy to permit real-time recording in a (circular) beam of electrons small enough to record the short wavelengths required. Recording has to be performed at a slower rate than real time (e.g., 200 times slower than real time) and requires some very ingenious solutions to achieve the necessary constancy of rotation speed [2, p. 19].

Electromechanical cutting of lacquer by a heated diamond stylus is a technique similar to that used in the recording of audio records. But the audio cutter has a frequency response of about 20,000 Hz, whereas the recording of a high-quality video signal, for example, would require a response in the megahertz range. The recording response could be extended only by a drastic size reduction of the cutter and tip, or by recording at a rate many times slower than the real-time rate. At such a slow rate the problems involved in maintaining a sufficiently constant rotation of the recording spindle (turntable) to prevent time-base errors in the recording are increased. Defects in the master obtained by electromechanical recorders are lower by at least one order of magnitude than in electron-beam masters processed under the cleanest conditions. Also, an electromechanically cut surface is smoother than the surface of an electron-beam resist, and the background noise level of signals from discs made from these masters is lower than for electron-beam recordings [2, p. 22].

Optical recording by means of a laser has the advantage that a sufficiently high density of energy can be obtained in a finely focused spot to permit the exposure of available photoresists, or melt holes in metal films, at real-time rates. Also, a direct-read-after-write (DRAW) possibility is a very important feature.

The master recorders vary in complexity and expense for each format, but most are similar in their functional aspects. The encoding electronics portion of the mastering system, for example, is common even to electromechanical systems. Most recording systems do the actual cutting of the master disc with a laser light beam. Without exception, all optical systems "master" with a laser, usually at from 15 to 500 mW, depending on the system. In each case, either an electro-optical or an acousto-optical modulator is used to control the beam.

The master disc itself is usually composed of a glass substrate upon which has been deposited a layer of positive photoresist material 1300-Å thick (or more). The cutting beam exposes minute portions of this photoresist, in accordance with the beam's modulation, as the disc rotates, generally at 1800 rpm. When the master disc is "developed," the exposed photoresist will be washed away, leaving a spiral track of pits. It is the length of these pits and the distance between them that carry the signal information. There are no data in the depth, or width, of the pits. The next step is usually to electroform nickel replicas of the photoresist surface, and from these replicas, produce the copies.

Some of master recording processes include a direct-read-after-write (DRAW) function. This permits monitoring of the program master as it is being written. This is a distinct advantage and in optical systems is accomplished by including a second, low-powered readout laser in the cutting head, or by beam splitting.

2.2 BASIC OPERATIONS

Various steps are involved in making a master recording and producing discs from it in quantity. The principal steps are depicted in Fig. 2–1.

The software for the recording can originate from various sources. Programs are received for mastering in the form of film (35mm or 16mm) or any professional form of magnetic tape. A video signal can be obtained from the camera, but can also be electronically generated, for example as a test signal or as a multilevel coding signal based on a digital signal. Video, chroma, and audio signals can be encoded in different ways. Composite video can be formed according to one of the existing standards (NTSC, PAL, SECAM) or in some other way, in order, for example, to get an extended playing time. Signal editing is performed to insert identification number codes in the vertical blanking interval, flags, and so on. The encoding method is used to produce a properly tailored signal for the videodisc channel.

The resulting fully encoded signal is then supplied to the recorder. Typically, recording is performed in real time, which requires the making of signal elements at a high average rate. Recording materials and technology should ensure reliable recording with sufficient resolution to make well-defined topographical signal elements approximately 1000 Å deep and between 0.25 and 0.8 μm long, with a high degree

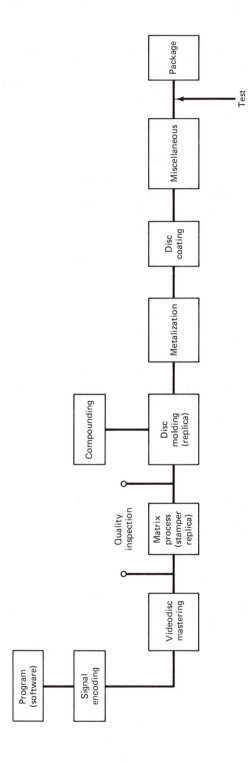

Figure 2-1 Recording steps.

of surface smoothness. Recording is performed in fine spiral tracks, maintaining uniformity of amplitude and cross-sectional shape of the signal elements. On the recording master, for example, a thin metal film or a photoresist-like organic material is the basis for developing processes for obtaining a flow-free metal master which is the starting point for the fan-out procedure to generate a number of metal stampers. The stampers are negative replicas of the original master, and they are used in pressing (molding) thousands of finished discs.

"Matrix" is a term used in the record industry for the process of electroforming thin metal replicas of the recorded sound master for use in the molding operation. The production of videodisc stampers from masters is the same as that used in audio record manufacture except that it is highly refined. The same basic processes for metal-to-metal duplication—utilizing nickel as the electroformed metal, and passivation of the nickel surfaces for duplication—are used. Electroforming of nickel metal molds and molding stampers from the recorded master is called matrix processing. An enclosure stamper of about 380 μm thickness takes 6 to 8 hours to manufacture. Quality inspection is performed after each step for possible defects.

The base material of the capacitive disc is a polyvinyl chloride (PVC) resin. The compounding process can also include additives to improve compound stability and flow characteristics or to form some desired characteristic. For example, carbon of extremely small and uniform particle size is mixed uniformly into the PVC base resin to achieve disc conductivity.

In general, both sides of a disc can be processed at once, or they can be processed separately and the two one-sided discs mounted together. Compression molding is the oldest method used for the production of audio records and remains the technique used almost exclusively for the production of 30-cm records. The stamper is precisely centered and mounted securely into the mold and force is used to form a disc with the recorded surface. Before the halves of the mold cavity are pressed together, forcing the warm PVC to assume the shape of the cavity, the mold is heated. Some time after the mold is closed, it is cooled to solidify the PVC, and then opened so that the disc can be removed. This technique, which is used to make thick videodiscs, has been used for many years to make audio LP records.

In injection molding, the molten compound is injected at high pressure into the cavity between the stampers (in a two-sided disc). The halves of the mold cavity are first clamped shut, then hot PVC is injected at high pressure to fill the cavity. The mold is cooled to solidify the PVC, opened, and the disc is removed. This technique is used to make 45-rpm audio discs and is being used to make thick videodiscs. One-sided discs may also be pressed by injection molding. Compression and injection molding are thermoforming processes commonly used to replicate thick videodiscs.

A third distinct thermoforming process is embossing. A preformed sheet of plastic is placed in a press. A heated stamper that bears the encoding is pressed against the surface of the sheet long enough to transfer the encoding without deforming the sheet. This technique, which is used to make thin, "mailable" audio discs or thin videodiscs, is a web press technique, similar to the method used to print newspapers.

Besides thermoforming, videodiscs may be replicated by casting and photopolymerization techniques. Thin videodiscs can also be made by a casting process. A thin film of liquid resin monomer is first applied to a preformed substrate, and then the liquid resin is pressed against an encoded mold surface at relatively low pressure. After contact, the resin polymerizes (either spontaneously, or in conjunction with heat or radiation) and the disc can be separated from the mold. Materials that are pigmented and difficult to cross-link by photographic methods may be polymerized with electron-beam radiation (M. Michalchik, personal communication). The mold is derived from the photoresist in the same way that the replica is made from the mold. A film of silicone (elastomeric) rubber monomer is applied to a glass disc and this is pressed against the photoresist surface. The elastomer is cured in contact with the photoresist and then the two are separated. The elastic properties of the cured mold make its separation from the photoresist, and the separation of the replica from the mold, relatively easy. A rigid mold would be more difficult to separate, and distortion of the encoding could result.

Because of the low temperatures, pressure, and shrinkage involved, the casting process introduces the least geometrical distortion of any of the replication techniques. The replicas are nearly as round as the masters, and the encoded detail is reproduced faithfully. The flatness, however, depends on the substrate.

In some systems, replicas can be made by means of fairly straightforward photographic contact-printing techniques. In such cases, the master disc serves the same purpose as the master negative, and is used to make copies in the same way that a photographic negative is used to produce photographs. This technique has the advantage of low cost per copy, plus the ability to make inexpensive copies one at a time. This permits replication at the local level, for very little cost in equipment. (This technique seems to have had limited success, possibly because of the poor contrast resolution.)

The reflectivity of the optical (reflective) disc is improved by deposition of a thin metallic film, usually made of aluminum. The metal film follows the shape of the information elements. To protect the reflective surface from dust and damage, a protective overcoating is added. The protective cover is a thin layer of transparent polymer. The original construction of capacitive discs consisted of a nonconductive compression-molded disc made from a polyvinyl chloride/polyvinyl acetate copolymer, the surface of which contained the modulating signal information and grooves. The required surface conductivity was obtained by vacuum deposition of 200 Å of a glow-discharge-deposited organic dielectric filter [2, p. 116].

After coating, there are a number of additional operations to be performed, depending on the purpose of the videodisc. The most important process is the formation of a hole, of a definite diameter corresponding to the adopted standards, in the middle of the videodisc. The disc shape is very important because, for example in the case of a color TV signal, any eccentricity may cause excessive time-lag errors, which would be reflected in the quality of the color reproduced.

For the pregrooved capacitive disc, a lubricant is required to keep wear of the disc surface, and especially the stylus, to a minimum. A thin coating of lubricant, approximately 200 Å thick, is applied by a relatively simple spray process to both sides of the disc simultaneously, after which it is respindled.

It is necessary to inspect the quality of the videodisc, and control is performed at each step in production. Some level of playback testing must be done to verify product quality for consumer acceptance criteria. The control tests are usually performed on a sample basis. Besides visual (subjective) inspection, some objective measurements must be performed, such as for dropouts, carrier level, video and audio signal-to-noise ratio, and intermodulation (IM) products.

The reflective disc does not need special protection and its packaging is simple. The capacitive disc employs a stylus in contact with the disc and requires protection. To ensure that the disc is shielded from surface contamination and handling damage, a protective package, or disc "caddy," is used. The disc leaves the protective package only during playback, when it is automatically extracted by a simple mechanism in the slot-loading player.

2.3 OPTICAL VIDEODISC

The main steps for optical videodisc production are:

- Mastering
- Electroforming (stamper replica)
- Molding
- Metallization
- Coating

The electrical signal intended for recording is amplified and used as the driving voltage of a light modulator. The modulator controls the intensity of a laser beam focused on the surface of the disc, which is covered with the recording material, metal film, or a photosensitive material, usually photoresist. The disc rotates with a fixed frequency (25 or 30 Hz for the video signal recording) and the writing head moves in a radial direction with constant speed. The disc can also rotate with a changeable speed instead of a fixed speed. After exposure and development (if photoresist is used) the surface of the disc shows depressions arranged along a spiral track and representing the (video) signal in their geometry. The encoding is reproduced by pressing in the surface of the plastic disc and, after metallization, is read out in the player. It is also possible to read a master disc without aluminizing its surface.

2.3.1 Mastering

There are several ways to produce a master disc, each of which has unique advantages. In addition to criteria that are strictly technical, the choice among mastering methods also depends on commercial considerations, such as capital and operating costs, yields, and process control. On the basis of both technical and commercial criteria, optical (laser) mastering is presently the method of choice. The principal advantage of this approach is real-time recording; it is possible to make signal elements at an average rate higher than 8 MHz.

The limits of an optical recording system are defined by the ability of the recording lens to direct the laser beam into a tightly focused spot. For a uniformly illuminated diffraction-limited lens, the focused spot width, w, between half-intensity points can be represented by

$$w = \frac{\lambda}{2NA} \qquad (1-1)$$

where λ is the wavelength of the light and NA is the lens numerical aperture. In practice, because high-numerical-aperture lenses are not perfect, the focused spot width is slightly larger than that predicted by Eq. (1-1). The finite size of the focused spot in an optical recorder for a videodisc affects the response of the system (i.e., the response tends to roll off at high spatial frequencies, the spatial frequency being a function of the temporal frequency of the recording signal, the frequency of rotation of the disc, and the radial position of the recording beam on the master disc).

2.3.1.1 Optomechanical system.   Figure 2–2 shows the optomechanical system for mastering [1]. It includes both a writing laser and a reading laser, providing direct-read-after-write (DRAW) capability. The argon laser of several watts generates the light beam for recording the information on the disc. Choice of laser power is, to a great extent, affected directly by the choice of mastering method. More specifically, it is dependent on the material used to coat the master disc substrate. The light beam is modulated in the electro-optical modulator or acousto-optical modulator, and then through the corresponding focusing lenses and mirrors is led to the disc. The cost and transmission efficiency are advantages of the acousto-optical modulator over the electro-optical modulator. The light of wavelength corresponding to the argon radiation passes through the microscopic lens with numerical aperture NA, which enables the metal or photoresist surface covering the disc to be illuminated in an area less than 1 μm in diameter. This ensures the formation of a spiral channel less then 1 μm wide. For the metal film, the laser intensity and the thickness of the disc metal cover film are adjusted such that at the maximal laser intensity corresponding to the higher level of electrical signal brought to the optomodulator, a hole is made in the metallic film. The light-beam intensity corresponding to a lower level of the electrical signal is not sufficient to make a hole in the metal film.

In order to keep the microscopic lens at the same, very small distance from the disc, it is supported by an air cushion obtained by use of an air bearing under relatively high pressure. The disc itself is in an enclosure that protects it against dust.

The light beam creates a spiral channel less then 1 μm wide; the contents of the channel aperture depend on the content of the signal led to the optomodulator. The spiral trajectory is obtained by a simultaneous rotation of the disc around a

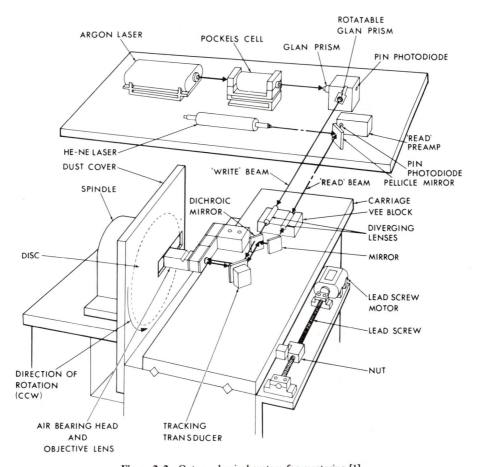

Figure 2–2 Optomechanical system for mastering [1].

fixed axis and horizontal shifting of the microscopic lens, realized through the lead-screw. Several factors are critical in master recording. These include hole-size tolerance, radial track-cutting accuracy, disc speed control, and lens focus control by closed servo loops.

2.3.1.2 Recording materials and processing.

The recording medium is one of the principal components of the optical disc system. Ideally, it should provide all of the following characteristics:

* High resolution: Since high-density optical recording systems require 1-μm or smaller spots, the material must have a resolution capability greater than 1000 line pairs per millimeter.
* High sensitivity.
* High smoothness.

- High signal-to-noise ratio.
- Real-time recording and instant playback: The material should have real-time recording characteristics and allow immediate retrieval of the information stored (i.e., there should be no processing steps required before reading).
- A high degree of freedom from defects: Material pinholes, dust, and other surface contaminants can prevent the recording of many signal elements, or obscure them after recording; thus the material must protect against this source of noise.

For the nonreplicated optical discs known as "direct-read-after-write" discs (more discussion in Chapter 7), some other characteristics of the recording materials should be provided. Probably the most important is archival storage. The material should allow permanent recording of data and should not degrade under ambient conditions or prolonged readout. A number of materials have been tested for such use, including thin metal films, ablative materials, silver halide materials (both evaporated and in emulsion), and both positive- and negative-working photoresists, but there is still room for other materials.

In ablative recording media, during recording, portions of the coating are melted away or vaporized by a modulated focused light beam, thereby exposing portions of the reflective layer, and recording (video) information as a reflective–antireflective pattern. Photoresists and metal films are most commonly used as optical recording media. Figure 2–3 shows typical signal element shapes in the photoresist and metal film. The depth of the photoresist, 1300 Å in the example shown, corresponds to the real depth of the pits on the plastic replica. The depth of the metal film is significantly less, 300 Å in the example. Thin metal films are needed because of the melting process involved in the recording process. The typical minimal hole size is approximately 0.4 μm, obtained after development of the photoresist. The corresponding diameter for metal film is 0.6 μm in tellurium (Te) to 0.8 μm in bismuth (Bi). In both cases the edges are distorted approximately 0.1 μm. This can cause both proportional and fixed-length asymmetry.

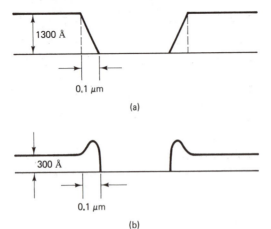

Figure 2–3 Typical signal element shapes in (a) photoresist, (b) metal film.

A glass substrate 365 mm in diameter and about 6 mm thick is generally used for the master disc. Substrate discs are cut from twin-ground plate glass, the surface of which contains hundreds of small pits per square millimeter. The glass substrate is therefore reground with a fine abrasive to eliminate the deepest pits. Finally, the surface is optically polished until the pit density has been reduced to less than 1 pit per square millimeter. The substrate disc is then chemically washed, thoroughly rinsed in deionized water, and spun dry. The glass substrate is then transferred to a thermal vacuum evaporator, where it receives a coating of metal film or a spin station for photoresist, depending on the mastering method to be employed. By regrinding and repolishing, glass masters can be reused several times.

Polymethyl methacrylate (PMMA) can be used as a substrate material. PMMA is a plastic material and is normally produced, at least for this purpose, by casting between two polished plates. Good gross flatness, minimal surface roughness, few surface defects, and long-term stability of the plexiglass material are some of the advantageous characteristics of these substrates.

If the substrate rotates at a constant angular velocity, the linear speed with which its coated surface moves under the beam increases as the recording proceeds from the inside radius to the outside radius. To maintain constant exposure, the beam power is changed in proportion to the radius.

2.3.1.3 Photoresist mastering.

Exposing a thin photoresist film [4] on a rotating substrate with a focused and intensity-modulated laser beam, and subsequent development, results in a pit pattern arranged in a spiral track on the disc. After development, the minute areas of exposed photoresist are washed away, leaving the series of holes, or pits if positive resist is used. Many parameters, including those of the modulator signal, the resist-film thickness, the laser power applied, and the substrate, determine the final dimensions of the pits. At this point, the developed disc is ready for galvanic processing.

On the basis of its high resolution and smooth surface, adequate sensitivity, and commercial availability, Shipley AZ 1350 positive-working photoresist has been found to be a good photosensitive material for optical exposure. An intermediate layer is necessary because the adhesion of this resist to glass is insufficient to survive subsequent etching and processing. A thin evaporated metal film on the glass disc can overcome this problem, since this resist was specially developed for application to metals. The metal film should be thick enough to accomplish good adhesion, but it should be as thin as possible to keep reflectivity low because highly reflective intermediate layers will generate undesired standing waves in the resist film (Appendix 2.1). A 5-nm-thick chromium layer satisfies these conditions well and even diminishes the reflectivity at the resist–substrate interface [3, p. 2004]. In this process great care should thus be taken in the choice of the substrate and also in the choice of intermediate layers. In Fig. 2–4, the effects of various intermediate layers on pit profiles are shown.

Dip coating and spin coating can be used to put on the desired 1100 to 1500 Å thickness of resist. In the dip-coating process, a plate is immersed into a tank of

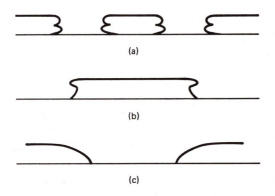

Figure 2–4 Profiles of pits in photoresist for different intermediate layers on the glass substrate: (a) reflectance about 50%; (b) reflectance about 10%; (c) reflectance about 2%.

1 part of Shipley AZ thinner and 4 parts of AZ 1350, for example. The rate at which the plate is withdrawn from the tank will determine the thickness of resist deposited. Typical withdrawal rates are 1 cm/min. In the spin-coating process, a mixture of 3 parts of thinnner and 1 part of photoresist is flooded onto a slowly spinning plate through a 0.2-μm filter. The plate is then spun at 300 to 400 rpm to remove excess resist. The thickness of the resist may be controlled by both the dilution and the spin rate. Spin coating, in general, gives better uniformity and more reproducible results than dip coating.

After coating, the plates are heat dried at a maximum of 90°F (or 65°C) for $\frac{1}{2}$ hour and are ready to be exposed. The effective sensitivity can be drastically reduced by baking at excessive temperature.

The optimum depth of the pits (i.e., the thickness of the resist film) is determined by the playback system. Reading out a replica in reflection through the bulk material yields an optimum modulation depth of the signal at a pit depth of

$$d_{\text{opt}} \geq \frac{\lambda}{4n}$$

For $\lambda = 633$ nm and the reflective index for a vinyl disc $n = 1.54$, the optimum thickness on a master disc is $d_{\text{opt}} \geq 103$ nm.

A carefully controlled process is necessary with regard to pit dimensions because playback of a disc is required to generate the original encoded signal and there must be no additional signals due to improper pit geometry. The exposure and development parameters (Appendix 2.2) have to be adjusted empirically. Optimum exposure energy levels can be obtained from the signal-quality control of test masters and replicas. Development of the exposed photoresist can be carried out in an aqueous solution (AZ 1350 developer). Spraying the solution on a slowly rotating disc has been shown to give good results. Since exposure energy, resist thickness, developer concentration, resist sensitivity, and other factors can vary from disc to disc, the development time should be reinvestigated for every master disc.

The generation of pit profiles in the photoresists film is shown in Fig. 2–5. The maximum and minimum width (γ_2 and γ_1) and the maximum and minimum length of a pit (β_2 and β_1) determine the steepness of the pit walls. This steepness

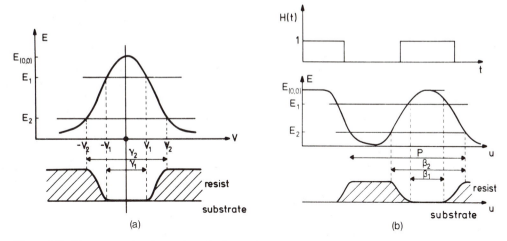

Figure 2–5 Pit structure in photoresist: (a) pit profile in the radial direction (v). (b) track profile in the tangential direction (u) in relation to the input encoded signal $H(t)$ [3].

must be small in replication techniques in order to achieve convenient separation of matrix and replica. On the other hand, for optimum detection of the pits the steepness must be high. The maximum and minimum lengths of a pit determine the important duty cycle of the signal. A compromise between these opposing requirements can be found experimentally by changing the exposure and development conditions.

2.3.1.4 Metal film mastering.

For mastering, the master disc, coated with a thin layer of metal film, is clamped to the spindle of the mastering machine and the cutting head is used for recording. Cutting is accomplished with a microscope objective lens of 0.75 NA, which focuses the 200-mW argon-ion laser beam to produce on the disc surface a spot of light of diameter less than 0.8 μm. The pulsed spot, whose duration is controlled by the electro-optical modulator, selectively records pits through a process of ablation or melting. The high NA of the lens, required to produce the small recording spot, makes the depth of focus very shallow (± 0.3 μm). This is accomplished by mounting the lens on a hydrostatic air bearing that rests on the disc surface. The bearing is loaded with enough force to make it follow disc runout as great as 25 μm. The cutting head is driven radially by a leadscrew, which advances it, for example, 1.66 μm per turn of the spindle.

The cutting process consists of melting holes in metal film which correspond to the positive half-cycles of the FM recording signal. The ends of the holes mark the zero crossings of the recording signal. Melting occurs when the power in the light spot exceeds a temperature threshold which is characteristic of the composition and the thickness of the metal film and the properties of the substrate. The cutting spot power is modulated by the Pockels cell, which is driven by the input signal, the FM-encoded video, for example. The on–off transitions are kept short to make the location of the hole ends precise in spite of variations in the melting threshold.

Although the thin-metal-films category sounds broad, functionally it means tellurium or bismuth, or, more precisely, alloys of the group 6A elements, in which a focused laser beam can easily produce small holes because of their 450°C melting point and low thermal conductivity. Pure tellurium is far over on the "sensitivity" side of the sensitivity/stability scale [5, 6]. It is rapidly oxidized and degraded by humidity; moreover, thin layers tend to flake off a substrate after vacuum deposition. Alloying tellurium with other elements adds stability. Selenium, for example, prevents the absorbtion of water, but it also reduces the material's infrared absorbtion. Another common alloying agent, antimony, makes the film amorphous—hence fine grained—but, like selenium, it reduces the medium's sensitivity.

One of the principal advantages of tellurium-based media is their fine resolution. The low thermal conductivity means that heat deposited by the laser stays in a small area—at least for the 100 ns or so that the laser pulse is on. Holes as small as 0.4 μm in diameter can be formed with an argon laser beam; smaller holes probably could be made with a more tightly focused laser.

Sensitivity, a material property, translates into recording speed, and thus is a systems property. Typically, 1-μm spots can be ablated at speeds of 10 million per second with the modulated output of a continuous-wave 10-mW laser. Fast recording requires short laser pulses, hence higher optical power to deposit the same amount of energy. The relationship is not purely reciprocal, however; in general, less energy is needed in a short pulse than in a long pulse because there is less time for the heat to dissipate.

Because the hole-forming process is a thermal, not an optical effect, the laser wavelength is not critical. That is significant because it allows writing (recording) with any laser that can deliver the required power at the required speed—a category that includes high-power, single-mode diode lasers, which are compact and easy to modulate.

The average power in the spot is of the order of 20 mW. Since the FM carrier frequency is about 8 MHz, 8×10^6 holes are cut per second, and the energy per hole is 2.5×10^{-3} J.

After the master disc has been cut, it must be transformed into a configuration from which replicas can be made. This is done by transforming the essentially "two-dimensional" master record into a "three-dimensional" configuration which can be used to stamp or form inexpensive, plastic replica discs.

The essentially flat hole pattern produced on the thin metal film could not be read if it were replicated by molding. The difference in height between the glass and the uncut metal (200 to 300 Å) is too small to produce much optical interference, or scattering, at visible wavelengths.

The master is therefore coated with a layer of negative photoresist material and is exposed through the rear (undersurface) of the disc. An ultraviolet-light source exposes (polymerizes) the photoresist through the information holes. The uncut metal film shields the photoresist where there are no holes. This exposure will polymerize negative resist at the hole sites and will leave bumps or mesas when the unexposed resist is developed away. (Positive resist will yield holes at the hole sites.) This results in an array of hardened areas that coincide with the initial array of information holes. The unpolymerized photoresist material is then washed away with an appropriate solvent, leaving bumps over the holes. This bumpy surface can then be reproduced in metal by the same galvanic technique as that used in audio-disc processing. Depending on the photoresist used, the hardening program, and other parameters, the height and profile of these bumps may be tailored to optimize the optical contrast between these bumps and the surrounding flat area when they are illuminated by the high-numerical-aperture, diffraction-limited, optical scanning system of the player. Next, a 500-Å layer of nickel is evaporated over the photoresist, and the discs are then played back on the test player with the 0.45-NA lens.

2.3.1.5 Direct read after write (DRAW).

It was noted previously that a DRAW function has certain desirable attributes in the master recording process. Among these are the ability to monitor signal quality during encoding, and the elimination of time lag between making the master and determining whether or not the master will be acceptable from a technical standpoint.

The DRAW optical system shown in Fig. 2–2 consists of a 1-mW He-Ne laser, a beam splitter, a second diverging lens, an adjustable mirror, and an adjustable dichroic mirror for combining the cutting and reading beams before they enter the microscope objective lens. The two adjustable mirrors are used to position the read spot about 10 μm downstream from the cutting spot, and directly on the center of the track that it has just cut. The 10-μm spacing ensures that the recorded surface has cooled to its final state at the time it is read.

As the system is reflective, the read beam returns from the surface of the disc in the same fashion as it was directed to the disc in the first place, until it reaches the antireflective-coated beam splitter, where it is diverted to a PIN photodiode for processing.

Because the metal film solidifies rapidly after melting, it is possible to monitor the mastering process through the cutting objective by directing the 1-mW He-Ne reading beam through the objective at an angle to form a reading spot 10 μm downstream from the cutting spot. The read spot is more strongly reflected by the metal than by the exposed glass, so the reflected beam can be used to measure the recorded signal, noise, distortion, and dropout frequency during cutting. This read-while-write feature is also used to check disc quality and machine adjustments while preliminary cuts are made inside and outside the area reserved for the video program.

Figure 2–6 shows a similar configuration for a master recorder with the DRAW function. As indicated, the optical system employs two lasers of the same wavelength: a 20-to-30-mW He-Ne laser for cutting and a 1-mW He-Ne laser for readout. Digital systems currently employ junction diode lasers operating at 830 nm. A power of 30 to 50 mW is used to write and 0.5 to 1.5 mW is used to read DRAW bits. The writing beam is modulated and then expanded by the spot lens to fill up the microscope objective. The half-wave plate rotates the polarization of the writing beam so that the writing and reading beams are polarized at 90°.

The reading, or playback, portion of the optics in this mastering recorder is

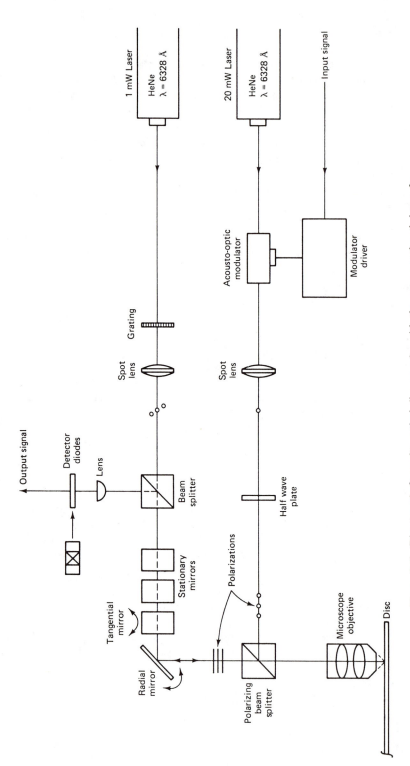

Figure 2-6 Direct-read-after-write optical disc recorder with the same-wavelength lasers for reading and writing.

similar to that of the videodisc player. After passing through the grating, the readout beam is split into the various diffraction orders; however, only the 0, +1, and −1 orders are used. The grating is such that about 82% of the readout beam's energy is concentrated almost equally among these three orders. The +1 and −1 diffraction orders are used for tracking. The 0 order is used for focusing and information reading. The spot lens is used for filling up the objective, the tangential mirror is used for time-base-error correction, and the radial mirror is used for radial tracking.

A lens-focusing method similar to that used in the videodisc player is shown. Photodiodes provide the information, tracking, and focusing signals. The radial mirror movement and the objective focusing servo are controlled by the tracking and focusing signals, respectively.

Separation between the reading and writing reflected beams is accomplished by angular separation and a 90° polarization between the reading and writing beams. The reading and writing beams form an angle θ, as seen in Fig. 2–7. Thus the reflected light from the writing beam will not interfere with the tracking and reading beams. Moreover, due to the 90° polarization between the two beams, only depolarized light from the writing beam will be reflected to the photodetectors. A preliminary measurement shows that only about 1/1000 of the light entering the writing spot lens is reflected to the photodetectors. This measurement was performed by using a mirror in front of the objective lens, thus providing about 96% reflection. In actual use, the reflection of the writing beam from the disc will probably be considerably

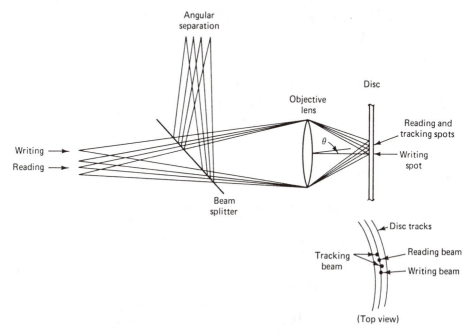

Figure 2–7 Angular separation between the reading and writing beams.

smaller than 96%, and thus the effect of the reflected writing beam in the photodetectors will be less noticeable. Good detectors with sharp cutoff after illumination (time-domain analysis) are required. Leakage current should be minimized (the state of the art is 1% edge resolution of the highest recorded frequency).

Photoresist mastering requires two fewer steps than are required for metal film mastering. The disc is coated only once, rather than twice; and photoresist exposure takes place during mastering, rather than in a separate operation. However, the read-while-write feature can be lost, so that the recorded master disc quality is unknown until the photoresist is developed and a metal film is deposited on it.

2.3.1.6 Quality control testing.   Quality control is an important aspect of the mastering process. The most direct and simple control would be to play back the disc. Measurement of signal-to-noise ratio (SNR), intermodulation products, dropouts, and so on, can be carried out easily. The nickel-coated photoresist master discs are played back on the test player. The master quality is then derived from the image on a TV screen, if a video signal is recorded.

In general, the duty cycle is one of the most important parameters of the recorded signal and should be carefully controlled. One way to check the duty cycle is to perform measurements on the specially recorded test electronic signal, consisting of a main carrier f_0 pulse-width modulated with an audio carrier signal f_a (e.g., $f_0 = 8.1$ MHz, $f_a = 2.5$ MHz). Intermodulation (IM) products will appear in the playback signal at frequency $f_0 \pm f_a$ whenever the average duty cycle differs from 50%. The ratio of the IM products and the main carrier must remain smaller than -29 dB, as was determined empirically to satisfy the requirements of TV picture quality. It can be shown that the duty cycle of the playback signal may vary between 40 and 60% (assuming that the average duty cycle ought to be 50%) and still satisfy the -29-dB criterion.

Typical values for the master at inner radius are:

- Carrier-to-noise ratio: 58 dB
- IM products due to asymmetrical duty cycle: -30 dB
- Dropouts: less than one per TV frame caused by 1-μm or larger defects on the disc

Visual inspection is normally performed. Track pitch regularity, dropouts, and other defects on a disc are investigated by visual inspection as well as by visible-light microscopy. Electron microscopy can also be very helpful in quality control. The latter method, however, generally needs small samples, which makes application of this method rather difficult.

A diffraction spectrograph provides a very rapid and accurate method of (indirect) measurement of pit geometry without a strong relation to the picture quality. The intensity ratio of the first and second radially diffracted orders gives the value of the radial duty cycle. If the resist thickness is known and hence the phase depth of the pits, the tangential duty cycle can be determined [3, p. 2005]. With a special setup, the phase difference between the first and zeroth radially diffracted orders can be measured, and the phase depth can then be carried out from this measurement.

Changes in track-to-track spacing are determined entirely by the mastering process; replication can neither improve nor degrade masters in this respect. Seismic vibrations during mastering, and stick-slip in the leadscrew, are the chief causes of changes in track spacing. As a precaution against a large source of vibration going unnoticed, or the leadscrew getting rough, pitch measurements on master discs should be made at least periodically (at at least three radii). Optical testing can also be performed, especially to detect possible track-to-track errors, called "track kissing." The preferred method is interferometric laser tracking control.

2.3.2 Galvanization: Stamper Production

After the metal-coated master is tested, it is nickel-plated to produce a mother, and then the mother is separated from the master. At this point, the mother can be used either directly as a stamper for injection molding, or it can be plated again to make an intermediate piece of tooling, which can in turn be plated to make additional submaster stampers. For experimental work, it is quicker to inject mold with the mother. In general, the number of stampers depends on the desired number of the replicas. To produce 20,000 to 30,000 or fewer replicas, the mother can be used as the only stamper. Otherwise, the fan-out process must be used to make additional stampers. Multiple metal masters can be made from the original master, so a large number of stampers, and better replicas, can be made from a single recording.

Figure 2–8a shows a cross section of a stamper made on a photoresist disc. First, thin metal (usually, nickel) film is vacuum deposited over the developed photoresist disc. Then electroforming processing is performed to form a stamper approximately

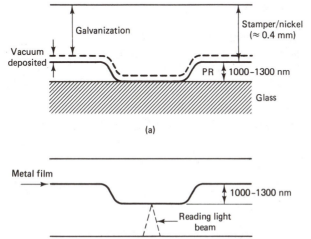

(a)

(b)

Figure 2–8 Stamper production: (a) cross section of the stamper (mother) before separation from the photoresist; (b) corresponding information element in the final replica (disc).

0.375 mm thick [7]. Following separation from the photoresist and inspection, the nickel stamper is prepared for use in the molding process.

Figure 2–8b shows a corresponding cross section of the final disc. The hole in a photoresist corresponds to a bump on the metalized surface, although the hole originally is made. This is because reading is performed through the injection-molded plastic layer.

To get a very large number of replicas per recording, metal-to-metal replication must be performed. This is illustrated in Fig. 2–9. The mother (Fig. 2–9a) is used for submother production (Fig. 2–9b). From one mother, 10 to 15 submothers can be obtained. In a similar process, stampers are made from the submother (Fig. 2–9c). Fan-out for this step is also 10 to 15. Thus more than 6,000,000 discs can be replicated from one recording.

For metal-to-metal duplication, nickel is electroformed, and passivation of the nickel surfaces for duplication is required. Normally, the information surface of the mother is passivated (basically, oxidized) and the electroforming process is continued. The nickel electroforming solutions are of the low chloride nickel sulfamate type at standard concentration. No additive agents are used. After electroforming, the parts are separated; the mother and submother (Fig. 2–9b) are cleaned in a mild alkaline cleaner, rinsed, sprayed (with anhydrous isoproponal), and dried.

2.3.3 Replication

2.3.3.1 Molding.   Once the stamper has been made, it is inserted into the injection-molding machine, usually a 300- to 500-ton machine. Liquid compound, PVC or acrylic for example, heated and under pressure, is forced into the mold cavity, where it duplicates the surface deformation pattern of the stamper. The compound is then cooled rapidly and extracted from the mold. This overall step requires from 20 to 45 seconds, depending on the molding compound, injection-molding machine being used, condition of the stamper, and predetermined cycling times (in automated systems).

Care must be taken in extracting the molded disc, to protect the sensitive surface of the replica from airborne dust, debris, and vapors, as well as from the accidents

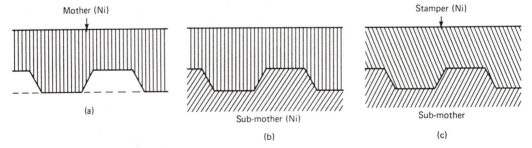

Figure 2–9 Stampers fan-out: (a) mother; (b) submother produced of the mother; (c) stamper produced of the submother.

of human handling. This step is therefore best accomplished as an automated process in a clean-room environment.

A standard injection process is characterized, among other things, by the following:

- Controlled mold temperature profiles and temperature cycles with care in selecting the molding resin
- Melt temperatures above 190.5°C (375°F)
- High mold pressure
- High material injection rates
- Small injection reservoir (shot reservoir)

A cycle for the molding starts with press closing and disc mold cavity evacuation. Next, the dry molding compound is injected into the closed, cooled mold through a center gate in about 1 s. The injection screw ram retracts and material is forced into the small shot reservoir for the next injection. After the disc is cooled, the press opens and simultaneously punches out the disc center hole and sprue. The disc is removed from the movable half of the mold by the pickup as the mold reaches full-open. After that, the press starts to close for the next cycle and the disc is removed from the mold area. Finally, the disc is loaded onto the handling system for further processing.

The key aspects of the process, which determine compound requirements, are the high shear rates involved during the injection cycle and the constant-low-temperature mold. Shear rates through the nozzle during the 1-s injection cycle are on the order of 10^4 s^{-1} with an injection pressure of 1500 psi. Higher shear forces are required for higher-viscosity compounds.

Since mold temperature is controlled constantly at a low platen temperature of approximately 57°C (135°F), the plastic is chilled instantly as it deposits against the stamper surface with very little lateral surface movement. The combination of fast injection time and cool mold surface require high compound melt flow and pressure in order to replicate the fine video signal element geometry. Melt temperatures for disc compounds are generally in the range 193 to 204°C (380 to 400°F). As a result of the high shear rates and temperatures, compound formulation is more critical than for compression molding. Additives must be chosen that are normally stable at the temperatures above and that will not separate out under high shear stress. Volatiles must also be minimized to avoid potential staining of the stamper and disc surface. The mold pellets are predried for a day or more before molding.

Good control of the total injection cycle is required to produce discs reproducibly. Control systems with pressure sensors would be helpful. Peak cavity pressure and fill control the material density and resultant weight, shrinkage, surface resolution, and dimensional stability of the molded disc.

Injection molding is the optimal process for thick-optical-disc molding. For thin discs, the casting method of replication is used.

2.3.3.2 Metallization.   After molding, the replica disc is transported to a vacuum coating machine, where a thin reflective layer of aluminum is applied over the surface containing the spiral pit pattern. This step requires 15 to 20 minutes per load of discs. A thin layer of metal, 700 to 1500 Å, is coated over the plastic videodisc to enhance the reflectivity, necessary for the reading process. The thickness of the layer does not affect the configuration indented in the plastic. The deposition parameters, however, must be controlled for optimum results.

2.3.3.3 Overcoating.   The metalized discs are then transported to another machine, where the aluminum coating is covered in a continuous conveyer with a thin layer of protective transparent plastic. The plastic coating cycle time is about 10 s.

Disc coating is performed to protect the aluminum cover against dust and various other mechanical damage. In such a way, freer handling of videodiscs than that for standard sound reproduction discs is feasible. In a double-sided disc, the overcoated disc, with metal side in, is bonded by adhesives to a second side, leaving the PMMA plastic as a protective surface through which the laser encoding is read.

2.3.3.4 Miscellaneous.   A hole of a definite diameter corresponding to the standards adopted must be formed in the middle of the videodisc. Centering is essential, because eccentricity could be reflected in poor reproduction quality. The application of labels and packing are among other final steps.

2.3.4 Cold Flow, Birefringence, and Other Problems

There are any number of real, and potential, problems associated with the injection-molding method of replicating videodiscs. These include cold flow, delamination, surface contamination, pinholes, birefringence, flatness, stamper wear, substrate materials, and cycle times. Most of these problems are a direct result of the thermal replication process itself.

Attempts to achieve rapid injection-molding cycle times have led to serious environmental stability problems with the replicated discs. The substrate material is formed, cooled, and withdrawn so quickly that stresses and strains which are set up within the material as a result of the thermal process do not have time to relieve themselves while the disc is still clamped securely inside the mold. Over a long period of time, these residual stresses in the disc undergo relaxation. Stress relief is accelerated through environmental conditions such as gradual temperature and humidity variations as the disc is slowly heated, cooled, exposed to humidity, dried out, and undergoes all of the other handling conditions at the end-user level. Warpage is one of the most common results.

By contrast, thin, flexible discs do not seem to experience many of these problems. Experiments indicate that thin discs can be made quite flat and remain quite stable even after undergoing rather severe environmental cycles. Very high and very low temperature changes in rapid succession, varying amounts of humidity and dryness, and similar conditions do not seem to affect the long-range usability of thin, flexible discs. However, in flexible discs the integrity of the aluminized surface is lost and video quality deteriorates rapidly. Also, no protection can be offered as they can with the relatively thick plastic surface.

Residual stresses are associated principally with thick discs. Simply because the discs are thicker, there is more bulk material to deal with, therefore that much more opportunity for the stresses and strains to be molded in. This is typically because the idea is to injection-mold the discs quickly, and as a consequence they do not have a chance to anneal, and completely relax, in the mold. The disc that comes out of the mold is fairly flat but not free of strain.

In extreme cases, these discs are then left to sit somewhere, at the user level. For example, where the room air temperature is reasonably warm for lengthy periods of time, the discs will soften. This assumes that no pressure whatever is applied to the discs, such as that which would be encountered if the discs were simply stacked one on top of the other. As the discs soften, the stresses and strains will relieve themselves. So if the discs started out flat and strained, after environmental cycling they are subsequently strain-free, but they are not flat anymore, either. The tolerance levels in terms of disc flatness may have been exceeded to the point at which the player's optics and electronics can no longer compensate for the out-of-flatness that now exists; consequently, the discs are unusable.

Cold flow is a particularly serious problem, and one that is inherent in every thick disc. It is not a product of the replication process itself, but rather a character defect of the polymer material used as the substrate material. When the thick disc is placed on an uneven surface—one that is not perfectly flat to within 3 to 4 μm— the surface of the disc assumes the characteristic shape of the surface on which it rests. Because of the memory of the material, once the disc has been removed from the uneven surface, it nevertheless retains the new shape, for a period. Again, any change from the original flatness of the disc, such as that which occurs with cold flow, may be enough to exceed the playable tolerance parameters, and the disc will not render satisfactory playback. At the user level, after these discs have been left lying about for a day or so on an uneven surface, they may not be playable.

Surface contamination can occur during any of the replication steps described previously. Generally, surface contamination takes the form of dust, airborne debris, and vaporized chemicals, including oils, which in anything other than a clean-room environment impinge on the exposed surface of the substrate or the aluminum coating. If the contamination is coated over, several problems may result. A "bump" or "bubble" may result, which causes a dropout on the screen. If enough contamination exists, whole sections of the programming will be unplayable.

As a practical matter, there is no reason to apply a "lacquer" layer or thin protective plastic coating over the aluminum coating, except for the fact that the aluminum layer has been applied in the first place. If a reflective player concept is being used, a reflective aluminum layer is required on the disc; otherwise, the reflective system will not read the disc. But the aluminum layer cannot remain unprotected, because experience has shown that, unprotected, it tends to flake or peel off, and is also responsive to airborne chemicals, which may weaken the adhesion of the metal coating. As the metal coating drops away, it fails to reflect. The aluminum coating must be applied perfectly; otherwise, pinholes result in the coating. Pinholes cause loss of reflected light, which causes loss of signal, which causes signal degradation.

Birefringence is a form of double refraction in which the light being reflected from the disc's surface is split into separate beams of different velocities. It is caused by stress in the substrate material (most commonly), usually resulting when (1) the substrate compound cannot be molded smoothly enough, or (2) there is a failure in uniformity of the flow modifier, which is added to the compound during the injection-molding process. The result is a surface that scatters the reflected light in a manner for which the player's optics cannot compensate, and the disc will therefore fail to deliver the playback quality required. The problem can easily be so severe that the disc will not play at all.

Stamper wear is yet another problem that must be dealt with in the replication process. For compound molding, the life of a videodisc stamper has proven to be exceedingly short compared to experience in the audio-disc field. Wear is caused by attempts to achieve rapid cycle times; higher and more rapid temperature variations in the mold; higher mold pressures; the type of compound being used; and so on. Quality control is much more critical here, because the signal information track is so much finer than that encountered with audio discs. Undetected stamper wear will very quickly cause extreme degradation of the molded signal track. In Table 2–1, PVC is included for the sake of comparison; PVC will be discussed later.

TABLE 2–1   DISC MATERIALS AND THEIR CHARACTERISTICS

Material	Cold flow	Birefringence	Stamper life
PMMA	Some problem	Problem	No serious problem
Clear PVC	Serious problem	Problem	Slight problem
PVC with carbon fill	Problem	Not applicable	Serious problem

2.3.5 Optical Transmissive Discs

The optical transmissive disc is usually made of clear PVC, 150 μm thick and 30.5 cm in diameter. Smaller-diameter discs can be replicated for shorter program lengths. Normal TV program length is 30 to 45 minutes. Disc can be single- or double-sided [8]. Signal information is in the form of pits, each having a constant width, and track spacing, say 0.8 μm and 1.66 μm, respectively. Pits vary in length from 1 to 2 μm. Figure 2–10 shows a videodisc with phase-relief modulation.

2.3.5.1 Mastering.   Programs are premastered in much the same way as that described for reflective discs. Signal input is from a 1-inch VTR.

Mastering of transmissive discs does not differ significantly from that for reflective

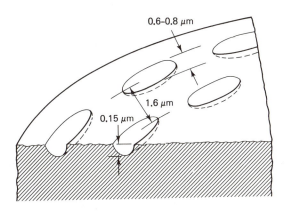

Figure 2–10 Cutaway view of an optical transmissive videodisc with relief modulation.

systems. For example, there is the same choice of photoresist mastering or metal film mastering. A 200-mW argon-ion laser can be used to write the information on the master. A glass master disc covered with AZ 1350 positive photoresist is a possible combination, although PMMA can be included. Because the system is transmissive, a DRAW function is permitted during mastering simply because readout occurs through the disc. This, in fact, means that the transmissive master does not have to be metallized to be played back. Stampers are produced in the fashion described previously.

2.3.5.2 Replication.   It is possible to perform replication by embossing a thermoplastic material with a metallic stamper obtained from the master. This process is similar to that used in audio-disc replication. More precisely, replication is accomplished by either embossing or by a web-roll-fed process, since the discs are thin and quite flexible and floppy. This process provides extremely rapid replication rates at very low cost. No coatings of layers or material are added to the copies once they have come off the embossing or web press. Instead, the discs are inserted in a protective cassette, for transport and use.

Primarily due to the thickness of the disc material and the fact that no subsequent steps are required, once the discs have been embossed, the transmissive discs have a high replication rate and concurrent low per disc cost. At 150 μm each, more than 110 discs may be obtained per kilogram of material.

Replication can be achieved extremely rapidly by either embossing or web press techniques. For the replication of millions of discs, however, the web process appears to be the only foreseeable practical approach.

2.3.5.3 Double-sided transmissive discs.   In general, double-sided transmissive videodiscs can be played without turning the record over. In principal, cross-talk between two sides can be kept low enough that there is no noticeable deformation in the displayed picture.

A cross section of a double-sided transmissive disc is shown in Fig. 2–11. The ratio of pit size to record thickness is 1 to 300, and beam is completely out of focus

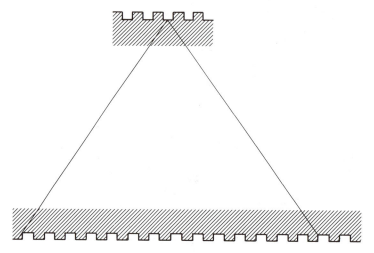

Figure 2–11 Cross section of the double-sided transmissive disc.

during its passage through the unused surface. One will have to record the opposite side backward, which should be no problem.

2.3.6 Comparison between Reflective and Transmissive Videodiscs

Although there is much to compare on the playback side, on the recording side, there is a great level of similarity between reflective and transmissive videodisc systems. The main advantages of transmissive disc systems are:

- Possible DRAW function during recording in the photoresist
- Single-step replication process

The main disadvantage is:

- Poor field life and quality for unprotected disc

To compensate for this, cassette disc protection should be added.

2.3.7 Film-Based Videodiscs

Besides reflection and phase modulation, optical videodiscs can, at least in principal, be made with absorption modulation. These are film-based videodiscs (silver halide emulsion and their equivalents).

Fine film structure, for example those originally created for holografic photograph, allows accurate recording of the high information densities associated with real-time video record/playback. Quoted resolutions for film are over 10,000 line pairs per millimeter. Track density exceeds 500 tracks per millimeter.

In addition to its capability for storing density-packed binary (on–off) signals, film can be used for the storage of multilevel continuous or pulse signals. For video-signal recording, FM or AM techniques can be used. Also, any carrier frequency can be omitted and direct analog recording of the video signal can be performed.

Figure 2–12 shows a simplified schematic of the recording process. The input signal is fed to an acousto-opticic modulator. The input signal is impressed on a low-power laser beam, and the laser beam is delivered to the film surface via a microscopic objective.

The film itself is on a rotating turntable and the microscope objective is on a moving stage keyed to the turntable rate. Thus the laser beam writes a spiral track as the storage moves in. The exact choice of laser power, numerical objective, turntable rate, spiral pitch, and other parameters are determined by the application and the particular film used. After passing through the film, the laser beam can be directed onto a photodiode, thus allowing real-time monitoring of the modulated laser beam throughout the recording process: a DRAW function.

Once the signal is recorded, the film can be developed at room temperature using standard darkroom techniques. The record may be reproduced using very straightforward and well-understood contact printing methods without degradation of the master recording. Duplication of the film master is by contact printing onto a silver halide or a diazo copy. The latter material appears to offer distinct cost advantages in large runs (Appendix 2.3). The process of production photocopying

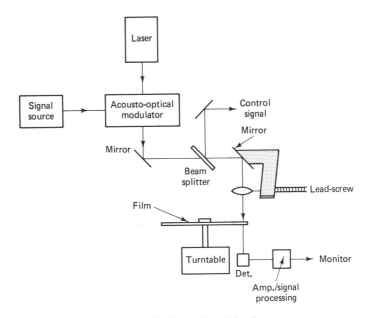

Figure 2–12 Recording of the film.

of the master disc is relatively expensive. Copies are negative images of the original; duplication time, including development, is measured in minutes rather than seconds; and there is an inherent loss of resolution from original to copy, irrespective of the quality of exposure optics, due to the grain size of silver particles. The processing steps for development are extremely sensitive, with highest-quality photographic emulsions required. The resulting cost can be prohibitive.

2.4 CAPACITIVE VIDEODISCS

This needle-in-groove approach is an extension of audio-disc technology and has tended to push that technology to its upper limits. The main steps for capacitive videodisc production are:

- Mastering
- Matrix (stamper replica)
- Molding
- Disc coating (lubricant; conductive-dielectric)
- Packaging

Encoded signal information is recorded on 1-inch video tape, which then becomes the input source for recording. Recorder can be mechanical or optical. The principal technical difficulties associated with optical recording stem from the fact that (1) the track must be wide enough to accept the mechanical contact from the stylus; (2) the configuration of the signal elements encoded on the surface of the disc is in the form of (relatively) wide slots rather than spots or pits; and (3) the smallest slot (0.25μm) is such that it cannot be recorded by laser methods because such a slot cannot be optically resolved.

In an electromechanical mastering, which is mainly used, the resulting fully encoded signal is supplied to the cutter head of the electromechanical recorder. A diamond cutting stylus driven by a piezoelectric element cuts the V-shaped groove in a smooth flat copper substrate and simultaneously cuts the signal elements in the bottom of the groove. The recording is made at real time, which requires the cutter head to respond at frequencies as high as 9.3 MHz, the upper limit of the upper sideband of the video carrier. Alternative approaches for recording (optical, electron beam, etc.) can be used. This original copper master has the surface relief pattern that will appear on the final disc. To permit large quantities of discs to be pressed, a matrix process—the process of electroforming thin metal replicas of the recorded master—must be performed. The copper master, after suitable passivation of the copper surface, is given a thin metal coating. This coating is then built up by electroplating and separated from the original, giving a submother ("metal master," mold, etc.) that is a negative replica of the original. This process is repeated to form a number of positive copies. Each submother can then be replicated to produce a number of negative replicas or stampers that are used to press discs.

The base material of the disc is a polyvinyl chloride (PVC) resin. Conductivity is achieved by loading this resin with carbon of extremely small and uniform particle size. Conductive disc can be pressed by compression or injection molding—both sides at the same time. The surface conductivity necessary for capacitive pickup can be obtained by coating discs pressed from conventional nonconducting PVC with thin coatings of metal, styrene, and oil as a continuous process in an automatic coating machine.

A thin coating of lubricant (approximately 200 Å thick) is extremely beneficial in prolonging the life of the stylus and disc. This lubricant is applied by a relatively simple spray process to both sides of the disc simultaneously, following which it is respindled.

To ensure that the disc is shielded from surface contamination and handling damage, a protective package, or disc "caddy," is used, which the disc never leaves except during play, when it is automatically extracted by a simple mechanism in the slot-loading player.

2.4.1 Mastering

In general, information elements can be recorded on the pregrooved surface, or both information elements and grooves can be recorded simultaneously. A 0.5-mm-thick bright copper layer electroplated on a smoothly machined flat, thick (approximately 12 mm) aluminum disc provides a substrate material into which to machine the groove with a sharp diamond cutting tool. Application of the resist to the surface of the grooved substrate can be done by a spinning process. For the 5555-groove per inch (gpi) format, trapezoidal-cross-section substrate grooves (Fig. 2–13) can be used to provide the desired slightly cusped groove shape. For the 9541-gpi format, triangular-cross-section substrate grooves are required to provide a desired final groove depth of 0.2 μm [2, p. 64].

To ensure an acceptable signal-to-noise ratio and an acceptable level of interference between sound and video when a TV signal is recorded, specifications for the signal geometry, applicable to the on–off format as well as to the linear format, were developed. These geometric signal specifications, defined in Fig. 2–14, are [9, p. 430]:

- *Signal amplitude* (SA): The signal amplitude at 4.3 MHz (peak of sync) must be larger than or equal to 70 nm:

$$SA_{4.3} \geq 70 \text{ nm}$$

- *Signal amplitude ratio* (R): The ratio of the signal amplitude at 6.3 MHz (white level) to the signal amplitude at 4.3 MHz (peak of sync) must be larger than 0.7:

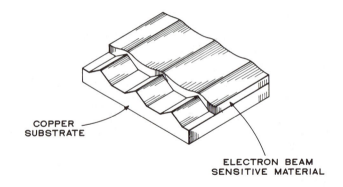

COPPER
SUBSTRATE

ELECTRON BEAM
SENSITIVE MATERIAL

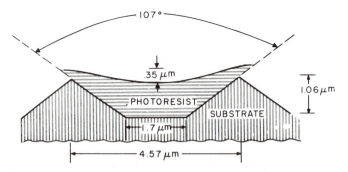

Figure 2–13 Videodisc grooved format: (a) sag-coated resist on grooved substrate; (b) cross section of photoresist-covered pregrooved substrate for the 5555-gpi format [2].

$$\frac{\mathrm{SA}_{6.3}}{\mathrm{SA}_{4.3}} > 0.7$$

- *Absolute track drop* (TD): The difference in the level of the bottom of an unexposed groove to the level of the top of a recorded 4.3-MHz carrier must not exceed 30 nm:

$$\mathrm{TD}_{4.3} \leq 30 \text{ nm}$$

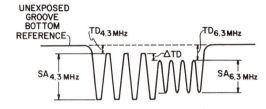

UNEXPOSED
GROOVE
BOTTOM
REFERENCE

$\mathrm{TD}_{4.3\,\mathrm{MHz}}$

$\mathrm{TD}_{6.3\,\mathrm{MHz}}$

$\Delta\mathrm{TD}$

$\mathrm{SA}_{4.3\,\mathrm{MHz}}$

$\mathrm{SA}_{6.3\,\mathrm{MHz}}$

TD = TRACK DROP
SA = SIGNAL AMPLITUDE
$\Delta\mathrm{TD} = \mathrm{TD}_{4.3\,\mathrm{MHz}} - \mathrm{TD}_{6.3\,\mathrm{MHz}}$

Figure 2–14 Physical specification and nomenclature of the capacitive videodisc signal geometry [9].

- *Differential track drop* (ΔTD): The difference between the track drop at 6.3 MHz and at 4.3 MHz must not exceed 5 nm:

$$\Delta TD = TD_{6.3} \leq 5 \text{ nm}$$

A test signal can be recorded near the outer and inner radii of the master. This signal included a few cycles each of 4.3-MHz and 6.3-MHz signals (TV signals recorded) of controlled, repeatable phase relationship. Specifications for signal depth (800 to 1000 Å) for the master were a compromise between a large enough depth to provide a good playback signal level and a small enough amount of "Δ track drop."

2.4.1.1 Electromechanical mastering. The principal elements of an electromechanical recorder (EMR) are shown in Fig. 2–15 [2]. A smoothly turning precision turntable is accurately locked to the signal source by a tachometer and speed servosystem (not shown). A cutterhead is mounted on a sturdy arm with a translation mechanism that moves the cutterhead smoothly a distance of 1 inch every 9524 turns. (Due to shrinkage in master replication and in disc-molding steps, the final disc has a slightly different number, nominally 9541 turns per inch.) On the turntable is a flat metal disc substrate covered by a layer of material that can be smoothly cut by sharp diamond cutting tools or styli. Before recording, the top surface of the material is carefully machined to be flat. The sharp diamond recording stylus has a tip-face shape that corresponds to the desired cross-sectional shape of the finished groove, and cuts the groove at the same time that it is recording the signal. In operation, the depth of the cut is determined by the position of the cutterhead support relative to the machined surface of the recording material. The cut is deeper than

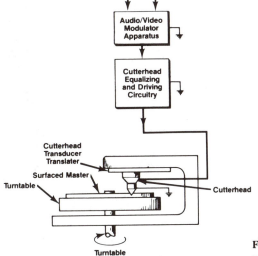

Figure 2–15 Simplified schematic diagram of an electromechanical recorder [2].

the groove, however, so groove depth is controlled by the shape of the recording stylus tip and the amount it has been translated between turns. The amplitude of the signal recorded is determined by the high-frequency motion imparted to the tip. Signals to drive the cutterhead are provided through an equalizing circuit, which is necessary to compensate for the way the amplitude and phase of the high-frequency cutting vary with frequency.

The basic cutterhead construction is shown in Fig. 2–16. A piezoelectric transducer (PZT) supports a diamond cutter tip that is bonded to it with an epoxy cement. In turn, the PZT element is bonded to a steel element, again with epoxy or a low-temperature metal alloy. The two, loaded by the diamond cutter, form a mechanically resonant unit that is damped by a layer of special hard elastomeric material used between the steel element and the support. The PZT element and the metal element have dimensions that are tapered and so chosen that the main resonance is broadened and the response is free of undesired resonance modes over a relatively wide frequency range. One of the functions of the steel element is to present, at the higher driving frequencies, a high mechanical impedance to the steel-element side of the PZT, so a given motion of the other side, upon which the diamond cutter is mounted, can be achieved at a lower driving voltage.

For real-time recording, the thickness of the PZT element and the steel element are each about 0.1 mm, and their widths and lengths somewhat greater. The diamond tip is made even smaller; typically, its greatest tip-to-base dimension is 0.1 mm. A back clearance angle of 28 to 35° is needed, so the tip-to-base dimension tapers to zero. Even for this small diamond tip mass, the dynamic stress on the bond between it and the PZT element is in excess of 1000 psi for ~1800 Å peak-to-peak motion at 5 MHz. Stresses in the PZT/steel bond are also high; if a bond fails in use, it is as likely to be the PZT/steel bond as the PZT/diamond bond.

Since no development process is required, electromechanical masters can be read out as soon as they are cut. Optical readers [9] are well suited for this purpose. They can be used to check quickly the equalization of the cutterhead, or can be used as monitors during the recording.

The main characteristics of electromechanical mastering are [2, p. 84] as follows:

- High yield of defect-free masters.
- Capability of the basic cutting process to cut signals with short spatial wavelength.

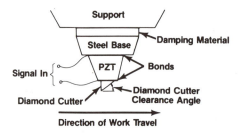

Figure 2–16 Schematic of electromechanical cutterhead [2].

- Least demanding of environment cleanliness in the mastering area.

- Smooth surfaces, resulting in high signal-to-noise ratios.

- Simple recording and processing facility needs.

- Immediate readout possible from original master during recording.

- Multiple copies can be made from a single original master.

- Deep grooves, which help stylus tracking during playback.

- Inherently linear recording.

- The groove is cut at the same time that the recording is made, simplifying the mastering procedure.

2.4.1.2 Optical mastering.

To overcome the principal technical difficulties associated with optical recording for capacitive videodiscs, let us review in brief the main disc characteristics. The videodisc signal is recorded as a "vertical" relief pattern in the grooves of a disc, which is either made out of conductive material or has a conductive coating. The disc rotates at 450 rpm and plays for 1 hour on each side. The recording starts at an outer radius of about 14.5 cm and proceeds inward at a groove pitch of about 10,000 grooves per inch to an inner radius of about 7 cm. The grooves are about 2.7 μm wide and 300 nm deep, and the peak-to-peak amplitude of the recorded signal is about 70 nm. At a radius of R inches the track velocity is $v = 1.2R$ m/s. Thus for a frequency of f MHz, the recorded wavelength is $\Lambda = 1.2R/f$ μm. The shortest wavelength of concern is $\Lambda = 400$ nm, which may occur at an inner radius of 7 cm for a maximum frequency of 8.3 MHz.

The optical master recorder is very similar in design and construction to the optical recorder system shown in Fig. 2–2 [9, p. 431]. Focus can be maintained by the use of an air-puck servo system that is monitored by a capacitive sensor: the air-bearing performance is monitored by a capacitance meter which monitors the capacitance between the (focusing) lens mount and the recording substrate. A fixed lens-to-substrate distance is maintained, despite slowly varying height changes of the substrate, by adjustment of the airflow to the air-bearing structure. For example, a lens with NA = 0.95 and a laser with $\lambda = 442$ nm can be used.

Usually, on–off recording is used. But linear recording is also possible. Whenever the beam width is not negligible compared with the wavelength of the recorded carrier, linear recording becomes attractive. In linear recording, the intensity of the recording beam varies continuously in linear response to the input signal applied.

In the nonlinear recording process discussed so far, the modulator turns the beam on and off as fast as possible in response to the on–off signal that emerges from the limiter. The locations of these on–off transitions on the disc convey the information. When the beam width is much smaller than the wavelength of the recorded signal, a reference "land" level on the disc is established when the beam is turned off, and a constant valley level is established when the beam is turned on. Under these ideal conditions, the nonlinear response of the photoresist causes no errors in the location of the transition. However, when the width of the beam is not negligible compared with the signal wavelength, distortion in the system causes errors in the apparent location of the on–off transitions as they are seen by the readout device. These errors are difficult to control or to compensate with predistortion of the input signal. In the player, the effect of these errors is distortion of the information, resulting in a variety of undesirable beats.

Figure 2–17 shows the transfer characteristic of the electro-optical modulator. The horizontal axis is the input voltage V volts from the driver, and the vertical axis is the output power P watts of the optical beam. The transfer characteristic is linear only over a limited range centered around a dc bias point, at which the input voltage is $\bar{V}$ and the output light power is $\bar{P}$. The output from the modulator is unavoidably attenuated by the optical system and intentionally by the adjustable attenuator with gain g_0.

The presence of a third-order harmonic is an indication of the introduced nonlinearities. In general, harmonics of the fundamental frequency can be detected and used for monitoring of the modulator during recording. The adaptive aperture correction can improve overall frequency characteristic of the recorder [10].

2.4.1.3 Electron-beam mastering.

For electron-beam recording, a metal disc containing a precut spiral groove is coated with an electron-beam-sensitive material, typically a positive-acting photoresist, and mounted on a turntable in a vacuum chamber. Smooth rotation of the turntable is achieved by driving it with jets of

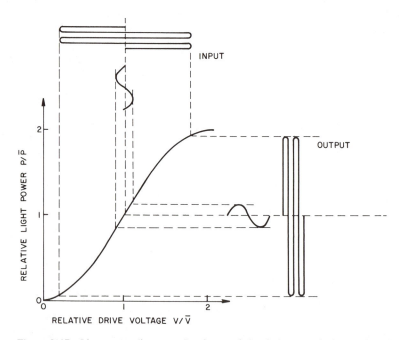

Figure 2–17   Linear recording: transfer characteristic of electro-optical modulator. [2].

vacuum pump oil. The rotational speed is measured by an optical tachometer, and is accurately locked to the signal-generating rate by means of servo control of the oil flow. The turntable rotates under an electron-optical column and translates under it at about 20 μm/s to keep the groove continuously in line with column axis, while the beam is deflected slightly so that it remains centered in the groove while it exposes the resist. The beam is blanked on and off according to the signals applied, which represent the video and sound information.

Following exposure, the resist is developed chemically to produce the desired relief-pattern track in the spiral groove. Evaporation of a thin conductive layer onto the recorded master, which is built up further by electroplating and separated from the original master, produces a negative metal master.

With a thin conformally coated substrate, an exposed signal element forms a hole after development, as shown in Fig. 2–18a. Three ways of forming a metal replica or master from it are shown in Fig. 2–18b–d [2, p. 62]:

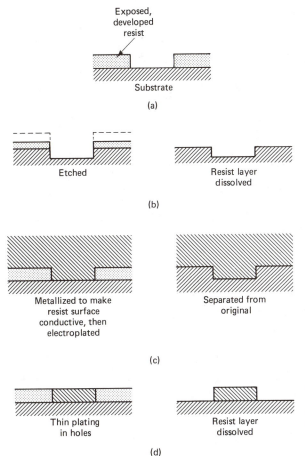

Figure 2–18 Methods for transferring to the metal substrate a signal element hole that has been developed in a thin layer of photoresist: (a) hole formed by exposure and development; (b) hole-to-hole transfer by ion etching; (c) hole-to-bump transfer by plating a thick overall layer; (d) hole-to-bump transfer by selectively plating in the hole [2].

1. Etching, preferably ion sputter etching to avoid undercutting, may be used to transfer the holes to the substrate, as shown in Fig. 2–18b. For this method, the substrate would be grooved and the metal master will be a "positive" (i.e., will be like the resist master).

2. Metallizing of the surface, for example by evaporating ~1500 Å of gold onto it, then electroplating to form a layer approximately 0.2 mm thick, followed by separation from the electron-beam substrate master as shown in Fig. 2–18c. For this method the substrate should also be grooved, but the metal master will be a "negative" of the resist master.

3. Selective electroplating into the holes in the resist coating, then dissolving away the resist material, to build up a signal element on the substrate as shown in Fig. 2–18d. For this method the substrate itself should be a negative, having ridges instead of grooves, and the resulting metal master will be a negative similar to that obtained by method 2 above, although made by a different process.

2.4.2 Stamper Production

The final process prior to the replication phase involves the preparation of disc stampers. This process is similar to that found in the preparation of stampers for audio LPs and similar to the process shown in Fig. 2–9. Electroplating of the master produces a negative metal master (0.038-cm-thick nickel matrix master). By further electroplating, the metal master is replicated to produce a positive metal master copy (variously called a mold or mother, or submatter) of the same thickness as the master. The submatter is replicated again to provide a 0.18- to 0.20-mm-thick stamper, which is finally what is used in the disc replication phase. One master recording will produce one metal master; approximately one metal master will produce 10 to 15 submatters (molds); each submatter will produce 10 to 15 stampers; and each stamper will produce 1000 to 1250 discs (this depends greatly on the molding process).

2.4.3 Replication

The principal steps involved in capacitive disc replication are compound processing, disc molding, and disc coating [11]. Thus the processes differ slightly for nonconductive and conductive discs.

2.4.3.1 Materials and compound processing.
To fulfill the requirements for videodisc applications, the materials used in disc processing must exhibit three main characteristics: (1) a uniform, homogeneous composition; (2) good melt flow with controlled rheology and process stability; and (3) dimensional and environmental stability of the molded disc.

The first two requirements imply a good process-control capability in practical manufacturing equipment, namely, good material blending in the compounding stage, controlled melt flow, and good thermal stability during molding. The third is necessary for good disc life under the environmental conditions encountered in normal use. Polyvinyl chloride/acetate copolymer compounds satisfy these requirements for audio discs, but considerable modification of these compound systems is necessary to meet videodisc needs. PVC compound technology is an established part of the plastics industry and is well documented in technical literature and industrial publications [12].

General considerations for the selection of additives for use in videodisc compounds are adequate performance of the specific additive, good thermal and process stability, and good environmental and life performance in the molded disc. The second consideration implies good compatibility of the additive material within the compound matrix under the thermal and shear stresses experienced during compound processing and disc molding. In this regard, choice of additive materials based on available product information and existing laboratory tests is a difficult task. In most cases, disc molding is the only true measure of the ultimate performance of additives within a compound.

Heat stabilizers are chemical additives required to retard the thermal degradation reaction of the base PVC resin. Lubricants are materials required to provide adequate processibility of PVC compounds during compound processing and disc molding. Such materials have a twofold requirement: to control shear effects between resin particles and to aid fusion, and second, to reduce interfacial frictional effects between the compound and process equipment surfaces. In the second role, they also act as a release during molding to prevent sticking of the molded disc to the mold surface. Lubricant materials are generally classified as internal and external according to their function in a compound. Internal lubricants are highly compatible and act like a plasticizer; external lubricants are less compatible and tend to migrate to the surface in processing. Both types are derived from similar materials, such as various waxes, oils, and low-molecular-weight polyolefins, and their functional behavior is determined by their chemical compatibility with the base resin.

Injection molding, with its increased shear rate and higher-temperature operation, places different requirements on the lubricant system. The high shear stresses experienced through the injection nozzle can produce a separation of lubricant additives because sensitivity to bleed-out of materials is increased. Selection and control of the lubricant compatibility and addition level is more critical than in compression molding. To provide acceptable performance, the lubricating action of other components must also be considered. Studies of various lubricants showed that organic acid and alcohol ester waxes used in combination with a lubricating processing-aid additive provided better control and a wider latitude of additive levels than did other lubricant types. These waxes also exhibit low volatility at the temperatures encountered during molding and produced no detrimental effect on the molding stamper or disc surface during injection runs on the order of 2000 discs.

Compression-molding-type compounds show little difference in performance with the addition of a process aid since shear rates encountered in compression molding are low. Injection-molding studies, however, show a marked difference in compound behavior as a function of process aid addition. PVC compounds without process aid exhibit banding due to nonuniform flow effects. Addition of an acrylic process aid improves flow uniformity and eliminates this problem, and is a necessary part of an injection-molding composition.

Other additives that may be used include a colorant to impart a certain surface appearance, and flow modifiers, such as plasticizer-type materials, to reduce melt viscosity. Selection of such materials must be guided as above, that is, according to their compatibility within the compound and effect on disc surface quality. These types of additives are not required for videodisc compounds of the nonconductive type in either compression or injection-molding compositions.

The material and process aspects of the conductive disc are slightly different from those mentioned above. Basically, the existing videodisc PVC base formulations are modified using highly conductive carbon. Plastics can be made conductive by the incorporation of conductive fillers such as carbon or fine metal particles. Carbon is preferred and has been used in the industry for the formulation and processing of flexible PVC compositions used for cable and microwave shielding and for static-free applications.

When high levels of carbon (in the order of 15% by weight) are added to a standard PVC formulation, three major effects occur: (1) a large increase in internal heat generation due to the frictional effects of the carbon particles during processing; (2) a large increase in shear stress, particularly at low shear rates; and (3) a lowering of lubricity. This results in a substantial increase in melt temperatures and melt viscosity of the compounds, thus making processing and molding more difficult.

2.4.3.2 Molding.   Videodisc molding by either compression or injection can be generally described in the process flow diagram of Fig. 2–19. Pelletized compound is transferred into the extruder hopper, plasticated through the extruder to a desired temperature, and a controlled shot of molten compound is delivered to the press. The molten compound is molded under controlled heat and pressure to form the disc; discs are transferred to a postprocess step (e.g., deflashing) and then stacked. Although process parameters differ between the methods, the major differences are in the actual disc-molding stage and the method of introducing the material into

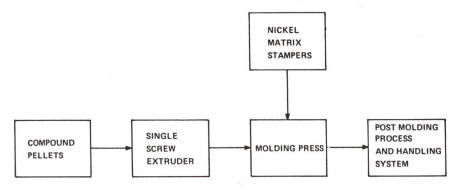

Figure 2–19 Capacitive videodisc molding process flow sequence [2].

the mold. Both methods utilize nickel-molding stampers containing the disc program information clamped to the surfaces of the press mold.

Nonconductive capacitive discs can be produced by either compression or injection molding. Compound requirements for compression molding are determined by available molding temperature and shear parameters of the process. Ideally, the mold temperature should be higher than the shot melt temperature to effect good surface replication but affect cycle times adversely. With lower mold temperature, cycle time must be increased to compensate for reduced melt flow. Disc compounds made with higher-temperature PVC homopolymers and copolymers are generally run at melt temperatures in the range 360 to 370°F (182–188°C) and produce good discs at cycle times on the order of 36 s.

Analysis of the flow rates involved in the compression process shows that maximum shear rates are low, on the order of 10 to 100 s^{-1}. Thus, only mild shear stress is placed on the material. Rheology tests show good, stable melt-flow behavior for disc PVC compounds under these conditions. The low shear rates encountered and time cycles of several seconds at high temperature result in good stress relaxation of the melt prior to cooling and low residual stress in the molded disc.

The low disc-stress levels and excellent signal-element definition are the main attributes of the compression process. The chief disadvantages are limited stamper life due to the movement, or working, of the metal stamper against the press mold during thermal cycling and greater change of contamination due to the open mold process and trimming operations.

Total cycle time for the molding of nonconductive discs is typically 26 s for most compounds. The key aspects of the process, which determine compound requirements, are the high shear rates involved during the injection cycle and the constant-low-temperature mold. Shear rates through the nozzle during the 1-s injection cycle are on the order of 10^4 s^{-1} with an injection pressure of 1500 psi. This is above the critical shear rate for most PVC compounds and unstable flow can occur. Also, high temperatures, on the order of 400°F (204°C), are generated in the nozzle area, making good compound thermal stability mandatory. Further analysis shows shear rates to decrease sharply as the melt flows out radially into the closed cavity.

Since mold temperature is controlled constantly at a low temperature of approximately 135°F, the plastic is chilled instantly as it deposits against the stamper surface with very little lateral surface movement. The combination of fast injection time and cool mold surfaces requires high compound melt flow and pressure in order to replicate the fine video signal element geometry. Melt temperatures for disc compounds are in the range 380–400°F (193–204°). As a result of the high shear rates and temperatures, compound formulation is more critical than with compression molding, as noted previously in the discussion of compound formulation. Additives must be chosen that are thermally stable at the temperatures above and that will not separate out under high shear stress. Volatiles must also be minimized to avoid potential staining of the stamper and disc surfaces. PVC compounds developed by RCA can be processed at the conditions above and show no adverse surface effects in press runs of 2000 discs, which is more than the typical life with soft electroformed nickel.

Characterization of injection-molded discs shows higher molded-in stress levels than in compression-molded discs, as a result of the high shear rates and rapid cooling of the material. The higher stress has not been found to affect disc performance. The process effects above also result in less fill at the surface. Groove and signal element depths generally measure 5 to 10% less than comparable compression discs from the same master. The overall advantages and limitations of injection molding compared to compression are summarized in Table 2–2. In general, the injection process produces discs with a better overall quality but puts greater demands on the compound materials. Compounds of the 53T type perform extremely well in injection and compression molding, producing disc yields on the order of 90% [2, p. 110].

Conductive discs are compression molded. Compression-molded conductive discs show good playback performance and little difference in visual playback quality compared to coated capacitive discs. Stylus- and disc-wear characteristics are also similar.

The physical properties of molded conductive discs are affected considerably by the added carbon. Impact strength is reduced and surface hardness is increased. Shock, vibration, and drop tests at room temperature of the as-packed disc show no apparent problems with the reduced-strength materials. Shrinkage in disc diameter at elevated temperature is also increased with the carbon but is controlled (by compound formulation) to an acceptable level. Disc warp at elevated temperature is decreased by the carbon, with discs exhibiting little static sag warp in tests at 130°F.

2.4.3.3 Coatings.

For nonconductive discs three layers are coated: metal, dielectric, and lubricant. From the standpoint of electronic signal recovery (playback),

TABLE 2–2   COMPARISON OF VIDEODISC MOLDING PROCESSES [2]

Advantages	Limitations
Compression molding	
Good signal element and groove definition (93–98%)	Limited stamper life
	Debris-generating secondary operations (deflashing)
Very low residual part stress	Increased material usage (180 g/disc)
Well-established process in record industry	Increased potential for surface contamination and blisters
Relatively low capital and maintenance costs	Limited control of disc consistency
	Limited to low-temperature materials (mainly PVC)
Injection molding	
Good control of part consistency and quality	Lower replication definition
Longer stamper life	More stringent material requirements (higher temperature and shear rates)
Less material usage (150 g/disc)	Higher residual part stress
Longer mold life (15 million cycles)	Higher capital and maintenance costs
No debris-generating secondary operations	
Wide choice of materials, including non-PVC resins	

the metal layer must be conductive, conformal, and continuous, and free of any features (point defects, pinholes, grain structure) larger than perhaps 100 Å [2, p. 137]. The dielectric must insulate the stylus electrode from contact (shorting) with the metal layer, and it should also be smooth, continuous, and defect-free. This layer must be as thin as possible because the signal strength and the system aperture response are both adversely affected in proportion to the separation between the stylus electrode and the disc conductive plane.

The requirements imposed by the mechanical properties of the system are considerably more severe. The stylus sliding rate is about 500 cm/s and its pressure is 1 to 2×10^8 Pa. The coatings must possess high cohesive strength and adhere tenaciously to one another and to the substrate in order to withstand those forces. The pressures and shear forces involved lead to the need for a lubricant. It must form a thin continuous film that inhibits intimate contact between the stylus and the dielectric layer to prevent wear of these two surfaces.

There are several sources of stylus-lifting material: small point defects in the molded disc surface or embedded in the coatings will be sheared off by the stylus; if the stylus is chipped, or if it experiences a sudden acceleration, it may dig into the coatings and shave off material; as a result of play-induced coatings stresses or because of chemically induced loss of cohesion/adhesion, particles of the coatings can flake off; dust particles of a variety of sizes, shapes, and chemical composition will settle on the disc surface. In many cases, the stylus will push the various particles aside, but in some cases, an accumulation of them will build up around the stylus, held together by mechanical or electrostatic forces and/or the wetting action of the lubricant. Such accumulations can become spontaneously detached or they may build up to the extent that they engulf the end of the stylus and lift it. In the most severe cases, particles can become wedged between the disc and the stylus "prow." They then have a high probability of becoming interposed between the stylus shoe and the surface, whereupon they lift the stylus (loss of signal strength) and because of the extremely high pressures, cut the coatings (generating more particles) as they are dragged along. The high pressures produced can literally weld the particle to the stylus shoe, resulting in extensive disc damage and disablement of the stylus. Particles 2 to 5 μm in size were more likely to cause stylus lifting than were particles 1 μm or smaller. Particles 20 to 30 μm in size cause locked and skipped grooves.

To protect the system from this mode of failure, the dielectric should have high cohesive strength and adhere well to the substrate. The lubricant should prevent adhesion of particulate matter to the stylus shoe and of dust particles to the disc surface. The dielectric layer should, if detached, break up into submicrometer-size particles.

The deposition processes and chemical composition of the coatings must be chosen to produce the properties discussed above. The metal layer must be amorphous. Its adhesion to the extremely smooth disc surface will be dominated by van der Waals and electrostatic forces and will require low intrinsic stresses. The dielectric surface energy and the surface tension of the lubricant must be compatible to ensure good spreading and retention of a permanent film. An optimum coating can be defined only with reference to the disc substrate to be coated and the stylus used to play the coated discs.

The conductive layer on the disc can be metallized, for example, by vacuum thermal evaporation of aluminium. The choice of a metal for the conductive layer is influenced to a great extent by the manufacturing process to be used for its deposition. The low materials cost, ease of application by sputtering, and the good adhesion characteristics of copper led its evaluation as the metal layer. Copper, however, has two drawbacks: the grain size is large, and of considerably greater significance, it is extremely prone to corrosion. "Trilayer" metal composition, or a "bilayer" (consisting of, e.g., 25 to 50 Å of copper under about 200 Å of Inconel/oxygen), is frequently used [2, pp. 146–147]. The dielectric layer should be thin, conformal, and defect-free. Among monomers, styrene is the most useful. The dielectric film is 200 to 250 Å thick. A lubricant is necessary for coated discs to prevent intimate contact between the stylus and the disc surface. With the most durable dielectric coatings, the principal benefit derived from lubrication is the reduction of stylus wear. Ideally, a lubricant will provide a nondestructive mechanism for dissipating the high shears involved as the stylus slides in the groove. With sufficiently thick layers of lubricant, the liquid film will undergo flow shear (hydrodynamic lubrication). On the other hand, if the film is very thin, rather more stringent requirements are imposed on the lubricant; ideally, it must form a tightly bound, densely packed, oriented molecular layer on both sliding surface in such a way that the new surfaces formed by the portions of the lubricant molecules that extend outward can slide past each other without destroying the integrity of the bound lubricant layers (boundary lubrication).

2.4.4 Flat Capacitive Disc

In general, special features—freeze frame, for example—had not been a property of the pregrooved videodiscs. These features can be included in a system with capacitive pickup if the disc surface is made flat. But tracking elements should now be included with the information elements (Fig. 1–9). The recording process for the flat capacitive videodisc (Fig. 2–20) is similar to that for the capacitive videodisc already discussed.

2.5 VIDEODISC TESTING

Evaluation of performance characteristics for any of the videodisc technologies is very important. Once initial development is complete, the primary task of testing operations is to ensure acceptable product performance for the consumer's application. However, testing only the finished product under actual consumer conditions would result in equipment, labor, and yield inefficiencies that would be economically prohibitive. Therefore, various process control tests are being used to establish product integrity and allow classical sampling plans to be used for all technologies mentioned.

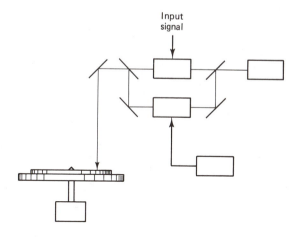

Figure 2–20   Recording of the flat capacitive videodisc.

Ideally, each replication step will be an exact reproduction of the physical signal and groove characteristics of the master. This also means, however, that any imperfections that may have been on the original part or that are created in an individual process will become permanent ("pattern") defects on all parts produced in following processes. Consequently, it is important to have test techniques and procedures that can identify defective metal parts prior to their utilization in molding operations.

Discrete defects can be formed in the electroplating process by solution problems and debris contamination. Many of the defects caused by solution and other plating problems can be detected by visual inspection using grazing high-intensity light. This technique provides a simple but effective process control test. Some of the defects that can be generated by debris contamination are very small and cannot be detected by visual techniques, yet cause significant playback problems.

The laser defect detector has the capability of testing a metal part of the standard LP size in only a few minutes. Any irregularity can be detected and its exact radius and angle location recorded for further analysis.

Using the tests above, a high confidence level is achieved for the metal parts produced in the master and matrix processes. One final test is made by making a sample press run to test the playback quality. Special test signals, located for example at the end of the program material, can be used to check disc/stylus system performance. Spectrum analyzer measurements can be made to determine important system characteristics, such as carrier level, frequency response, signal-to-noise ratio of the video carrier, and noise spectrum. This type of test provides technical information on how well the original signal information is transmitted through the replication medium and detected by the stylus pickup. In the disc-molding operation it is important that the signal elements and possible tracking elements be replicated as closely as possible to the information recorded on the master.

The efficiency of this process is affected by the setup and stabilization of various process conditions. During the molding operation, the disc is also subject to a number of effects that may produce defects; and as with all plastic materials, stress variations within the disc can cause physical distortions to occur. In addition, defects can be caused by compound contaminants, melt-flow variations, dispersion, and nonuniformity.

A simple visual inspection is effective for identifying many molding problems, such as stains, voids, and stamper damage. Also, special instruments can be designed to monitor the vertical displacement of the disc surface of predefined radii, measure entering, measure the replication efficiency of the molding operation, and so on.

All of the test techniques discussed thus far have been designed to check various physical characteristics of the major processes at low-volume points or with high-speed systems in order to produce high-quality product at a minimum cost of testing. Since none of these tests actually simulate the playback system to be used by the customer, some level of playback testing must be done to verify product quality for consumer acceptance criteria. However, the process control tests make it possible for playback tests requiring 1 to 2 hours per disc to be made with a small number of samples and still maintain a high confidence level in product quality. Special instruments and techniques can be developed to measure dropout characteristics, carrier(s) level, video and audio signal-to-noise ratio, and so on.

Environmental testing under a variety of environmental conditions should also be considered. The disc should be capable of surviving all reasonable environmental conditions without suffering significant damage to itself or adversely affecting playback performance. Those environmental conditions include conditions normally encountered in distributor warehouse, handling, shipping, retail outlets, and the consumer's home or user's place. This should also include the typical environments of different geographic locations.

Due to system requirements of all playback tests, a majority of testing must be conducted under carefully controlled conditions to prevent extraneous effects from dominating test results. At the same time, the wide diversity of consumer environments must be considered. Test design is therefore composed of two major divisions. One is concerned with tightly controlled tests to isolate specific problem areas so that they may be corrected (e.g., the determination of disc warpage limits from a player standpoint, environmental ranges, and long-term stability). Second, system tests are used in determining the product limitations when it is subject to the broad span of storage/shipping/handling and long-term consumer environment and usage conditions likely to be encountered.

Proper selection of sample size is as important as meticulous control of test conditions. As disc manufacturing processes are brought into better control, larger quantities of discs must be tested to eliminate statistical errors in the interpretation of the result.

2.6 SERVO SYSTEMS FOR MASTERING

The size of the information elements on the disc surface requires accurate control of the recording parameters. Some of the typical control systems will be discussed here.

In the mastering process some compensation circuits can be included. Their use depends on both the particular videodisc system and the particular application. For example, in the video signal recording a group delay predistortion should be included: namely, the video reconstruction filter following the FM demodulator in the playback unit has a nonlinear phase versus frequency characteristic due to the steep attenuation slope. It is easier to compensate for that once during recording than in each player. Also, if optical recording is performed using a focusing lens, a simple aperture corrector can be designed to keep (1) the peak-to-peak exposure of the (video) carrier at a frequency corresponding to black, level, constant, and independent of the recording radius; (2) the signal amplitude of frequency corresponding to white, level equal to the signal amplitude of frequency corresponding to the sync tip and independent of radius; and (3) a linear phase versus frequency characteristic.

2.6.1 Laser Control

Profile and intensity of the laser beam are important parameters: Smaller changes can cause harmonic distortion, for example by changing effective duty cycle or increasing the cross-track. Large changes in the beam power can cause severe distortion.

Write laser beam power is controlled by adjusting the current to the laser tube, as shown in Fig. 2–21. The laser tube current is controlled by an error signal derived from the difference between the output of a photodetector, whose input is a sample of the write beam, and a specified power profile. The power profile increases with radius in the CAV mode and is constant in the CLV mode since it is proportional to the tangential velocity of the write head. A profile of the laser current for on–off recording in the time domain is not necessarily of pulse form, but rather a smoother signal. This is because of the threshold of the photoresist, which acts as a (hard) limiter. In general, minimization of even harmonic distortion (the second harmonic of the carrier, in particular) can be performed with this feedback system.

A closed loop laser noise-reduction system is shown in Fig. 2–22 (U.S. Patent No. 4,114,180). A modulating signal modulates a laser beam through light modulator 1, and the recording process is performed as usual. A half-silvered mirror is disposed in the path of the beam to transmit a greater part of the light energy to light modulator 1 and to reflect some of the light to a photoelectrical transducer to convert the reflected light into the corresponding electrical signal. Two filters are included to monitor noise in the corresponding spectral domains. Amplified signals are combined, clamped, and used as a control signal for light modulator 2. This closed loop consists of the mirror, transducer, electronic circuits, and light modulator 2.

2.6.2 Focus Servo

In the mastering process the write laser beam is focused to a very small spot (<0.5 μm) by an objective lens with a high numerical aperture (~ 0.75). To maintain this small spot requires that the focus servo keep the lens positioned to achieve within ± 0.1 μm of focus. A block diagram of the servo loop to do this is shown in Fig. 2–23.

The compensator for the loop has a transfer function with one zero and one pole:

$$F(s) = K_c \frac{s + z}{s + p} \qquad \text{where } z < p$$

This results in the response curve shown in Fig. 2–23. This is a very stable loop with a phase margin of about 66°. There is a constant-acceleration slope for the lower frequencies where the maximum acceleration is a function of the peak vertical displacement and vertical frequency.

In the optical videodisc recording, a portion of the wire beam is split off-axis and is imaged on a split photodetector, all this giving a differential indication of focus. In the capacitive videodisc recording, focus was maintained by the use of an air-puck servo system that was monitored by a capacitive sensor [2, p. 75]. The output of this sensor was used for making coarse adjustments that kept the servo system within its operating range. Accurate determination of focus could be made by observing the frequency spectrum of the beam reflected back from the master surface. The relative amount of high-frequency content, due to resolution of micro-

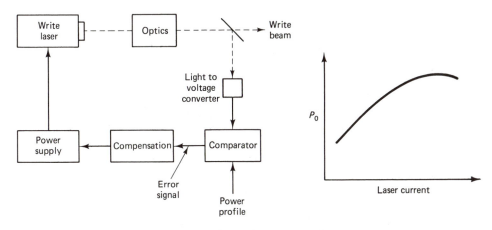

Figure 2–21   Laser power stabilization.

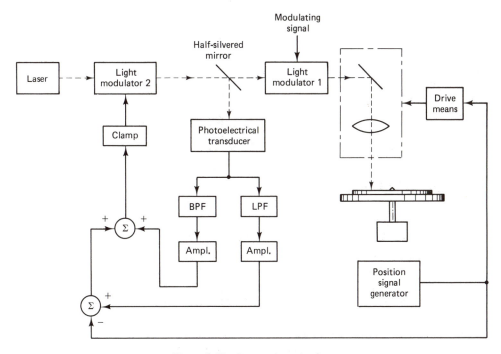

Figure 2–22   Laser noise reduction.

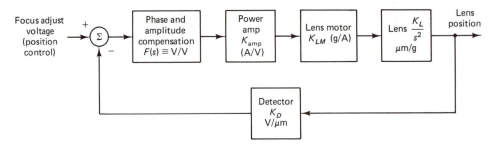

Figure 2–23 Focus servo loop.

roughness of the master surface, indicated the condition of the spot focus. A collinear beam from an He-Ne laser was found to be well suited for tracking.

The optical system of the focus detector for the optical videodisc master is shown in Fig. 2–25. Collimated, linearly polarized laser light is incident on condensing lens L_1 and continues on through beam splitter B_1 to quarter wave plate Q. Now circularly polarized by Q, the beam is imaged by objective lens L_2 onto videodisc D. Upon reflection from D, the light (its sense of circular polarization now reversed) returns back through lens L_2 to quarter-wave plate Q. The beam emerging from Q

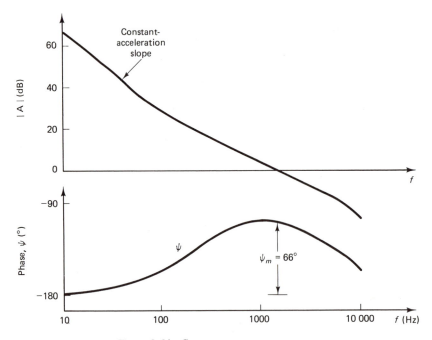

Figure 2–24 Compensator response curve.

(now linearly polarized in a plane orthogonal to the polarization plane of the original beam) is reflected by beam splitter B_1 to beam splitter B_2. Half the beam intensity is diverted by B_2 to a focus behind aperture A. The remainder is transmitted by B_2 to mirror M, and then comes to a focus in front of aperture A. The power of each beam is monitored by the split photodiode P. A change in distance between objective L_2 and videodisc D will move the focal points f_1 and f_2 closer to or farther from aperture A. The aperture will then intercept more of one light cone, less of the other. Accordingly, the power falling on each diode element will change. The apparatus may be adjusted so that disc D lies at the focus of objective L_2 when the potentials appearing across the diode elements are equal.

A plot of the difference of the split photodiode outputs as a function of objective-to-disc distance (flying height) is shown in Fig. 2–26. The range, defined as the change in flying height between maximum and minimum potentials, is dependent on aperture size—increasing the aperture diameter results in a narrower range, and vice versa. The slope in the region of interest is reasonably linear, and flying height changes of 0.1 μm are easily detected. The sensitivity is dependent on laser power and disc reflectivity; the range is independent of these.

The response is not adversely affected by the presence of encoded information on the disc. Since the device operates in a differential mode, power fluctuations such as laser ripple do not result in erroneous focus indication.

2.6.3 Spindle Servo

To maintain a zero static time error on the disc, the angular velocity of the master recording disc must have a zero mean error. This requires that the spindle be phase locked to a master timing source for entire system (a submultiple of the color subcarrier

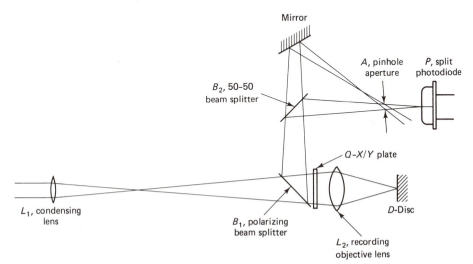

Figure 2–25 Optical system of the focus detector.

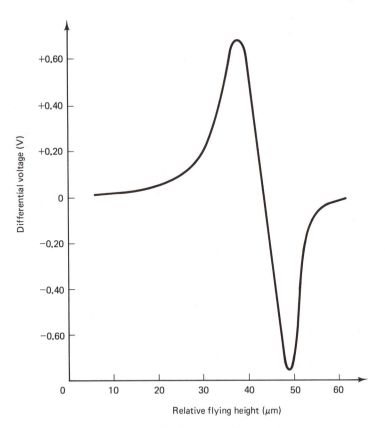

Figure 2–26 Transfer function of the focus detector.

frequency when a TV signal is recorded). A block diagram of the mastering spindle servo is shown in Fig. 2–27.

The reference input to the phase detector is the color subcarrier frequency, f_{sc}, divided by the integer. The value of the radius function R is fixed for recording in the CAV. But for the CLV mode, the radius function R is a function of radius and keeps the tangential velocity of the write spot constant. The feedback input to the phase detector is the angular velocity of the spindle times N and divided by T. A saturating type of phase detector is again used to obtain faster lock. The same loop can be used for more than one TV frame per revolution (extended play).

2.6.4 Carriage Drive

The carriage drive is the system that translates the lens (writing spot) radially across the disc. For the CAV mode of recording, the radial translation rate is uniform with time, thus producing a constant pitch. To produce a constant pitch in the CLV mode, the translation rate decreases with increasing radius and in fact is proportional

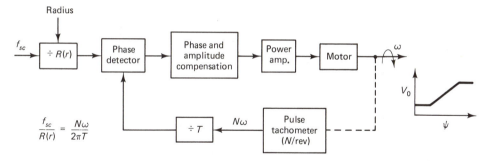

Figure 2–27   Block diagram of the mastering spindle servo.

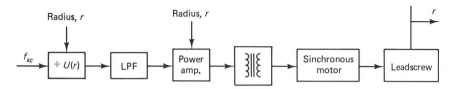

Figure 2–28   Block diagram of the carriage drive.

to the disc angular velocity. A block diagram of the carriage drive is shown in Fig. 2–28.

The translation is accomplished by a threaded nut on a leadscrew driven by a synchronous motor. The frequency to drive the motor is derived by dividing the color subcarrier frequency by $V(r)$, which is constant in the CAV mode, and a function of radius in the CLV mode. The low-pass filter removes square-wave harmonics out of V. The peak-to-peak voltage to the motor is varied as the frequency varies in the CLV mode due to the motor characteristics.

REFERENCES

1. K. D. Broadbent, "A review of the MCA DiscoVision systems," 115th SMPTE Tech. Conf. Equip. Exhibit, Los Angeles, Apr. 26, 1974.

2. *RCA Review*, Vol. 39, No 1, Mar. 1978: special issue on videodiscs.

3. *Applied Optics*, Vol. 17, No. 13, July 1, 1978: seven papers on video long play systems.

4. W. S. DeForest, *Photoresist: Materials and Processes*, McGraw-Hill, New York, 1975.

5. H. Brody, "Materials for optical storage: a state-of-the-art survey," *Laser Focus*, Vol. 17, No. 8, Aug. 1981, pp. 47–52.

6. T. H. Allen, and G. S. Ash, "Optical properties of tellurium used for data recordings," *Opt. Eng.*, Vol. 20, No. 3, May–June 1981, pp. 373–375.

7. J. V. Alfriend, Jr., *Electroplating*, International Textbook Company, New York, 1935.

8. G. Broussaud, "Le vidéodisque," *Rev. Tech. Thomson-CSF*, Vol. 10, No. 4, Dec. 1978.

9. *RCA Review*, Vol. 39, No. 3, Sept. 1978: special issue on videodisc optics.

10. J. Isailović, "MTF compensator for digital videodisc recording systems," Optica '80, Budapest, Nov. 18–21, 1980.

11. J. J. Brandinger, "The RCA CED videodisc system—an overview," *RCA Rev.* Vol. 42, Sept. 1981, pp. 333–366.

12. *Encyclopedia of PVC*, Vols. I and II, ed. L. I. Nass, Marcel Dekker, New York, 1977.

APPENDIX 2.1 STANDING WAVES

Superposition of two waves of the same frequency traveling in opposite directions causes standing waves. This means that when monochromatic light source is used for recording, reflection of a part of the incident beam at the resist–substrate interface changes the energy distribution in the resist film; this affects the pit dimensions. The incident and reflective waves are

$$\phi_i = A_i \sin k(ct - z)$$
$$\phi_r = A_r \sin [k(ct + z) + \alpha]$$

where α is the phase jump due to reflection at the resist–substrate interface, k the wave number, c the velocity of light, and z the position along the direction of progagation. The intensity distribution along z is given by

$$A^2 = |A_i|^2 + |A_r|^2 + 2|A_i A_r|\cos(2kz + \alpha)$$

The extremes of the intensity are:

$$\text{Max.: at } 2kz + \alpha = 2p\pi \qquad p \text{ an integer}$$
$$\text{Min.: at } 2kz + \alpha = (2p + 1)\pi$$

The distance between adjacent maxima and minima in the photoresist film is

$$\Delta z = \frac{\lambda}{4 n_{\text{resist}}}$$

The modulation depth of standing waves in a photoresist film is a function of the reflectance of the incident beam at the resist–substrate interface; it becomes high for the A_r of the same order of magnitude as A_i.

APPENDIX 2.2 EXPOSURE AND DEVELOPMENT

By electro-optical modulation the input signal is converted into a variation in laser intensity, $I(t)$. The energy distribution per unit surface area (E) on the disc is given by

$$E(u, v) = \int_{-\infty}^{\infty} I(t)h(u - st, v) \, dt$$

$E(u, v)$ is the normalized energy distribution within the focused laser spot, where u and v are Cartesian coordinates in the plane of the disc and along the perpendicular to the track, respectively. s is the tangential speed of the disc, $s = 2\pi nR$, where n is the number of disc revolutions per second and R the radius on the disc.

The equation for E can be rewritten as

$$E(u, v) = \frac{1}{s} \int_{-\infty}^{\infty} I(u')h(u - u', v) \, du'$$

It is obvious that E should vary inversely proportional to R.

In the case of continuous tracks, for $I(t) = I_0$,

$$E(u, v) = \frac{I_0}{s} \int_{-\infty}^{\infty} h(u, v) \, du$$

assuming that $h(u, v)$ is an even function. The laser power P on the disc is

$$P = I_0 \int_{-\infty}^{\infty} \int h(u, v) \, du \, dv$$

Energy distribution per unit surface area at the center of the track is

$$E(u, 0) = 4P(\pi d_{\text{eff}} s)$$

where the effective spot diameter d_{eff} is

$$d_{\text{eff}} = \frac{4}{\pi} \frac{\int_{-\infty}^{\infty} h(u, v) \, du \, dv}{\int_{-\infty}^{\infty} h(u, 0) \, du}$$

The required exposure energy for AZ 1350 photoresist at a wavelength of 350 nm krypton laser is 60 mJ/cm² for $d_{\text{eff}} = 0.5$ μm and for exposure at the outer radius the minimum laser power on the disc should be 6 mW. Assuming a 10% transmissive optical path of the laser-beam recorder [3, p. 2002], we need a 60-mW laser for exposure.

The maximum and minimum pit widths (γ_2 and γ_1, respectively) are determined by the required energy level (E_1) to etch the exposed photoresist away to desired pit depth (d) and the threshold energy at which etching will just start (E_2):

$$E_i = \frac{I_0}{s} \int_{-\infty}^{\infty} h(u, v_i) \, du \qquad i = 1, 2$$

so that $\gamma_1 = 2v_1$ and $\gamma_2 = 2v_2$.

If the time-dependent recording signal is

$$I(t) = I_0 H(t)$$

where $H(t)$ represents the (encoded) input signal, the maximum and minimum pit lengths (β_2 and β_1) can be found by tracking E at the center of a track:

$$E_i = \frac{I_0}{s} \int_{-\infty}^{\infty} H(u')h(u_i - u', 0)\, du' \qquad i = 1, 2$$

where $\beta_1 = 2u_1$ and $\beta_2 = 2u_2$.

APPENDIX 2.3 THE DIAZO PROCESS

The original diazo technology was developed over 50 years ago and was first used for microfilm copying in the late 1930s. Since that time it has gained a prominent place in the microfilm industry.

The diazo coating consists of diazonium salts, couplers, and azo dies of various colors. The coating is sensitive to light in the ultraviolet (UV) range. During exposure to a UV source, such as a mercury vapor lamp, the diazonium salts are decomposed by the loss of nitrogen gas. The decomposed salts will not couple and thus form the clear or white areas of an image. The coating may be compounded for continuous low-contrast tones, whereas partially exposed and decomposed salts form halftones. The contrast ratio of a particular diazo coating is referred to as the gamma ratio. Ratios greater than 1 yield higher-contrast copies, and ratios less than 1 yield lower-contrast copies (with more gray tones).

The coating is inherently acidic and pH sensitive. When subjected to a strong alkali, such as ammonia vapor, the salts react with the couplers and the image darkens, assuming the color of the dye.

A high-resolution, direct positive copy is made from transparent masters such as a microfilm. The microfilm master is placed in contact with the unexposed diazo copy film, forming a sandwich. Ultraviolet light is introduced from the master film side, which decomposes the diazo salts, selectively matching the transparent areas of the master. After exposure, the master and copy are separated and the copy is then developed by ammonia vapor.

The "speeds" of diazo coating (or emulsions) are extremely slow compared to speeds of silver halide emulsions used in camera films. This is why diazo films can be so conveniently handled in ambient room light. Their very low speed, however, demands very high energy levels in the UV range to create the latent image in the exposure process.

Chapter 3

OPTICAL PLAYBACK

3.1 INTRODUCTION

Having constructed pits in the form of a spiral track upon the disc surface, an accurate means of reading the information is necessary. First, when a video signal is recorded, the videodisc must be rotated at a speed compatible to the frequency at which the television system is operating (i.e., either 50 or 60 Hz). The 50-Hz PAL system is designed to give 25 pictures per second on the TV screen, or 1500 per minute. A PAL-system videodisc must therefore be rotated at 1500 rpm. The 60-Hz NTSC system gives 30 TV pictures per second, requiring that an NTSC-system videodisc be rotated at 1800 rpm. We will now consider optical reading of the signal information by means of a light beam—reflected or transmitted in a method similar to that used to write the information on the master videodisc. First, reflective systems will be considered.

The most important advantages of the optical system are contact-free readout of the information, as a result of which wear (disc or readout device) is nonexistent, and the possibility of effective protection of the information on the disc against the influence of dust, fingerprints, and so on.

The optical principles used are based on light-ray diffraction, a phenomenon that occurs if the object dimensions are the same order of magnitude as the wavelength of light. Well known is light diffraction by means of a narrow slit (Fig. 3–1a).

An analogous situation occurs if a light beam impinges on a reflecting surface with depressions of the shape and size of the afore-mentioned pits. Particularly strong is the effect if a light spot comparable in size to the wavelength of light is concentrated on a pit with accurately chosen depth (Fig. 3–1b). In the case of a flat surface,

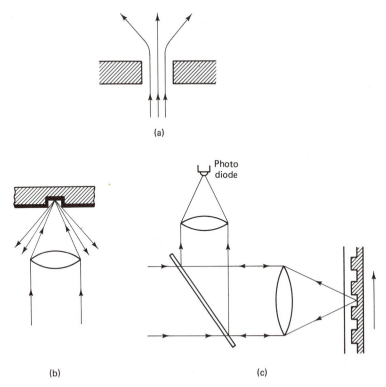

Figure 3–1 (a) Diffraction in a narrow slit; (b) deflection of light when the spot is focused on a pit of well-defined depth; (c) modulation of the reflected light by means of "pits" passed by the light spot, resulting in a corresponding signal on the photodiode.

nearly all the light is reflected and can be picked up by a photodiode fitted in the beam path, whereas if a depression is present, the major part of the light is deflected and accordingly less light is detected by the photodiode (Fig. 3–1c).

In order to read the signal information represented by the individual pits along the spiral track, the laser beam must be focused, and reduced, to a spot of light 1 μm in diameter. There is, however, a fundamental lower limit to the size of the details that can be read, depending on the wavelength of the light and the numerical aperture (NA) of the objective lens. The numerical aperture is defined as the product of the refraction index and the sine of the angle between the optical axis and the outermost light ray contributing to the imaging: $NA = n \sin \alpha$ (Fig. 3–2).

Due to diffraction at the lens aperture, the light "spot" in reality is a spot with around it annuli of decreasing brightness. If spot diameter is defined as the half-intensity diameter, it is found that for $NA = 0.4$ and a light wavelength of 0.63 μm, the minimum (half-intensity) spot diameter is 0.9 μm (Fig. 3–3). The resolv-

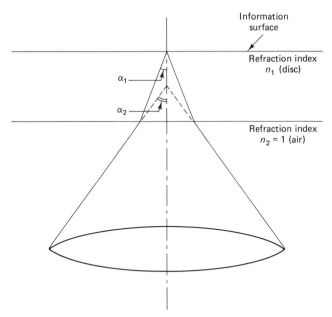

Information surface

Refraction index n_1 (disc)

α_1

α_2

Refraction index $n_2 = 1$ (air)

Figure 3–2   Numerical aperture NA $= n_1$ $\sin \alpha_1 = n_2 \sin \alpha_2$.

ing power of a system of lenses can be expressed by Abbe's law giving the minimum distance Δ between two points that still can be distinguished separately:

$$\Delta = \frac{0.6\lambda}{\text{NA}}$$

With the formula

$$f_{\text{max}} = \frac{2\text{NA}}{\lambda}\omega R_i$$

the cutoff frequency at the inner radius of the disc can be calculated by substituting the following values for NTSC:

$$\text{NA} = 0.4$$
$$\lambda = 0.63 \ \mu\text{m}$$
$$\omega = 2\pi \times 30$$
$$R_i = 55 \ \text{mm}$$

resulting in $f_{\text{max}} = 13.2$ **MHz**.

Depth of focus of the lens is about 3 μm and the beam is concentrated at the surface of the material between the pits (Fig. 3–4), not at the surface of the video disc itself. Since the video disc is 1.1 mm thick (± 0.1 mm), and most of this thickness is represented by the thickness of the clear plastic protective coating over the reflective side of the disc read by the laser, this means that the active depth-of-focus area

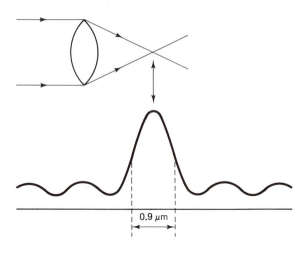

Figure 3–3 The focused spot has a half-intensity diameter of approximately 0.9 μm and an objective lens with NA = 0.4.

0.9 μm

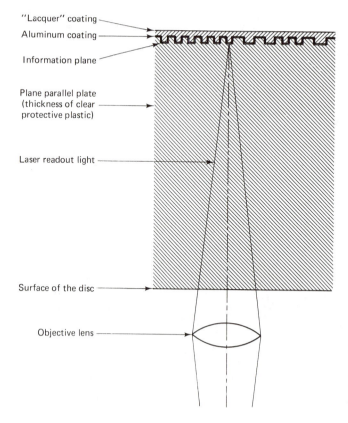

"Lacquer" coating

Aluminum coating

Information plane

Plane parallel plate (thickness of clear protective plastic)

Laser readout light

Surface of the disc

Objective lens

Figure 3–4 Configuration of laminated video-type disc. (Scale is approximate. Actual focus-to-surface ratio is 0.1:1000.)

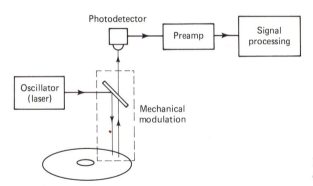

Figure 3–5 Model for the reflective optical videodisc reading process.

read by the lens is about 900 μm below the actual surface of the disc itself. This method of focusing, combined with the type of disc construction (i.e., laminated), results in a definite advantage, when disc handling and durability are considered.

Reflective video discs show their greatest utility when they are read through the body of the disc; this mode of use puts the information-bearing surface out of harm's way, a distance many times the depth of focus from the nearest optically effective exposed surface. This has the benefit of holding dirt and scratches out of focus to reduce or remove their effect on signal playback; it has the drawback of requiring compensation elsewhere in the optical system. Focusing a beam of light through a slab of refracting material makes the point of focus move axially and introduces spherical aberation. Above a certain tolerance limit, this aberration degrades the focused spot enough so that the signal is no longer useful.

A simplified model for the reflective videodisc reading process is shown in Fig. 3–5. The oscillator, usually a laser, generates a reading carrier light beam which is then modulated by the pit pattern recorded on the disc surface. A photodetector converts the light to electrical signals which are then amplified and possibly equalized in the low-range preamplifier. After signal processing, the original information is recovered. The same model is useful for transmissive (and absorbtive) optical systems, except that the photodetector should be placed below the disc.

The basic elements of the reproduction system are:

- An optical system serving to focus the light beam of a low-intensity laser on the videodisc, to gather the reflected light energy from the videodisc surface and to direct it to photodiode
- A system for rotating the videodisc
- A system enabling the reading light beam to follow the channel with grooves on the disc
- A servosystem for maintenance of focus of the reading light beam on the surface of the videodisc
- A control system and associated interfaces for the user
- (For TV signals) a suitable electronic system for placing the video signal into a suitable frequency channel so that reproduction on the TV set can be realized

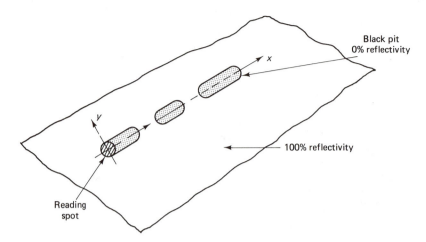

Figure 3–6 Ideal optical disc.

3.2 OPTICAL PICKUP: OPTICAL PRINCIPLES

Two questions should be answered here:

1. What is an ideal optical disc?
2. How can an (ideal) optical disc be approximated?

Part of the ideal optical disc is shown in Fig. 3–6. The reflectivity of the disc surface is 100%; for the transmissive disc, transmitivity is 100%. Information pits are black holes: 0% reflectivity, 0% transmitivity, that is, 100% absorptivity. The best contrast is obtained if the diameter of the reading spot is equal to the width of the black pit or less. Greater contrast means greater noise immunity. The closest approximation to the ideal optical disc is obtained by

- Scattering effect, or
- Grating phase modulation

In Fig. 3–7 it is shown that a black pit can be obtained if focused light is scattering from a properly shaped bump. Again, final contrast depends on the relative sizes of the reading spot and bump.

3.2.1 Diffraction Grating

Let us idealize a signal waveform on the optical disc as a periodic relief structure of spatial period L (Fig. 3–8a). Within the approximations of color diffraction theory, transmissive and reflective phase gratings are equivalent when they impart the same spatial phase variations to the readout beam. For the equivalent phenomena, a pit

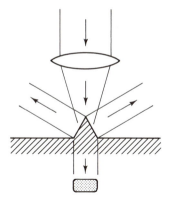

Figure 3–7 Scattering.

depth of the transmissive disc is twice that of the reflective one. For example, the equivalent pit depths are:

$$\lambda/2 \text{ (transmissive disc)} \Longleftrightarrow \lambda/4 \text{ (reflective disc)}.$$

In the following discussion, the diffraction pattern will be sketched for a reflective phase grating. The diffraction orders (-1, 0, and $+1$) are marked in Fig. 3–8b.

3.2.1.1 A pit depth of $\lambda/4$.   Two facts can be pointed out. First, the pit width (w) is smaller than the reading spot diameter. Second, there is a 180° phase difference between reflected (0 order) rays from the pit (c) and from the surface (Fig. 3–9a). For the sake of simplicity, only one recorded track will be considered.

To get good contrast power, equality should exist in rays with two different phases; that is,

$$P_a + P_b = P_c$$

where P_a is the power contained in light reflected from the surface with width a, and similarly for the other symbols.

As one example, in Fig. 3–9b a triangle profile is shown. Total power, when the pit is scanned, will be zero (this is a hypothetical case, of course) if

$$w = \left(1 - \frac{1}{\sqrt{2}}\right) d$$

That is,

$$w \approx 0.3d$$

Thus, from the pit, zero signal is detected, and from the flat area maximal signal is detected.

In Fig. 3–10, the output signal for a periodical pit pattern is shown. Only 0-order diffraction is assumed. The output signal is not square because the finite size of the reading spot in the tangential direction (x) is assumed. Thus the output signal is obtained by convolving pit pattern with the reading spot profile.

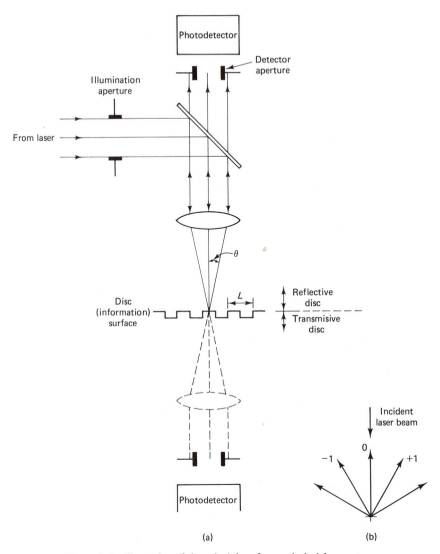

Figure 3–8   Illustration of the principles of an optical pickup system.

3.2.1.2 A pit depth of λ/8.   Three facts can be pointed out:

1. The reading spot diameter is not larger than the pit width.
2. There is a 90° phase difference between 0-order diffracted rays from the pit and from the surface.
3. There is a 180° phase difference for −1-order diffraction and 0° for +1-order diffraction (from the pit and from the surface).

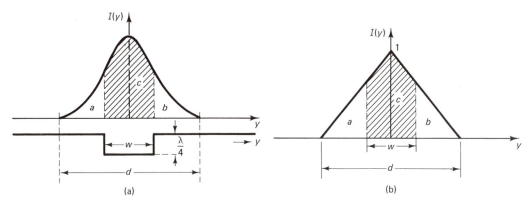

Figure 3–9 Cross section of the pit and the reading beam profile: (a) beam profile relative to pit width; (b) triangle beam profile.

For the reading beam focused on the pit edge, the power (intensity) carried by three diffracted orders is illustrated in Fig. 3–11. When the reading spot is focused either on the pit or on the surface, that is, when the edge is not included, output power is the same and pit and surface cannot be distinguished.

In this simplified example, the intensity of the 0-order diffracted light is practically constant; thus a differential detector should be used. The output signal is proportional to the difference of −1- and +1-order diffracted light.

In Fig. 3–12, the output signal, when differential detection is used, is shown for the periodical pattern. Marked points are obtained from Fig. 3–11. The output signal is the convolution of the pit pattern with the reading spot profile.

Usually, four detectors are integrated on one chip (Fig. 3–13). The output signal is then

$$I_s = (I_a + I_d) - (I_b + I_c)$$

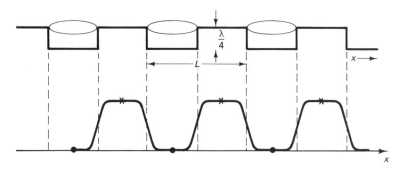

Figure 3–10 Output signal for the periodical pit pattern ($\lambda/4$ pit).

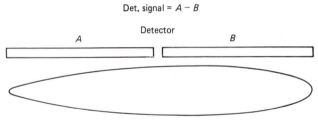

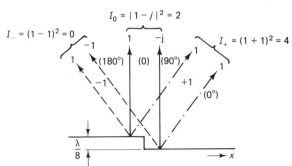

Figure 3–11 Edge diffractor for the $\lambda/8$ pit.

where I_a is the current of detector (segment) A, and similarly for the other symbols. The radial difference signal is expressed by

$$I_{\mathrm{DR}} = (I_a + I_b) - (I_c + I_d)$$

The same detector arrangement can be used for the $\lambda/4$ pits. The output signal is equal to the sum of the four currents.

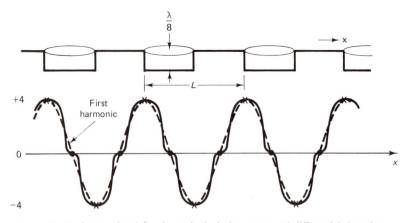

Figure 3–12 Output signal for the periodical pit pattern and differential detection.

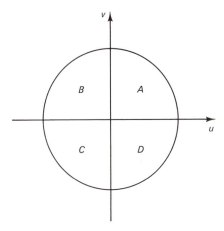

Figure 3–13 Quadrant sensor geometry.

3.3 PLAYER OPTICS (OPTICAL SYSTEM)

The playback optical system is shown in Fig. 3–14 [1]. The light source used is a low-power He-Ne laser tube. The laser tube, which has a nominal output power of 1 mW, is single mode (TEM_{00}) and linearly polarized. The laser beam is initially directed by two mirrors that are adjustable for alignment purposes. The laser beam is then expanded to fill the back of the objective lens by a planoconvex lens. This

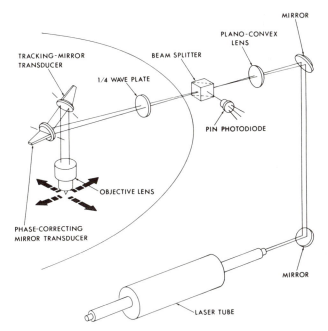

Figure 3–14   Player optics [1].

beam-expanding lens is adjustable along its axis to provide for fine focus of the optical system. The beam is then transmitted through a specially coated beam splitter. The direction of the laser beam polarization is such that most of the beam will pass through the beam splitter. A quarter-wave plate changes the beam polarization from plane to circular. The beam is then directed into the back of the objective by two mirror transducers, which consist of mirrors that can be rotated by piezoelectric or electromagnetic bender motors. These rotations produce a corresponding motion of the read spot on the disc surface. One transducer is used to move the read spot in a radial direction to provide the high-speed tracking corrections required to follow the data track. The other mirror transducer causes the read spot to move in a tangential direction on the videodisc to provide the time-base corrections. These high-speed tracking and time-base corrections are required because of videodisc eccentricity, mechanical vibrations, and so on.

The objective lens which focuses the laser beam to a small spot on the surface of the videodisc has a numerical aperture of 0.45 and an effective focal length of approximately 12 mm. The laser light that is reflected from the surface of the videodisc is collected by the objective lens and returned along substantially the same path that the incoming beam traveled. When the reflected beam passes through the quarter-wave plate, it is changed again to plane-polarized light, but it is polarized at a right angle to the direction of the incoming laser beam. The reflected beam is then reflected by the beam splitter to the PIN photodiode detector, where the optical signal is converted to an electrical signal for processing by the electronics. The use of the plane-polarized laser tube, the specially coated beam splitter, and the quarter-wave plate results in a high-efficiency optical system that minimizes the reflected signal that is fed back into the laser cavity.

When a large flat videodisc area is illuminated, about 10% of light is reflected, 3% returns from the videodisc grooves, and 5.5% returns from the area in between the grooves. These data, naturally, change with the frequency of the signal and with groove size, as well as with the size of channel radii at the spot where the reading from the videodisc is made. But on the basis of these data, an insight into the functioning of the system can be obtained. For this case, at a laser intensity of about 1 mW, the intensity of the signal returned reaching the photodiode is about 0.025 mW, generating a photocurrent of a mean value of about 2×10^{-6} A. The current of the photodiode noise is about 2×10^{-8} A, which yields a signal-to-noise ratio of about 100 : 1.

3.4 MODULATION TRANSFER FUNCTION AND ITS COMPENSATION

The implementation of optical readout requires an optical lens capable of forming a focused spot whose length along the track is less than the spatial wavelength of the signal to be read out. The fundamental resolution limit and readout frequency response depend on the numerical aperture of the focusing lens. Obviously, the resolution

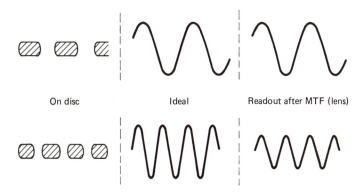

On disc   Ideal   Readout after MTF (lens)

Figure 3–15 Output signal (fundamental component) as function of spatial frequency on the disc.

limit and the highest frequency in the system should be considered together with the system's signal-to-noise ratio.

The amplitude of the readout signal differs with a corresponding spatial frequency on the disc surface (Fig. 3–15). A smaller amplitude is obtained for the higher frequencies. The phase shift is not included in Fig. 3–15.

The focused spot size can be measured in one dimension using what is called a knife-edge scanning technique [2]. Although microscope objectives with numerical apertures approaching unity are available, their actual performance is equivalent to substantially smaller numerical apertures. Figure 3–16 shows the effective one-dimensional focal intensity distribution of 633-nm and 442-nm wavelength for a lens with NA = 0.95 [3,4]. Figure 3–17 presents these data in the form of a one-dimensional modulation transfer function (MTF). These are the MTFs of an incoherent optical system. The broken curve is the modulation transfer of a uniformly illuminated, perfect, diffraction-limited lens with NA = 0.8. These measurements are indicative of the effective numerical aperture of the lens.

Aperture or MTF compensation is a technique for boosting the high-frequency

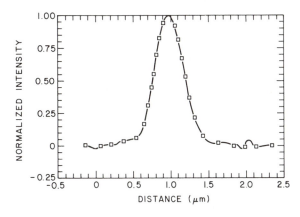

Figure 3–16 One-dimensional focused intensity profile of microscope objective with NA = 0.95 of a wavelength of (a) 442 nm, (b) 633 nm [3,4].

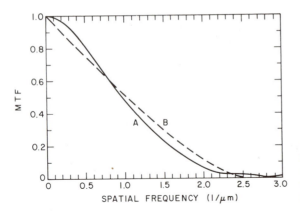

Figure 3–17 Modulation transfer function for intensity profile depicted in Fig. 3–16. Dashed curve corresponds to a uniformly illuminated diffraction-limited lens with NA = 0.8 [3,4].

components of a (recorded) signal without affecting their phase. The technique is a desirable one because the limited resolving power of lenses causes the effective frequency response characteristic of readout signals to have a low-pass characteristic. Even though it is possible to resolve more than 1000 pits per millimeter at the inside radius, the contrast for the fine detail is usually very low. The output signal can be made much sharper by boosting the amplitude of the high-frequency components so that the overall response curve is more nearly flat in the frequency band of interest.

Theoretical analysis of MTF compensation systems, as well as their experimental testing, are frequently reported in literature [5–7]. The best solution suggested so far has been the cosine equalizer. But the cosine curve does not match the MTF curve very well; and in addition, after compensation with the cosine compensator, the total transfer characteristic is radius dependent, so that its compensation also has to be allowed for. This technique is described below, together with a compensator system consisting of an optical feedback loop.

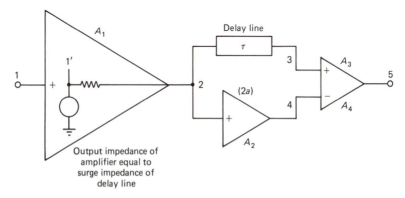

Figure 3–18   Cosine equalizer.

3.4.1 Cosine Equalizer

A block diagram of an aperture compensation circuit [5] is shown in Fig. 3–18. This circuit employs a small section of artificial transmission line which is sending-end terminated but open-circuited at the receiving end. The electrical length of the line is adjusted to correspond to 180° at the frequency for which maximum boost is desired.

A theoretical analysis will be carried out for an unterminated receiving end of the delay line. For this case (Fig. 3–18), assuming a delay line without loss, it is

$$V_1(t) = V_1 \sin \omega t$$

$$V_2(t) = V_2 \cos \omega \tau \sin \omega (t - \tau)$$

$$V_3(t) = V_3 \sin \omega (t - \tau)$$

$$V_4(t) = V_4 \cos \omega \tau \sin \omega (t - \tau)$$

$$V_5(t) = V_5(1 - 2a \cos \omega \tau) \sin \omega (t - \tau)$$

where $2a = V_4/V_3$ and $V = V_2 = V_3$ [8]. Then the normalized envelope of the output signal is

$$A(\omega) = 1 - 2a \cos \omega \tau$$

In Appendix 3.1 another circuit is analyzed. It is an interesting property of open-circuited transmission lines that the signal components at points (2) and (3) are always in phase or exactly out of phase with each other, but the slopes of the amplitude versus frequency curves at the two points are quite different.

In Fig. 3–19, output envelopes, $A(f)$, are shown for a as a parameter. The cosine-wave characteristic would continue indefinitely through the spectrum except for losses in the line and the finite bandwidth of the rest of the circuit.

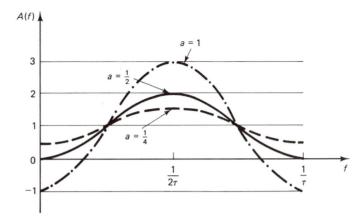

Figure 3–19 Frequency dependence of the compensator transfer function.

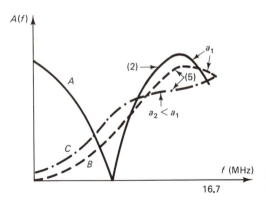

Figure 3–20 Experimental results for the transfer function.

From Fig. 3–19 it can be seen that the circuit should be so adjusted that cancellation (in a differential amplifier) never reaches zero. It can be seen that there is no phase shift in the output signal if

$$0 < a < 1$$

For $a = 0$, the normalized envelope of the output signal is flat. If $a > \frac{1}{2}$, there is a phase shift of 180° for some frequencies. For $a < 0$, the output envelope has a slope similar to the MTF curve for frequencies $0 < f < \frac{1}{2}\tau$.

In Fig. 3–20 experimental results are shown: (a) a spectrum in point (2) in Fig. 3–18, (b) a corresponding spectrum in point (5), and (c) a spectrum in point (5) for a different a ($a_2 < a_1$). The delay line used has $\tau \simeq 30$ ns.

3.4.1.1 Modeled response.
The next pattern (in rectangular form) considered was: two cycles of 7.2 MHz, one cycle of 1.8 MHz, two cycles of 7.2 MHz. The Fourier series is

$$f(t) = \sum_{n}^{\infty} b_n \sin n \left(2\pi \frac{f_1}{2}\right)t \qquad f_1 = 1.8 \text{ MHz}$$

or

$$f(t) = b_1 \sin \omega_{0.9}t + b_2 \sin \omega_{1.8}t + b_3 \sin \omega_{2.7}t + \cdots$$

If $F(\omega)$ is the spectra of the input signal $f(t)$, then the next three cases are of interest:

1. For reference only, the signal that would exist in the case when MTF is flat [or no MTF influence, that is, $H_1(\omega) = 1$] is

$$S_a(\omega) = F(\omega)H_2(\omega)$$

   where $H_2(\omega)$ is the amplitude characteristic of the preamp.

2. The output signal (spectra) following the lens (MTF) and the preamp is

$$S_b(\omega) = S_a(\omega)H_1(\omega) = F(\omega)H_1(\omega)H_2(\omega)$$

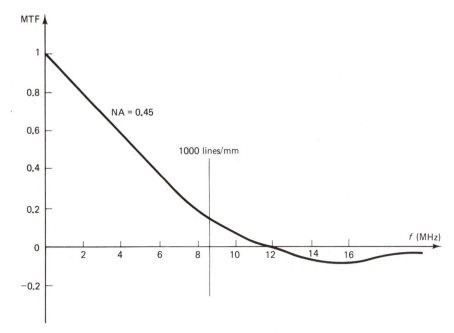

Figure 3–21 Approximated MTF.

3. The output signal (spectra) after MTF compensation is

$$S_c(\omega) = S_b(\omega)H_3(\omega) = F(\omega)H_1(\omega)H_3(\omega)$$

In this simulation, an approximation for the MTF was chosen as shown in Fig. 3–21. For the MTF compensation circuit, a parameter a was chosen such that $a = 0.25$, so $H_3(\omega)$ is

$$H_3(\omega) = 1 - 0.5 \cos \omega\tau$$

and τ has a value which boosts signals at 14.4 MHz ($\tau \simeq 35$ ns).

For the preamp an idealized characteristic was assumed: linear phase (no phase shift) and flat amplitude characteristic. In the first case, the bandwidth of the preamp was 11 MHz and in the second case it was 23.5 MHz. In Fig. 3–22 three corresponding curves are shown when the bandwidth is 11 MHz, while in Fig. 3–23 three corresponding curves are shown when the bandwidth is 23.5 MHz. In Fig. 3–24 the output signal is shown when the bandwidth is 11 MHz and $a = 0.5$. It is obvious that the circuit should be adjusted such that a is less than 0.5, so that cancellation never reaches zero (see Fig. 3–19).

3.4.1.2 Influence of the instant radius on the output signal.   Figure 3–25 is a simplified block diagram for the optical readout; H_1 represents the amplitude characteristic of the lens and H_2, the response of the preamplifier. The amplitude characteristic of the compensating circuitry is H_3 and the overall amplitude character-

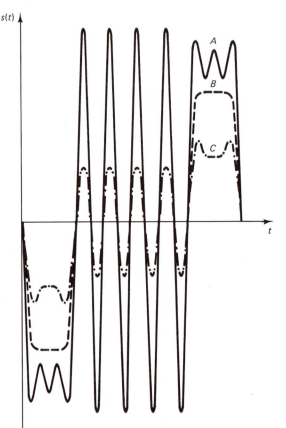

Figure 3–22 Waveforms for the bandwidth of 11 MHz.

istic is H. The model shown in Fig. 3–25 is a general one and holds true for a videodisc as an analog channel medium as well as for a videodisc as a digital channel medium.

Figure 3–26 shows the MTF curve for the actual lens. The horizontal axis is a number of black and white lines of the test pattern grid (per millimeter) and the vertical axis represents a corresponding value of the MTF. It is advantageous, however, to express MTF as a function of frequency. The frequency of the readout signal, f, is

$$f = n(2\pi R)L \tag{3–1}$$

where n = number of rotations of the disc per second
       R = radius at the readout point
       L = number of lines per millimeter

The corresponding scales for the inside radius ($R_1 = 70$ mm) and the outside radius ($R_2 = 140$ mm) are shown in Fig. 3–26b and c, respectively. For any other radius between the foregoing two values the scale would be between the foregoing two scales.

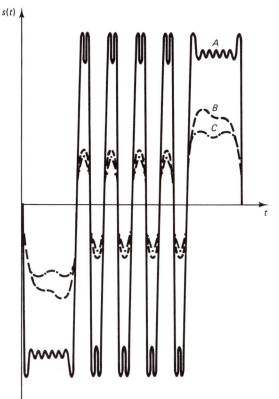

Figure 3–23   Waveforms for the band-width of 23.5 MHz.

The circuit transfer function is

$$H_3(\omega) = 1 - 2a \cos \omega\tau \tag{3-2}$$

Thus (1) for the same lens, the curve representing MTF versus frequency varies with changes in the disc radius (CAV disc), and (2) the transfer function of the compensation circuit does not change with a change in the disc radius. This means that if additional compensation is not made, the total transfer function would depend on the instantaneous radius at which the readout from the disc is performed. Therefore, it is necessary to make a quantitative analysis of the influence of the radius on the overall characteristic of the system.

According to Fig. 3–25, the transfer function of the readout system, $H(\omega)$, is

$$H(\omega) = H_1(\omega)H_2(\omega)H_3(\omega) \tag{3-3}$$

The preamplifier transfer function, $H_2(\omega)$, can be approximated with a transfer function of a low-pass filter. Then it is within the passband B:

$$H(\omega) = H_1(\omega)H_3(\omega) \tag{3-4}$$

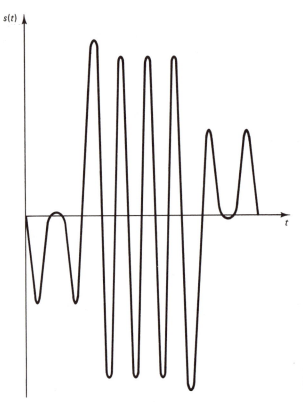

s(t)

t

Figure 3–24   Waveforms for the
bandwidth of 11 MHz and $a = 0.5$.

For the lens with the characteristics shown in Fig. 3–21, the frequency responses
were determined for several values of the parameter a. Figure 3–27 shows a corresponding system characteristic for the inner radius, $R_1 = 70$ mm. It is assumed that a
maximum delay is $\tau = 35$ ns, corresponding to the frequency maximum of 14.4
MHz.

To illustrate the influence of the readout radius R on the transfer function of
the entire system, Fig. 3–28 shows corresponding characteristics of the system during
disc readout at the outside radius $R_2 = 2R_1 = 140$ mm. For comparison, three
curves from Fig. 3–27 are also shown.

The change in the overall frequency response of the system due to changes in

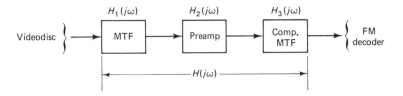

Figure 3–25   Simplified block diagram of the optical readout.

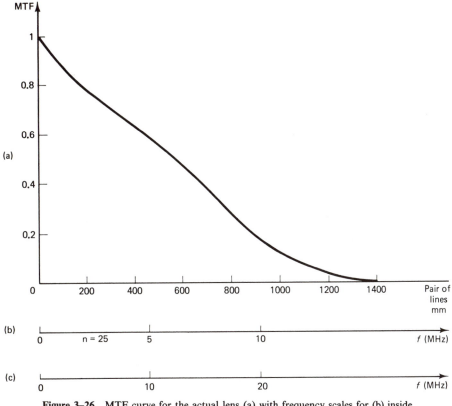

Figure 3–26 MTF curve for the actual lens (a) with frequency scales for (b) inside and (c) outside radius.

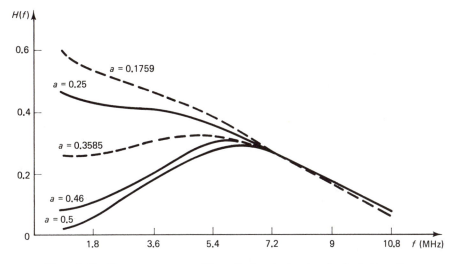

Figure 3–27 Frequency response of the optical readout system for the inside radius.

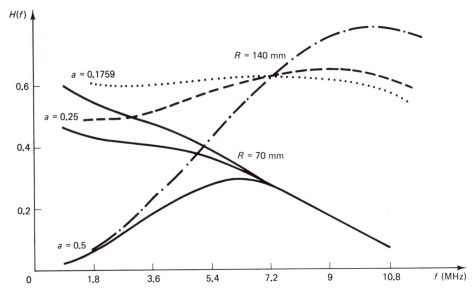

Figure 3–28   Frequency response of the optical readout system for the outside radius.

radius is very evident. The reasons for such strong dependence on radius R are clear.

In Eq. (3–4) H_1 is a transfer function of the readout lens, and H_3 is a transfer function of the compensating circuit. Figure 3–29 shows these responses (for simplicity, H_1 is shown by a straight segment, but this does not change the generality of the

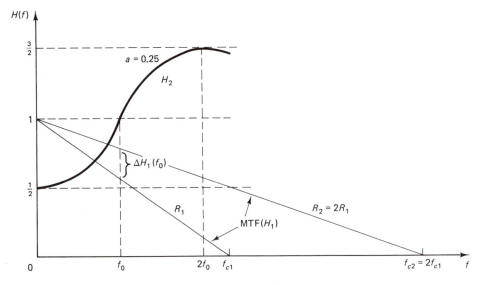

Figure 3–29   Idealized characteristics of the optical readout system.

case). In this example, $H_3(f)$, given for $a = 1$, does not change as a function of the readout radius, whereas H_1 (i.e., MTF) does. If at inside radius R_1, the cutoff frequency is f_{c1}, then at outside radius $R_2 = 2R_1$, the cutoff frequency will be $f_{c2} = 2f_{c1}$. This follows directly from Eq. (3–1).

Since the lens characteristic is dependent on the instantaneous readout radius, whereas the characteristic of the compensation circuit is not, the overall characteristic of the system does change with a change in the radius. The previous considerations refer to the case of constant angular velocity. If the circumferential velocity is constant, the foregoing problem does not exist. In Appendix 3.2, a simple circuit is shown that will compensate for this problem.

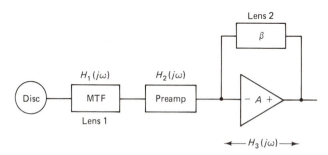

Figure 3–30 Optical feedback for the MTF compensator.

3.4.2 Optical Feedback for the MTF Compensator [9]

This method, although possible theoretically, is impractical to implement with the present level of technology. The idea is very simple: If one lens is bad, use two lenses to make it good! You have to knock a wedge out with a wedge—old Serbian proverb. (You have to fight fire with fire.) The corresponding block diagram is shown in Fig. 3–30, and

$$H_3(j\omega) = \frac{A}{1 + \beta A} \tag{3-5}$$

If

$$\beta = H_1(j\omega) \tag{3-6}$$

the product $H_1 \cdot H_3$ is frequency independent:

$$H_1 \cdot H_3 = \text{const.} \tag{3-7}$$

for frequencies where

$$\beta A \gg 1$$

If the lenses are well matched, this method of compensation is radius independent.

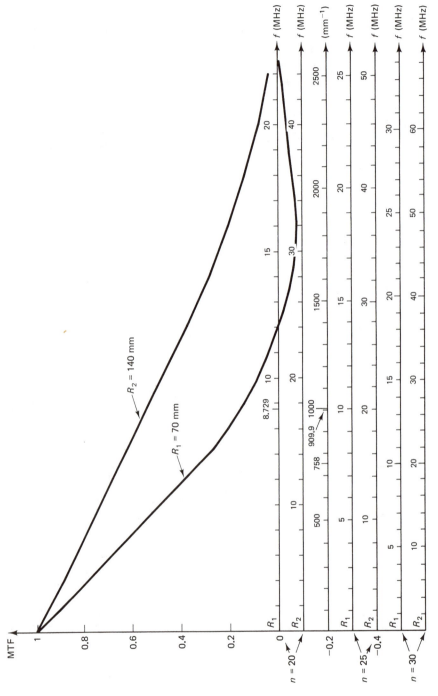

Figure 3-31 MTF for empirical measurement: $n = 20$ rps, $n = 25$ rps, and $n = 30$ rps.

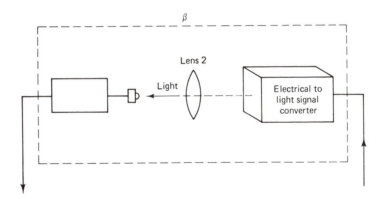

Figure 3–32   Principal solution for the optical feedback.

3.4.3 Comparison: Cosine Equalizer Versus Optical Feedback

Usually, the MTF is given as a function of L (lines/mm or cycles/mm). With n revolutions per second, rescaling to get MTF as a function of frequency can be done using the expression

$$f = n(2\pi R)L$$

This is illustrated in Fig. 3–31. The aperture compensator can change the frequency response of the (reading) system from low-pass (linear or flat) to bandpass. The latter is obviously unwanted. If the aperture compensator is adjusted to compensate for a certain radius, then for other radius the MTF would not be properly compensated. This is because the frequency response of the compensator is radius independent, but MTF is radius dependent for a CAV disc.

    Figure 3–32 shows the principal solution for the β part of the method proposed. It can be seen that realization is impractical with the present level of technology. The main problem is to transfer a one-dimensional electronic signal into a two-dimensional light signal.

    For frequencies where $\beta A \gg 1$, the overall transfer characteristic is frequency and radius independent. But from the stability point of view, the cosine equalizer has no stability problem. On the other hand, in the method proposed, attention should be paid to the stability problem, because the system has feedback. Additionally, when the reading lens (1) is out of focus, the phase shift can be 180° for some frequencies.

3.5 TIME-BASE ERROR

In the case when the center of the disc is shifted by ΔR with respect to the desired center, due to a failure in hole punching, for example, this eccentricity will provoke an undesired modulation of the readout signal. The modulated radius is

$$R(t) = R + \Delta R \cos \omega_d t$$

where $\omega_d = 2\pi f_d$ is the angular frequency of the disc rotation ($f_d = 25$ Hz in Europe, $f_d = 30$ Hz in the United States).

It can be shown that, for example, for a single-tone recorded signal, $s(t) = S_m \sin \omega_s t$, the readout signal would be

$$s*(t) = S_m \sin (\omega_s t + \beta_d \sin \omega_d t)$$

where $\beta_d = (f_s/f_d)(\Delta R/R)$, or

$$s*(t) = S_m \sum_{n=\infty}^{\infty} J_n \beta_d \sin (\omega_s + n\omega_d)t$$

Thus an angular modulation is obtained with ω_s as a carrier. This causes a time-base error, which can be compensated for electronically during reading. In Appendix 3.3, an analysis is made of the influence of disc eccentricity on the signal recorded on a videodisc.

A television picture consists of lines that have been "written" in an exactly defined time interval (63.49 μs for NTSC). Deviations result in a distorted picture and in phase errors that cause color aberrations. In view of this, broadcast TV transmitters are crystal controlled for stability and the receivers are equipped with synchronization circuits to ensure that even with deteriorated signals (e.g., in fringe areas) perfect synchronization is maintained. In a videodisc player, however, we are faced with a number of problems as a result of which the linear speed of the track, as seen by the objective, is not constant. The causes of this difficulty are to be found in deviations of the rotational speed of the motor, imperfections in the disc, and unavoidable tolerances in the centering of the disc on the turntable.

The eccentricity of the track is the main cause of timing errors. Since the player/disc combination allows for a maximum eccentricity of 100 μm, it can be calculated that the resulting timing error

$$\Delta t = \frac{\Delta R}{\omega R} = 10 \ \mu\text{s}$$

where ΔR = eccentricity
$\quad \omega = 2\pi f \ (f = 30$ Hz)
$\quad R$ = inner radius = 55 mm

A maximum timing error of 5 ns can be permitted to ensure satisfactory performance in combination with any TV receiver, so a reduction of 66 dB is required. Higher-frequency timing errors occur, but it has been found that their statistical amplitude decreases at a rate of approximately 12 dB per octave.

To minimize timing errors it is necessary in the first place to keep the rotational speed of the disc as close as possible to 1798.2 rpm. To this end the phase of the line synchronization pulses is compared with that of a crystal-controlled oscillator and the signal derived is fed to the motor-speed control circuit. However, using this method it is not possible to obtain an acceptable reduction in timing errors for frequencies of 30 Hz and above. For effective reduction, a second pivoting mirror, scanning the track tangentially, is used. A signal derived from the burst is used to control the movements of the tangential mirror. This type of time-base error correction makes it possible to connect a player to any type of NTSC television receiver without the need to adapt the flywheel synchronization circuit as required for video tape recorders.

There exists another type of timing error that has nothing to do with the above-mentioned phenomena but originates from imperfections in the mastering of the disc. If corresponding synchronization pulses on adjacent tracks are not positioned on a straight radial line, difficulties may occur with still pictures and other playing modes in which track jumping is involved. For this reason the disc specification requires a track-to-track timing error of no more than ± 25 ns.

3.6 DROPOUT DETECTOR

Monitoring dropouts is very important for almost any application of the videodisc. When a video FM signal is recorded, the usual effect of a disc dropout is one or more missing half-cycles. This would appear as a large reduction in instantaneous frequency and would demodulate as a "blacker than black" video signal. This can be, for example, misunderstood as a sync pulse. A circuit for detecting the missing half-cycles can be implemented using a retriggerable one-shot that is triggered by every positive and negative transition of the FM signal, with duration set to just exceed the maximum half-cycle period. If the Q output of the one-shot falls, a dropout has been detected.

Protection from the dropout influence depends on the particular application. For example, if a video signal is being recorded, a one-line delay is used; the line in which dropout is detected can be substituted in part or in whole for the preceding line. If digital data are being recorded, rewriting or an error correction technique can be used, separately or in combination.

3.7 CONTROL AND SERVO SYSTEMS IN THE PLAYER

To maintain the actual performance of a system (player) to the desired performance creates a control problem. The necessary basic equipment is assembled into a system to perform the desired control function.

The basic optical platform layout is shown in Fig. 3–33. The system utilizes three beams formed from one: a grating is used to form those three beams (Fig. 3–34). The outside beams, E and F, are used for tracking, and the middle beam is used for reading and focusing. After reflection from the disc surface, point P (Fig. 3–33) is imaged into point P' on the surface of the cylindrical lens. Whenever the disc surface is moved vertically at the reading point, the position of point P' will move. This can be used for focus error signal generation.

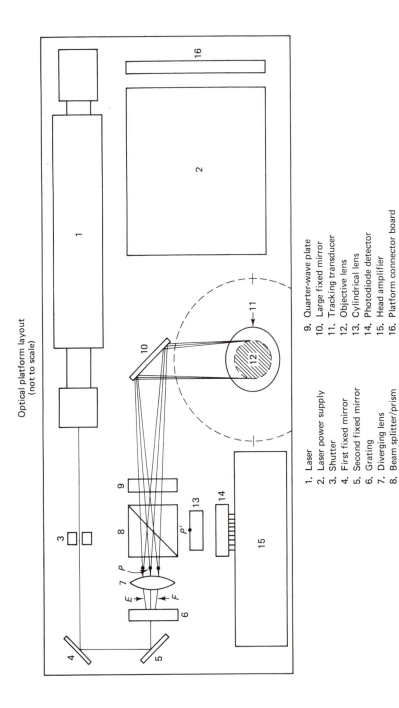

Optical platform layout
(not to scale)

1. Laser
2. Laser power supply
3. Shutter
4. First fixed mirror
5. Second fixed mirror
6. Grating
7. Diverging lens
8. Beam splitter/prism

9. Quarter-wave plate
10. Large fixed mirror
11. Tracking transducer
12. Objective lens
13. Cylindrical lens
14. Photodiode detector
15. Head amplifier
16. Platform connector board

Figure 3-33 Optical platform layout (not to scale).

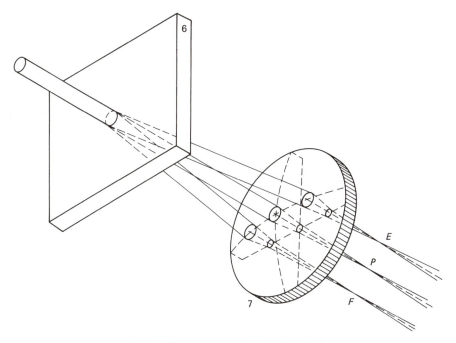

Figure 3–34   Grating: three-beam formation.

Very fine details on the videodisc require very precise control of the optical system so that the reading incident ray will be always within the channel. This cannot be obtained by mechanical systems; the use of electronic control systems is indispensable.

Next, the main servo systems in the player will be considered: spindle, focus, and tracking.

3.7.1 Spindle Servo

The videodisc is set on an axis driven by an electrical motor (with, e.g., 1798.2 rpm). Variations in rotating speed are reflected in phase and frequency change in the signal that is read. This is why the number of rotations should be stabilized to a tolerance of 0.1% and a mean of error of 0. There are several ways of doing this, as discussed below.

3.7.1.1 System with tachometer.   On the axis that serves to provide rotation of the disc, an auxiliary disc with an LED is mounted; a phototransistor is used as a sensor. In the phase-locked loop (PLL) circuit, a comparison of the detected frequency from the tachometer (at the output of phototransistor) with the frequency from the reference quartz oscillator is made. The error signal passes through suitable

RC circuits to the receiver controlling the number of rotations of the operational electromotor.

3.7.1.2 System with recorded reference signal.

In this system, the read tachometric signal is replaced by the signal from the disc itself. At the mastering stage, a periodical signal of definite frequency is added to the information that is to be stored on the disc. At the reproduction stage, this signal is detected, separated by a passband filter, and applied as input to a phase detector. Another input signal is generated by the quartz oscillator. The error signal is processed further similar to the method used in the previous case.

Figure 3–35 shows two such signals at the output of their respective detectors. The ratio of their frequencies is about 10 (pilot end color subcarrier). Both signals are sinusoidal and their shapes on the figure are the consequence of the use of the particular detectors.

In Fig. 3–36 a functional block diagram for the spindle servo is shown; a standard video signal (NTSC) is assumed to be recorded on the disc. The burst signal of horizontal rate is separated from the video signal off the disc, before time-base corrector (TBC). Signals obtained from the disc and from the reference color subcarrier oscillator are compared in the phase detector and an error signal is generated. This signal is used as a horizontal error to the TBC and as a control signal for the spindle motor.

The compensation for the player servo loop has the same form as for the mastering spindle servo. However, the loop bandwidth must be less than 20 Hz. Otherwise, there would be destructive interference with the TBC.

When the spindle is activated from rest, there are no horizontal sync pulses since no video is being read. Thus an error signal must be generated to bring the spindle speed to approximately the correct range. This is accomplished using a bridge circuit to measure the motor back electromotive force (EMF) (a voltage proportional

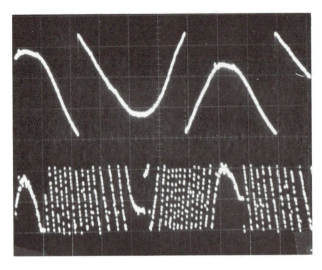

Figure 3–35   Reference signals at the outputs of detectors.

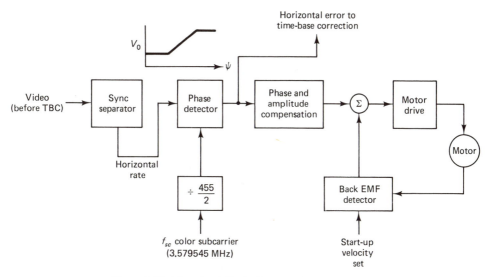

Figure 3–36 Functional block diagram of the spindle servo.

to angular velocity) to form a low gain loop to start the motor. When video is read, the high-gain PLL is dominant, and it tacks over.

3.7.2 Focus Servo

A thick videodisc with a thickness of 1.1 mm rotating in free air at 30 Hz (NTSC TV signal) is in general stable but not critically damped. A stabilization system can be used to restrict the action of transverse waves in the frequency range 0 to 15 Hz and to compensate for the "umbrella" form deflection of the disc under its own weight.

Stabilization is achieved in the player by means of an air bearing, which is created by the centrifugal action of the rotating disc on its surrounding air. A plate with a central hole is positioned in the vicinity of the rotating disc's surface, which creates an air bearing having both the damping and stiffness desired. Disc stabilization obtained in this way is by no means sufficient to guarantee that the distance between the disc and the read objective has the accuracy required.

To read the microscopically small information details on the disc, an objective lens is required with a large numerical aperture and thus with a small depth of focus. With NA = 0.4 the maximum out-of-focus allowance is 2 μm. In view of the tolerances in the disc and in player construction, this accuracy can be realised only by means of a servo control system, including a moving read objective that can follow the undulations of the disc.

As can be seen in Fig. 3–37, the objective lens is mounted in a system similar to that of a loudspeaker voice coil and thus operates according to electrodynamic principles. Depending on the direction and magnitude of the current through the

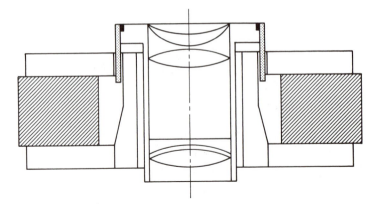

Figure 3–37   Principle sketch of move objective.

coil, it makes well-controlled vertical movements in order to follow the information on the disc.

To understand the way in which the control signal is obtained, it is necessary first to have a closer look at the photodiode on which the light beam reflected by the disc falls. In fact, this diode consists of two segments, *E* and *F*, with a third segment between the other two composed of four quadrants *A*, *B*, *C*, and *D* (Fig. 3–38). The light beam focused on the central segment will normally create a circular spot, and thus all four quadrants will receive equal amounts of light. The sum of the electrical signals over these four quadrants is the radio-frequency (RF) signal or modulated video information.

However, on its way to the diode, the reflected beam passes a cylindrical (astigmatic) lens. As a result, the light spot on the diode, which normally is circular, will suffer elliptical deformation if the information plane of the disc changes its distance from the objective. Depending on whether the disc is too close or too far, the main axis of this elliptical spot has positions perpendicular to each other. The amounts of light impinging on the four quadrants are no longer equal and a difference signal can be derived. After amplification and processing, this signal can be used to correct the position of the objective lens so that the distance with respect to the disc is correct.

It appears that the largest amplitudes in vertical movement of the disc occur at the speed of rotation (30 Hz) and decrease rapidly for higher frequencies. Disc specification sets the maximum value at 10*g*, but the player can cope with accelerations as high as 14*g* and still remain below an out-of-focus value of 2 μm. Figure 3–39 illustrates the servo open-loop gain versus frequency.

When the encoded surface of the disc is in focus, point *P* in Figs. 3–33 and 3–40 is imaged in point *P′* on the surface of the cylindrical lens—13 in Fig. 3–33. When the disc is far, point *P* will be imaged before beam reaches the lens surface—point *M′* in Fig. 3–40. When the disc is near, point *P* is imaged in point *N′*—in the cylindric lens (Fig. 3–40).

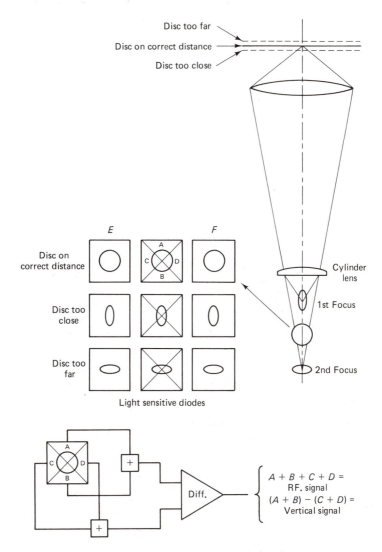

Figure 3–38 Principle of astigmatic focusing.

When the beam is focused on the astigmatic lens surface (Fig. 3–41), a circular spot is created on the detector surface. When the beam is focused before or after that point, elliptic spots are created on the detector surface. Thus three courses can easily be distinguished by the four-segment photodetector. (Fig. 3–41 is on p. 118.)

The signal "in focus" is generated when the focus servo (Fig. 3–42) is activated. This signal does not mean that the system is really in focus, just that it works on focusing. Loop gain and focus offset can be adjusted separately. (Fig. 3–42 is on p. 119.)

The focus servo in the player has much the same form as the focus servo

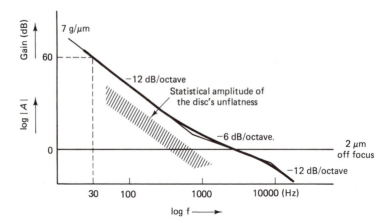

Figure 3–39 Focus servo open-loop gain as a function of frequency (A is the disc amplitude).

used for mastering. The differences are due mainly to specification changes. The NA (numerical aperture) of the reading lens in the player is 0.45, which allows the servo an error of ± 1 μm from true focus. The other difference comes from the disc specification, which allows a vertical acceleration of $10g$. Thus the ratio of open-loop gains at 30 Hz for the two servos would be

$$\frac{A_{\text{player}}}{A_{\text{mastering}}} = \frac{10g \times 0.1\ \mu\text{m}}{1g \times 1\ \mu\text{m}} = 1$$

Therefore, the open-loop response curve for both servos would be the same. The focus error signal is shown in Fig. 3–43. Normally, the system operates in the linear region. The error signal could also be generated from the auxiliary beam specifically generated from the main (reading) beam for this purpose.

3.7.3 Tracking (Radial) Servo

The information on the disc is contained in a spiral groove that is read from the inside to the outside. For this purpose the read objective and other optical elements are mounted on a sled, driven by a small dc motor, moving radially under the disc. With an average track pitch of 1.6 μm and a speed of rotation of the disc of 1800 rpm, this means an average linear speed of the sled of 3 mm/min. The scanning light beam has to remain focused on the track with a radial accuracy of 0.1 μm, a requirement that cannot be met by a purely mechanical guidance system. By varying the speed of the drive motor by incorporating it in a servo control system, certain slow corrections are possible. However, to cope with the effects caused by eccentricity of the spinning disc, additional measures are required and use is made of a pivoting mirror by means of which the light spot can move radially over the disc. This mirror is mounted in an assembly resembling the construction of a moving coil, where the

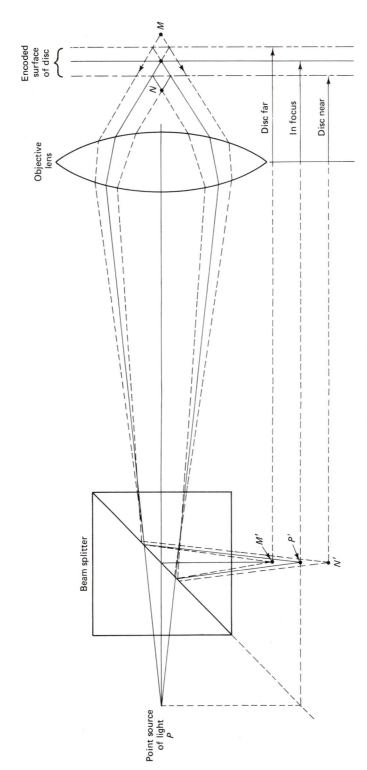

Figure 3–40 Principle of focusing.

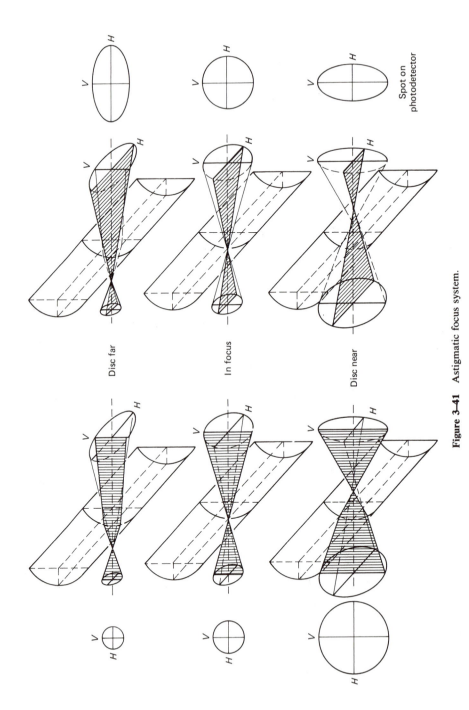

Spot on photodetector

Disc far

In focus

Disc near

Figure 3–41 Astigmatic focus system.

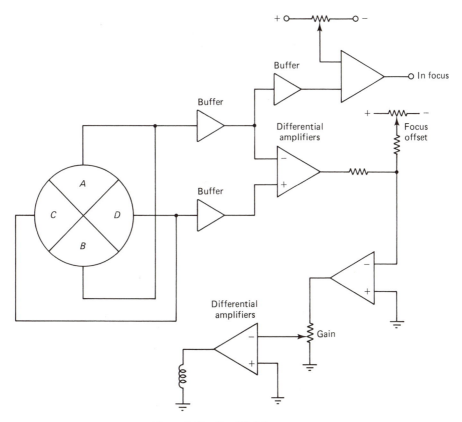

Figure 3–42   Simplified focus servo.

coil is part of the radial servo control circuit. The information in the track can only be read out optically, and thus the deviation of the beam from the center of the track can also only be measured optically. For this purpose two auxiliary beams of light are used which are slightly displaced from the center line of the track, in opposite directions, so that they are partly on and partly alongside the track (Fig. 3–44).

After reflection at the disc each of the two auxiliary beams falls on its own photodiode (E and F in Fig. 3–44) and the average current through the diodes depends on the amount of reflected light, thus on the position of the auxiliary beam relative to the track. In fact, the difference signal of the two diodes after amplification passes a low-pass filter, with a cutoff frequency of, for example, 20 kHz, and is then used as an error signal in the control system.

Unroundness of the track and unavoidable tolerances in player construction may result in a total eccentricity of 100 μm for the combination, which means that at 30 Hz, which is the rotational frequency, a reduction of at least 60 dB is required to keep the read spot within a tenth of a micrometer from the track. The amplitudes

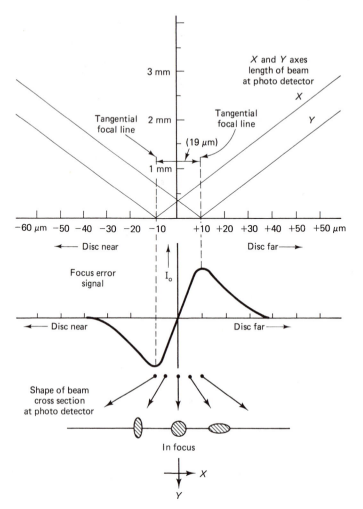

Figure 3–43 Focus error signal.

of radial tracking deviations at higher frequencies again decrease rapidly, and Fig. 3–45 shows the relation between open-loop gain and frequency.

A simplified block diagram of the tracking servo is shown in Fig. 3–46. The tracking mirror transducer has a transfer function which is closely approximated by a pure second-order system up to about 10 kHz. It has a transfer function defined by

$$H_T(s) = \frac{\omega_n^2}{s^2 + 2\xi\omega_n^s + \omega_n^2}$$

For a typical transducer, $\xi \simeq 0.35$ and $\omega_n = 250$ rad/s.

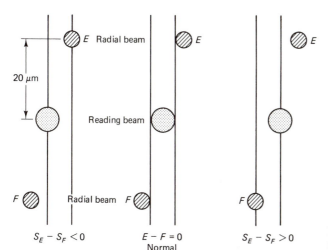

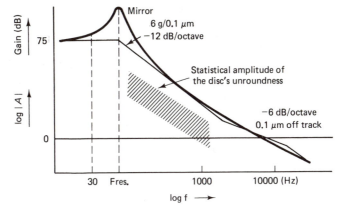

Figure 3–44   Principle of track following.

A compensator of the form

$$H_c(s) = K \frac{s+z}{s+p} \qquad z < p$$

would provide a stable loop that would very closely match the specification limits. If the average position of the mirror deviates from its zero position, the average current is used to control the slide motor for correction of the position of the sled.

When special effects are realized, for instance slowed-down registration requiring "a jump back," the radial system is used on the former channel of the light beam for reading. The jump of the light beam cannot, for this purpose, be assured by a mechanical system for the tangential moving of a videodisc. The respective shifting back of the light beam by a track (1.6 μm) is obtained by incorporating an additional

Figure 3–45   Radial servo open-loop gain as a function of frequency.

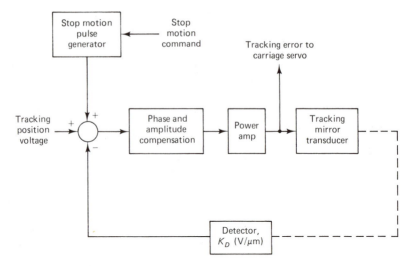

Figure 3–46   Tracking servo: functional block diagram.

pulse into the feedback loop of this system. Having in mind the dynamic characteristics of the system, several compensational circuits are included into the feedback loop so that the light beam can be controlled to remain in the channel after the "jump."

3.7.4. Tangential (Time-Base Correction) Servo

Correction of timing errors can be achieved by a second pivoting mirror positioned at right angles to the radial mirror (Fig. 3–46), with an open-loop gain as shown in Fig. 3–47. This can be confirmed by electronic compensation [charge-coupled device

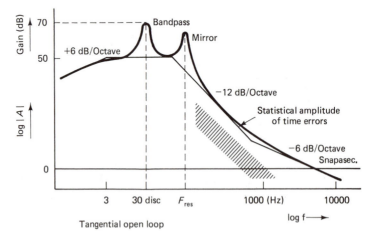

Figure 3–47   Tangential open loop gain.

(CCD), for example]. The choice of a low resonant frequency for the mirror suspension is such that it approximates a ballistic galvanometer, which has desirable characteristics, especially for the application of a pulse used for operating features. Because the stiffness of the tangential mirror is not compatible with the required gain at 30 Hz, it is necessary to simulate a resonant peak by means of an active filter and thereby achieve higher gain reduction at the disc's rotation frequency. Since the repetition frequency of the burst that is usually used to obtain an error signal is that of the line frequency (approximately 16 kHz), the bandwidth of the system is limited in practice to approximately 1.6 kHz in order to obtain sufficient stability.

3.7.5 Carriage Servo

The carriage servo-drive motor must operate over a large dynamic range, from high speed in the scan mode to very low speed in play or slow motion. In low-speed operation the servo loop must prevent stalling or intermittent motion.

In the play mode the input from the tracking servo (Fig. 3–48) contains a dc component when the read beams are off center. The dc component will alter the carriage velocity to bring the beam back to the center of the track. The control commands provide the appropriate input to the summer to generate the necessary carriage velocity.

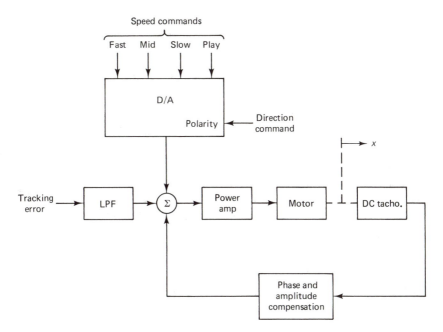

Figure 3–48 Carriage servo.

3.7.6 Signal-to-Noise Ratio of the Error Signal

In spite of the small detector area used for the extraction of the error signals, the signal-to-noise ratio (SNR) of these signals is better than 40 dB because of their small bandwidth [10]. The different noise sources, such as shot noise of the photons, preamplified noise, and noise due to the surface roughness of the disc, are included in this number.

The amplitude modulation of the light source is another source of noise, but it can be canceled out by electronic means. A distortion of the light distribution across the beam causes imbalance of the detectors. This imbalance effect can determine the reliability of the error signals.

3.8 OPTICAL TRANSMISSIVE DISCS

Optical readout of the transmissive discs is accomplished by means of a 1-mW He-Ne laser beam focused to a spot of the order of 1 μm. Transmitted light through the disc is modulated by the recorded pattern and detected with photocells. The original electrical signal is then restored [11].

A precise positioning of the information with respect to the focal point of the light beam is necessary. This light beam acts as a "reading stylus" without any mechanical contact, and its position has to be controlled very accurately to obtain a good readout:

- ± 2 μm in the vertical direction
- ± 0.1 μm in the radial direction

For vertical tracking (focusing) two approaches have been stated:

1. Aerodynamic stabilization
2. Vertical servoing of the readout objective

The focus servo for the floppy (transmissive) optical disc is in principal of the same type as that for reflective discs; an astigmatic sensor is used.

Aerodynamic stabilization is well adapted to flexible discs. The two systems that are significant in this type of stabilization are discussed next.

3.8.1 U-Shaped Stabilizer

This stabilizer (known as a Thomson stabilizer) consists of a U-shaped guide in which the rotating disc is introduced (Fig. 3–49). When the disc is rotating, air is forced between the upper and lower parts of the guide, creating a combination of aerodynamic forces. This rather simple and efficient solution has as its major drawback

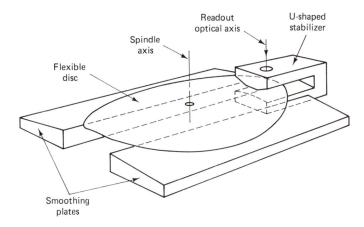

Figure 3–49   U-shaped stabilizer.

the fact that it is necessary to use a narrow gap between the two guides to obtain the desired stabilization for all discs (typically 50 μm in addition to the disc thickness). In a dusty ambiance this may cause severe damage to the disc surface when large dust particles are passing through the gap. For this reason the U-shaped stabilizer has been abandoned, and new solutions have been proposed to overcome this drawback.

3.8.2 Stabilization by Bernouilli Effect

In place of a passive aerodynamic stabilizer, an active aerodynamic stabilizer can be used (Fig. 3–50). A properly distributed airflow is blown onto the rotating disc through a nozzle placed under the readout objective. According to the Bernouilli effect, the laminar flow created between the nozzle and the disc provides local stabilization of the disc surface. This type of stabilization is more efficient than that obtained with the U-shaped stabilizer. Furthermore, dust particles, which would otherwise pass under the readout lens and affect the quality of the video signal, are blown away by the airflow. This solution, however, demands the addition of a compact air pump.

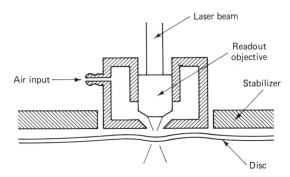

Figure 3–50   Bernouilli stabilizer.

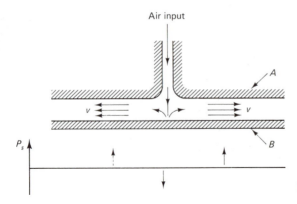

Figure 3–51 Bernouilli's formula.

Figure 3–51 illustrates Bernouilli's formula. When a fluid with a specific mass ρ is submitted to a laminar flow of velocity v, static pressure P_s inherent to the fluid on walls A and B is related to parameters ρ and v as follows:

$$P_s + \tfrac{1}{2}\rho v^2 = \text{const.} \qquad \text{(Bernouilli's formula)}$$

Fast flow thus means a local relative depression which can be put to use to "tame" mobile wall *A* to fixed wall *B*. Figure 3–50 shows the adaptation of this principle to the specific case of a flexible videodisc spinning under its stabilizer.

For the purpose described, a nozzle is placed under the microscope objective. Both laser beam and an air jet under pressure flow through it. The air flows out between the nozzle and the disc in a space whose straight section increases as the distance from the hole gets longer. A depression is thus created in the vicinity of the hole, which yields excellent stabilization of the disc at the exact spot where readout takes place.

Furthermore, the disc, at this very spot, takes a slightly convex shape, which lends itself perfectly to "ironing" out the unwanted microwrinkles. The characteristics of the airflow are as follows:

- Overpressure (above atmospheric pressure): 600 g/cm^2
- Flow: 6 liters/min

In these conditions thickness of flow is close to 100 μm and original air speed is close to 1 Mach. This means that any dust particle that would tend to creep in between the disc and its stabilizer would immediately be blown away. In fact, emptying an ashtray on the spinning disc proves totally harmless to the disc as well as to the recorded information.

In principle, radial tracking control for transmissive discs can be obtained the same way as for reflective discs: using asymmetry in the light beam intensity distribution which occurs when the light spot is not well centered on the track ("single light spot"), or using two or three tracking spots.

3.9 FILM-BASED VIDEODISCS

Playback of the (video) information stored in the spiral track of the film disc can be accomplished by use of either a laser or an incandescent light source [12]. Use of the laser provides a higher bandwidth and an increased signal-to-noise ratio, at the expense of the requirement for more sophisticated optical and electronic servo systems. Also, the introduction of the laser playback operated at its higher bandwidth introduces coherent noise at the detector, resulting from residual random film structure. However, careful optical alignment can produce a high-quality video signal compatible with a standard television display. The laser recorder can be used to replay a previously exposed and developed film disc.

Many systems applications require a less critical configuration involving an incandescent light source rather than the coherent laser source. Figure 3–52 shows a standard optomechanical design appropriate for use with an incandescent source. Playback device development has been, in general, directed toward maximum ease and maximum stability in use for such applications. Basic to this capability is the

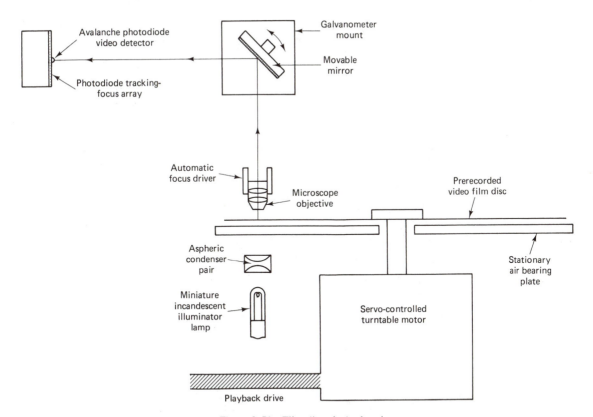

Figure 3–52 Film disc playback unit.

inherent flexibility of the noncontact optical read method, the utilization of an incandescent source for illumination, and careful exploitation of the advantages of flying the disc in the floppy disc mode. This approach has two major advantages: (1) an inexpensive white light bulb may be used as a source of illumination, and (2) the image of the area of interest on the disc surface is presented at increased magnification for subsequent processing. It is an easier task to extract the desired information from the appropriate video track in this magnified image than to extract it at the disc surface itself, at least in an ideal noise-free case.

A principal element in the optical system of the video player is the microscopic objective being used. This objective must be capable of transferring information recorded on the film to the video detectors with sufficient modulation and minimum phase inversion. The MTF declines toward the center of the recorded disc; this occurs since the disc spins at a constant rate, requiring signal elements to be recorded closer together at the shorter disc radii. In addition, the MTF also declines as the lens moves from optimum focus for any fixed radius, which is why automatic focusing is essential to preserve signal quality.

The least expensive light source available to provide illumination for the system is a simple miniature incandescent light bulb; for higher intensities, a 25-W quartz halogen lamp has also been used. By using an aspheric condenser system in the playback system, incandescence levels of approximately 1 mW/cm² have been achieved at the detector. By the nature of blackbody radiation, the output of the lamps is peaked in the same spectral region as that in which silicon video detectors are most sensitive (0.9 μm). As an alternative to the direct illumination of the light bulb, simple fiber optics can be used to illuminate the disc surface, thereby simplifying the playback configuration.

Given a perfect videodisc recording, the signal-to-noise ratio in playback is determined by the intensity of the light source used. For an incandescent source, a calculation of the expected signal-to-noise ratio must include consideration of the temperature of the filament, the transfer efficiency of the condensing optics, the maximum density of the film, the transmission efficiency of the viewing optics and its associated mirrors, and the sensitivity of the detector averaged over the emission wavelengths of the filaments. Higher values of signal-to-noise ratio can be obtained by turning up the intensity of the light source, but at a considerable penalty in terms of the lifetime of the bulb.

For the video detector, one can use a masked photomultiplier tube, a channeltron, or a silicon photodiode operated in the avalanche mode. The most important criterion is that a suitably small active detector area be used. Successful video operation has been attained by limiting detector areas to a maximum of 0.1 mm² with typical magnifications of 20 to 100 times, at numerical apertures of 0.4 to 0.7.

There are many tracks imaged into the vicinity of the video detector, on the order of 10 to 100 cm⁻¹. The appropriate track is maintained on the detector against excursions due to mechanical tolerances by using an oscillating mirror in the optical path between the microscope objective and the detector. The servomechanism may be either a wideband closed-loop configuration or may be run open loop, tuned to the basic rotation rate of the disc. The position of the desired track is sensed by a linear detector array (in the simplest case a split photodiode), and an error signal is generated to drive the mirror to the appropriate angular position.

3.10 OPTICAL READOUT OF THE GROOVED DISC

The primary factors in considering the optical readout of the grooved disc are possible contact-free playback and high-speed measurement techniques suitable for quality and process control evaluation of master discs, replicas, and intermediate parts [13]. The primary technical difficulty is that the geometric features of interest on the capacitive disc surface are substantially smaller than the wavelength of the reading light. The highest-frequency signal elements recorded on the disc can be as small as 0.2 μm.

As seen in Chapter 2, capacitive discs can be mastered electromechanically, optically, or by means of an electron beam. The finished master disc produced by any of these methods is highly reflective. The electromechanical master is cut directly into the metal substrate, while the optical and electron-beam-exposed masters are overcoated with evaporated gold prior to an electrochemical plating operation. Diffractive scattering results when, upon reflection, the smooth wavefronts of an incident beam are altered in phase by the pattern of signal features on the surface of the master disc. Since there is negligible absorption of light by the features, an optical reader must respond to the angular variations of the scattered light. Similarly, typical optical disc, both reflective and transmissive types, also act as phase objects.

In the case of disc systems designed specifically for optical readout, there is diffraction not only from the leading and trailing edges of the signal pits, but from the sides of the pits as well. The comparatively wide flat areas between tracks on these discs influences the zero diffraction order so that the effective reference surface is quite close to the actual surface. There is thus a strong in-phase component in the cyclic interference between the zero order and the two first orders having directional components along the track, a component easily sensed with a single-centered detector. In effect, the flat surface of the optical disc can be viewed as providing a stable interferometric reference mirror. Resolution of the surface features up to the fundamental limit specified by the cutoff period ($= \lambda/2\text{NA}$) is obtained directly with a single-centered detector for these optical videodiscs.

A capacitive disc format, which acts as an edgeless diffraction grating in an optical reader, requires the use of special phase-contrast methods for full resolution. The differential detector provides a form of differential phase contrast that is simple to implement. Although differential or off-centered detectors can also be used successfully for typical optical videodisc formats, the reasons for such a choice involve considerations such as pit depth and surface noise, rather than resolution. Typical readers for optical videodiscs use a numerical aperture of about 0.4 and a He-Ne laser with $\lambda = 0.633$ μm to detect minimum signal periods near 1.8 μm. Choosing NA = 0.8

and $\lambda = 0.442$ μm provides resolution of equal quality for signal periods more than three times smaller.

In general, to go from optical readout in principle to optical readout in practice, for capacitive discs, requires the solution to a series of design problems, similar to the design problems for typical optical discs. Optical setup is similar. Tracking and focusing control are basically the same for optical and capacitative discs, because the nature of the reading process is the same.

REFERENCES

1. K. D. Broadbent, "A review of the MCA DiscoVision system," 115th SMPTE Tech. Conf. Equip. Exhibit, Los Angeles, Apr. 26, 1974.

2. A. H. Firester,, M. E. Heller, and P. Sleng, "Knife-edge scanning measurements of subwavelength focused light beams," *Appl. Opt.*, Vol. 16, 1977, p. 1971.

3. A. H. Firester, C. B. Corroll, I. Gorog, M. E. Heller, J. P. Russell, and W. C. Stewart, "Optical readout of the RCA videodisc," *RCA Rev.*, Vol 39, No. 3, Sept. 1978, pp. 392–426.

4. Firester A. H. et al., "Optical recording techniques for RCA videodisc," *RCA Rev.*, Vol. 39, No. 3, Sept. 1978, pp. 427–472.

5. J. W. Wentworth, *Color Television Engineering*, McGraw-Hill, New York, 1955, pp. 305–307.

6. J. Isailović, "MTF compensator for digital videodisc recording systems," Optica '80, Budapest, Nov. 18–21, 1980.

7. J. Isailović, "The influence of the radius on the output signal in the videodisc system with the optical read out" (in Serbian), Proc. 25th ETAN Conf., Mostar, June 8–12, 1981, Vol. II, pp. 333–340.

8. R. A. Chipman, *Theory and Problems of Transmission Lines*, Schaum's Outline Series, McGraw-Hill, New York, 1968, p. 157.

9. J. Isailović, "Optic feedback for the MTF compensator," ICO-12, GRAC, Sept. 1981.

10. J. J. M. Brost and G. Bouwhis, "Position sensing in videodisc readout," *Appl. Opt.*, Vol. 17, No. 13, July 1, 1978, pp. 2013–2021.

11. G. Brossaud, "Le vidéodisque," *Rev. Tech. Thompson-CSF*, Vol. 10, No. 4, Dec. 1978, pp. 655–679.

12. J. A. Jerome, "Systems application of film-based optical videodiscs," *Inf. Display*, Vol. 12, No. 2, 1976, pp. 6–11.

13. I. Gorog, "Videodisc optics," *RCA Rev.*, Vol. 33, No. 3, Sept. 1978, pp. 389–391.

APPENDIX 3.1 COSINE EQUILIZER

Figure A3.1–1a shows one practical realization, for which a simplified block diagram is shown in Fig. A3.1–1b. Assuming a loss-less delay line, it is

$$v_i(t) = V_1 \sin \omega t \qquad\qquad\qquad\qquad \text{(A3.1–1)}$$

$$v_o(t) = -k_1 \sin \omega(t - \tau) + k_2 a_1[\sin \omega t + \sin \omega(t - 2\tau)] \qquad \text{(A3.1–2)}$$

This can be simplified as

$$v_o(t) = -k_1(1 - 2a \cos \omega\tau) \sin \omega(t - \tau) \qquad\qquad \text{(A3.1–3)}$$

or

$$v_o(t) = -k_1 A(\omega) \sin \omega(t - \tau) \qquad\qquad\qquad \text{(A3.1–4)}$$

Thus the normalized envelope of the output signal is also of the form

$$A(\omega) = 1 - k \cos \omega\tau \qquad\qquad\qquad\qquad \text{(A3.1–5)}$$

where k is a constant defined by amplifier in the circuit, and τ is a line delay.

Equation (A3.1–5) can be obtained by direct analysis of the circuit in Fig. A3.1–1a. Suppose that the transistor's output impedance is ∞, and that the load

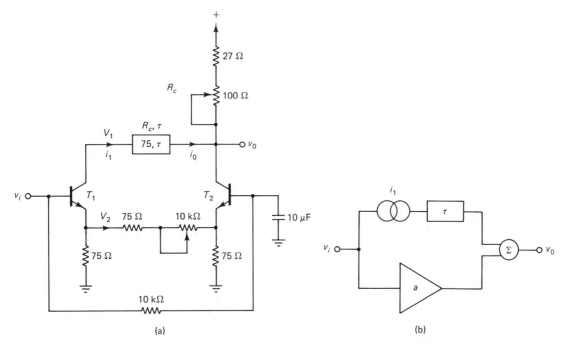

Figure A3.1–1 Cosine equalizer: (a) Practical realization and (b) its simplified block diagram.

resister of transistor T_2, R_c, is equal to the characteristic impedance of the delay line. The parameters of the delay line are:

- Characteristic impedance:

$$Z_c = R_c$$

- Characteristic transfer function:

$$\Gamma_c = A_c + jB_c = 0 + j\omega\tau$$

$$A_c = \text{amplitude function}$$

$$B_c = \text{phase function}$$

Equations for the delay line are (all variables are complex)

$$V_1 = a_{11}V_o + a_{12}I_o$$

$$I_1 = a_{21}V_o + a_{22}I_o$$

where

$$a_{11} = a_{22} = \cosh \Gamma_c = \cos \omega\tau$$

$$a_{12} = Z_c \sinh \Gamma_c = jR_c \sin \omega\tau$$

$$a_{21} = \frac{1}{Z_c} \sinh \Gamma_c = -j \frac{\sin \omega\tau}{R_c}$$

The output signal V_o can be obtained by superposition of the signal V_o' obtained through the delay line (from T_1) and the signal V_o'' through the transistor T_2:

$$V_o = V_o' + V_o''$$

where

$$V_o' = V_1 e^{-j\omega\tau} \qquad V_1 = A_1 V_i = K_1 R_c V_i$$

$$A_1 = \text{voltage gain of the stage with } T_1$$

$$V_o'' = A_2 V_2 \qquad V_2 = A_{12} V_i \approx V_i$$

$$A_{12} = \text{gain of the emitter follower } (T_1)$$

$$A_2 = K_1 Z$$

$$Z = R_c \parallel Z_{in2} \qquad Z_{in2} = \text{input impedance of the delay line}$$

$$Z_{in2} = -jR_c \cot \omega\tau \qquad \text{or}$$

$$Z = \frac{R_c}{1 + j \tan \omega\tau}$$

Then

$$V_o = (A_1 e^{-j\omega\tau} + A_2) V_i = A V_i$$

where

$$A = A_1 e^{-j\omega\tau} + K_1 Z$$

or

$$A = R_c (K_2 + K_1 \cos \omega\tau) e^{-j\omega\tau}$$

for

$$K_1 = -K_2 = K$$

It follows that

$$A = K R_c (1 - \cos \omega\tau)$$

APPENDIX 3.2 LC RESONANT CIRCUIT OF THE RADIUS-COMPENSATED MTF COMPENSATOR FOR THE VIDEO FM SIGNAL

The MTF of the (reading) lens acts like a radius-dependent low-pass filter to the FM signal (CAV mode). Although the video signal can be recovered without compensation, there may be an amplitude imbalance between the chroma and luma signals at the inner radius. This can be seen from the simplified spectrum shown in Fig. A3.2–1.

If the magnitude of J_0 represents the luma signal and J_1 the chroma signal, it is obvious that the ratio J_1/J_0 is larger at the inner radius than at the outer radius. This will then demodulate as an increased chroma signal (the luma signal is normalized).

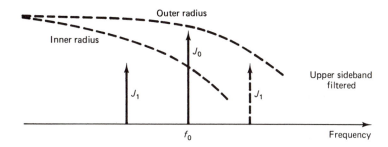

Figure A3.2–1 Influence of the MTF on the FM carrier and first chroma sidebands for the inner and outer radii.

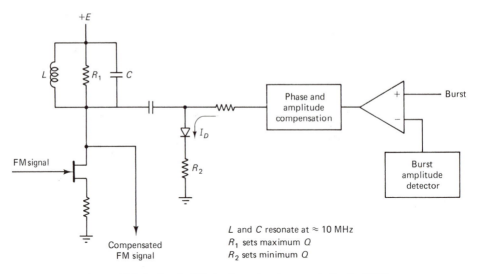

L and C resonate at $\approx$ 10 MHz
R_1 sets maximum Q
R_2 sets minimum Q

Figure A3.2–2   Simple LC circuit used to compensate for the MTF.

A circuit that will compensate for this problem uses deviation of the amplitude of the demodulated burst signal to generate an error signal to control a frequency-selective circuit. A simplified circuit is shown in Fig. A3.2–2.

The current I_0 through diode D is proportional to the amplitude of the burst signal from the disc. The diode resistance for the ac is inversely proportional to I_D. Thus the larger the detected burst, the smaller will be the equivalent Q factor of the LC circuit. This is illustrated in Fig. A3.2–3. The radius influence will be compensated, but same nonlinearity will be introduced because of the relatively large frequency deviation (typically, frequency deviates from 8.1 MHz to 9.3 MHz).

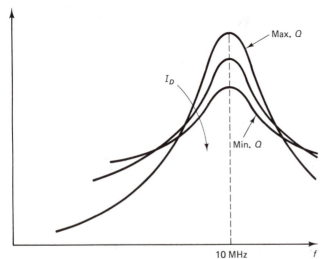

Figure A3.2–3   Resonant curves for different I_D.

APPENDIX 3.3 THE INFLUENCE OF ECCENTRICITY ON THE SIGNAL RECORDED ON A VIDEODISC

The influence of disc eccentricity on the signal recorded on a videodisc is examined in this appendix. The following cases are considered: sine wave, sine FM, pulse train, pulse FM, and sum signal.

A3.3–1 SINE WAVE

Suppose that we have a sine signal (Fig. A3.3–1a):

$$s(t) = S_m \sin \omega_s t \qquad (A3.3\text{–}1)$$

In the disc the sine signal is (a linear model is assumed)

$$s(d) = D_m \sin 2\pi\alpha d = D_m \sin \Omega_s d \qquad (A3.3\text{–}2)$$

where $\alpha = 1/L$ is the space frequency (Fig. A3.3–1b). If the writing velocity is constant, we have

$$L = v_o T_s = R\omega_o T_s \qquad (A3.3\text{–}3)$$

where R is the current (writing) radius and $T_s = 1/f_s$.

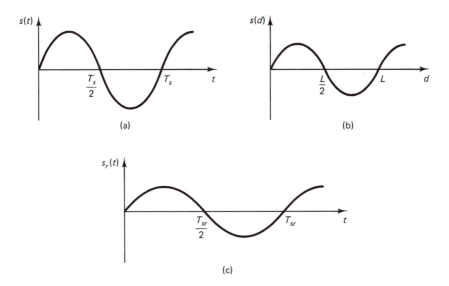

Figure A3.3–1   Sine-wave recording: (a) input; (b) recorded; (c) output signal.

A3.3–1a The Influence of Disc Velocity

If the reading velocity is not equal to the writing velocity, the output signal is (Fig. A3.3–1c)

$$s_r(t) = S_{rm} \sin\left(\frac{\omega_d}{\omega_o} \omega_s t\right) \tag{A3.3–4}$$

because

$$L = v_r T_{sr} = R\omega_d T_{sr} \tag{A3.3–5}$$

and

$$\omega_r = \frac{2\pi}{T_{sr}} = \frac{\omega_d}{\omega_o} \omega_s \tag{A3.3–6}$$

According to Eq. (A3.3–6), the signal frequency after reading from the disc is

$$f_{sr} = f_s \frac{f_d}{f_o} \tag{A3.3–7}$$

or

$$\frac{f_{sr}}{f_s} = \frac{f_d}{f_o} \tag{A3.3–7'}$$

Table A3.3–1 contains some results.

TABLE A3.3–1

		28	29	29.5	30	30.5	31
	f_d	28	29	29.5	30	30.5	31
	f_{sr}/f_s	0.93	0.9667	0.983	1	1.0167	1.033
f_{sr} (MHz)	$f_s = 0.4$ MHz	0.373	0.3867	0.393	0.4	0.407	0.413
	$f_s = 8$ MHz	7.467	7.733	7.87	8	8.13	8.267

A3.3–1b The Influence of Disc Eccentricity

Suppose that

$$\omega_o = \omega_d = \text{const.} \tag{A3.3–8}$$

that is,

$$L = R\omega_d T_s \tag{A3.3–3'}$$

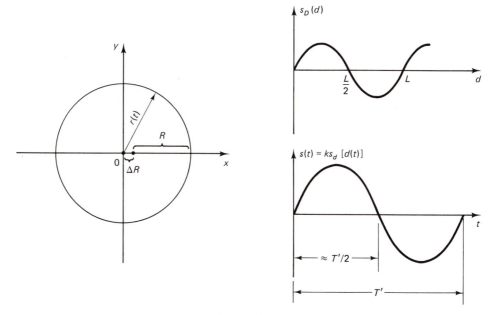

Figure A3.3–2   Disc eccentricity.

The output signal, $s(t)$, is

$$s(t) = ks[d(t)] = S_m \sin \phi(t) \tag{A3.3–9}$$

According to Fig. A3.3–2, the modulated radius, $r(t)$, is

$$r(t) = R + \Delta R \cos \omega_d t \tag{A3.3–10}$$

The tangential velocity, $v(t)$, is

$$v(t) = \omega_d r(t) = \omega_d R(1 + \delta \cos \omega_d t) \tag{A3.3–11}$$

where $\delta = \Delta R / R$. For one period, T', it is

$$v(t) \approx \text{const.} \tag{A3.3–12}$$

so that

$$L = \int_t^{t+T'} v(t)\, dt = T' v(t) \tag{A3.3–13}$$

and

$$T' = \frac{T_s}{1 + \delta \cos \omega_d t} \tag{A3.3–14}$$

or

$$f' = f_s(1 + \delta \cos \omega_d t) \tag{A3.3–15}$$

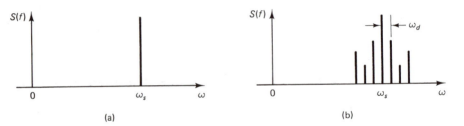

Figure A3.3–3   (a) Input and (b) output spectra.

Finally, according to Eqs. (A3.3–15) and (A3.3–9), the output signal is

$$s(t) = S_m \sin (\omega_s t + \beta_d \sin \omega_d t) \tag{A3.3–16}$$

where

$$\beta_d = \frac{\omega_s}{\omega_o} \delta = \frac{f_s}{f_o} \frac{\Delta R}{R} \tag{A3.3–17}$$

or

$$s(t) = S_m \sum_{n=-\infty}^{\infty} J_n(\beta_d) \sin (\omega_s + n\omega_d t) \tag{A3.3–18}$$

Figure A3.3–3 illustrates input and output spectra.

A3.3–2 FM SIGNAL

The frequency-modulated carrier is given by

$$s(t) = S_m \cos (\omega_c t + \beta \sin \omega_m t) \tag{A3.3–19}$$

We may express it in the form

$$s(t) = S_m \sum_{n=-\infty}^{\infty} J_n(\beta) \cos (\omega_c + n\omega_m)t \tag{A3.3–20}$$

In the disc, the signal is function of the distance d:

$$f(d) = ks[t(d)] \tag{A3.3–21}$$

During the writing process, the tangential velocity is

$$v = \omega_d R \tag{A3.3–22}$$

and

$$d = \int_0^t v(t) \, dt = \omega_d R t \tag{A3.3–23}$$

so

$$f(d) = D \cos{(2\pi\alpha_c d + \beta \sin{2\pi\alpha_m d})} \qquad (A3.3-24)$$

where α_c and α_m are spatial frequencies.

If during the reading process the tangential velocity is constant [Eq. (13.3–22)], then after substitution,

$$d = \frac{L_m}{T_m} t = \frac{L_c}{T_c} t = \omega_d Rt \qquad (A3.3-25)$$

and the output signal is

$$s_r(t) = k_1 f[d(t)] = S_{rm} \cos{(\omega_c t + \beta \sin{\omega_m t})} \qquad (A3.3-26)$$

That means that the output signal has no distortion. In case disc eccentricity exists, the tangential velocity is given by Eq. (A3.3–11). The distance d is

$$d = d(t) = R\omega_d(t + \tau_d \sin{\omega_d t}) \qquad (A3.3-27)$$

where

$$\tau_d = \frac{\delta}{\omega_d} = \frac{1}{\omega_d} \frac{\Delta R}{R} \qquad (A3.3-28)$$

The output signal is

$$s_r(t) = k_1 f(d) = k_1 f[d(t)] \qquad (A3.3-29)$$

or

$$s_r(t) = S_{rm} \cos{[\omega_c t + \beta_{dc} \sin{\omega_d t} + \beta \sin{(\omega_m t + \beta_{dm} \sin{\omega_{dt}})}]} \qquad (A3.3-30)$$

where

$$\beta_{dc} = \frac{\omega_c}{\omega_d} \delta = \frac{f_c}{f_d} \frac{\Delta R}{R} \qquad (A3.3-31)$$

and

$$\beta_{dm} = \frac{\omega_m}{\omega_d} \delta = \frac{f_m}{f_d} \frac{\Delta R}{R} \qquad (A3.3-32)$$

For example, for $f_c = 8$ MHz, $f_m = 3.6$ MHz, $f_d = 30$ Hz, and $\Delta R = 2 \times 10^{-3}$ in.:

R (in.)	β_{dc}	β_{dm}
6	89	40
3	178	80

Starting from Eq. (A3.3–30), we can write

$$s_r(t) = S_{rm} \operatorname{Re}\{e^{j\omega_c t} e^{j\beta_{dc} \sin{\omega_d t}} e^{j\beta \sin{(\omega_m t + \beta_{dm} \sin{\omega_{dt}})}}\} \qquad (A3.3-33)$$

Using the identity

$$e^{j\beta \sin \omega t} = \sum_{n=-\infty}^{\infty} J_n(\beta)e^{jn\omega t} \qquad \text{(A3.3–34)}$$

we have

$$s_r(t) = S_{rm} \sum_{p=-\infty}^{\infty} \sum_{n=-\infty}^{\infty} \sum_{q=-\infty}^{\infty} J_p(\beta_{dc})J_n(\beta)J_q(p\beta_{dm})$$

$$\times \cos[\omega_c + n\omega_m + (p+q)\omega_d]t \qquad \text{(A3.3–35)}$$

or

$$s_r(t) = S_{rm} \sum_{n=-\infty}^{\infty} \sum_{k=-\infty}^{\infty} J_n(\beta)J_k(\beta_{dc} + \beta_{dm}) \cos (\omega_c$$

$$+ n\omega_m + k\omega d)t \qquad \text{(A3.3–36)}$$

Figure A3.3–4 illustrates input and output spectra. For $\Delta R = 0$, is $\beta_{dc} = \beta_{dm} = 0$ and Eqs. (A3.3–35) and (A3.3–36) become equivalent with Eq. (A3.3–20).

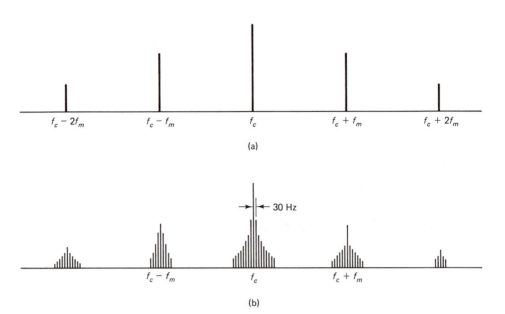

Figure A3.3–4   Eccentricity influence on the recorded FM signal: (a) input, (b) output spectra.

A3.3-3 PULSE SIGNAL

Consider an infinite periodic pulse train as shown in Fig. A3.3–5 of width τ_s, period T_s, and constant amplitude A_s. The exponential form of the Fourier series is given by [1, p. 23]

$$s(t) = \frac{A_s \tau_s}{T_s} \sum_{n=-\infty}^{\infty} \frac{\sin (n\omega_s \tau_s/2)}{n\omega_s \tau_s/2} e^{jn\omega_s t} \qquad (\text{A3.3–37})$$

The time of occurrence of the positive and negative steps is given by [1, p. 535–536]

$$\omega_s \left(t + \frac{\tau_s}{2} \right) = 2n\pi \qquad (\text{A3.3–38})$$

$$\omega_s \left(t - \frac{\tau_s}{2} \right) = 2n\pi \qquad (\text{A3.3–39})$$

If the writing velocity is constant, we have

$$L_s = R\omega_o T_s \qquad (\text{A3.3–40})$$

$$l_s = R\omega_o \tau_s \qquad (\text{A3.3–41})$$

$$d = \int_0^t v(t)\, dt = R\omega_o t \qquad (\text{A3.3–42})$$

so

$$\omega_s \tau_s = 2\pi \alpha_s l_s = \Omega_s l_s \qquad (\text{A3.3–43})$$

$$\omega_s t = 2\pi \alpha_s d = \Omega_s d \qquad (\text{A3.3–44})$$

$$\frac{\tau_s}{T_s} = \frac{l_s}{L_s} \qquad (\text{A3.3–45})$$

In the space domain the signal is given by (Fig. A3.3–5b)

$$\phi(d) = ks[t(d)] = \frac{D l_s}{L_s} \sum_{k=-\infty}^{\infty} \frac{\sin [k\Omega_s (l_s/2)]}{k\Omega_s (l_s/2)} e^{jn\Omega_s d} \qquad (\text{A3.3–46})$$

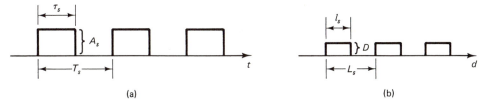

(a)   (b)

Figure A3.3–5   Pulse signal recording: (a) input, (b) recorded pulse train.

The positions of occurrence of the steps are given by

$$\Omega_s\left(d + \frac{l_s}{2}\right) = 2n\pi \tag{A3.3-47}$$

$$\Omega_s\left(d - \frac{l_s}{2}\right) = 2n\pi \tag{A3.3-48}$$

A3.3–3a The Influence of Disc Velocity

If the reading velocity, v, is not equal to the writing velocity, we can write for the output signal

$$L_s = R\omega_d T' \tag{A3.3-49}$$

$$l_s = R\omega_d \tau' \tag{A3.3-50}$$

$$d = R\omega_d t \tag{A3.3-51}$$

or

$$\Omega_s l_s = \omega_s \frac{\tau}{2} = \omega' \frac{\tau'}{2} \tag{A3.3-52}$$

$$\Omega_s d = \omega' t = \frac{\omega_d}{\omega_o}\omega_s t \tag{A3.3-53}$$

The output signal, $s_r(t)$, is

$$s_r(t) = k_1\phi[d(t)] = \frac{A_r T_s}{T_s}\sum_{k=-\infty}^{\infty}\frac{\sin[k\omega_s(\tau/2)]}{k\omega_s(\tau/2)}e^{jn\omega' t} \tag{A3.3-54}$$

where

$$\omega' = \frac{\omega_d}{\omega_o}\omega_s = \frac{f_d}{f_o}\omega_s \tag{A3.3-55}$$

$$\frac{f'}{f_s} = \frac{f_d}{f_o} \tag{A3.3-55'}$$

Equation (A3.3–55) is equivalent to Eq. (A3.3–7) and Table A3.3–1 is valid for this case.

The time of occurrence of the steps is given by

$$\omega'\left(t + \frac{\tau'}{2}\right) = 2n\pi \tag{A3.3-56}$$

$$\omega'\left(t - \frac{\tau'}{2}\right) = 2n\pi \tag{A3.3-57}$$

A3.3–3b The Influence of Disc Eccentricity

Suppose that

$$\omega_0 = \omega_d = \text{const.} \tag{A3.3-58}$$

and tangential velocity, $v(t)$, is given by Eq. (A3.3–11); then distance d is given by Eq. (A3.3–27). For one period, T', it is

$$L_s \simeq T' R \omega_d (1 + \delta \cos \omega_d t) \tag{A3.3-59}$$

$$l_s \simeq \tau' R \omega_d (1 + \delta \cos \omega_d t) \tag{A3.3-60}$$

According to Eqs. (A3.3–47) and (A3.3–48), the time of occurrence of the steps is given by

$$\omega_s \left[\left(t + \frac{\tau}{2} \right) + \tau_d \sin \omega_d t \right] = 2n\pi \tag{A3.3-61}$$

$$\omega_s \left[\left(t - \frac{\tau}{2} \right) + \tau_d \sin \omega_d t \right] = 2n\pi \tag{A3.3-62}$$

Panter suggests a procedure for evaluating the spectral components of a pulse train for this case. The spectrum is given by

$$F(t) = k_1 \phi[d(t)] = \frac{A_r}{2\pi j} \sum_{k=-\infty}^{\infty} \frac{1}{k} \left\{ e^{j[k\omega_s(t_s/2) + k\omega_s \tau_d \sin \omega_d t]} \right.$$

$$\left. - e^{-j[k\omega_s(\tau_S/2) - k\omega_s \tau_d \sin \omega_d t]} \right\} e^{jk\omega_s t} \tag{A3.3-63}$$

This equation can be simplified to

$$F(t) = A_r \frac{\omega_s \tau_s}{2\pi} + A_r \frac{\omega_s \tau_s}{\pi} \sum_{k=1}^{\infty} \frac{\sin [k\omega_s(\tau_S/2)]}{k\omega_s \tau_S/2} (J_0(k\omega_s \tau_d) \cos k\omega_s t$$

$$+ \sum_{n=1}^{\infty} J_n(k\omega_s \tau_d) \{ \cos [(k\omega_s + n\omega_d)t] + (-1)^n \cos (k\omega_s + n\omega_d)t \}) \tag{A3.3-64}$$

Evidently, each of the pulse-repetition frequency harmonics is phase modulated, the maximum deviation being $k\omega_s \tau_d$. The amplitudes of the pulse-repetition frequency harmonics and their sidebands decrease with k according to the term

$$\frac{\sin [k\omega_s(\tau_s/2)]}{k\omega_s(\tau_s/2)}$$

A3.3–4 PULSE FM SIGNAL

With modulation, the time of occurrence of the rising and falling edges of the resulting
pulse train is given by

$$\omega_c\left(t + \frac{\tau_c}{2}\right) + \beta \sin \omega_m t = 2n\pi \qquad \text{(A3.3–65)}$$

$$\omega_c\left(t - \frac{\tau_c}{2}\right) + \beta \sin \omega_m t = 2n\pi \qquad \text{(A3.3–66)}$$

The expression for the modulation pulse train is

$$F(t) = A\frac{\tau_c}{T_c}$$

$$+ 2A\frac{\tau_c}{T_c} \sum_{k=1}^{\infty} \sum_{n=-\infty}^{\infty} \frac{\sin\left[k\omega_c(\tau_c/2)\right]}{k\omega_c(\tau_c/2)} J_n(k\beta) \cos\left(k\omega_c + n\omega_m t\right) \qquad \text{(A3.3–67)}$$

If the writing velocity is constant, we have (Fig. A3.3–5) the relation given by Eq.
(A3.3–40) ÷ Eq. (A3.3–42).
In the space domain signal is given by

$$\phi(d) = kF[t(d)]$$

$$= D\frac{l_c}{L_c}$$

$$+ 2D\frac{l_c}{L_c} \sum_{k=1}^{\infty} \sum_{n=-\infty}^{\infty} \frac{\sin\left[k\Omega_c(l_c/2)\right]}{k\Omega_c(l_c/2)} J_n(k\beta) \cos\left(k\Omega_c + n\Omega_m\right)t \qquad \text{(A3.3–68)}$$

The positions of occurrence of the steps are given by

$$\Omega_c\left(d + \frac{l_c}{2}\right) + \beta \sin \Omega_m d = 2n\pi \qquad \text{(A3.3–69)}$$

$$\Omega_c\left(d - \frac{l_c}{2}\right) + \beta \sin \Omega_m d = 2n\pi \qquad \text{(A3.3–70)}$$

If during the reading process the tangential velocity is constant, after substitution,

$$d = \frac{L_m}{T_m}t = \frac{L_c}{T_c}t = \omega_d Rt = \frac{l_c}{\tau_c}t \qquad \text{(A3.3–71)}$$

where $\omega_d \equiv \omega_o$, the output signal is

$$F_r(t) \equiv F(t) \qquad \text{(A3.3–72)}$$

and it has no distortion.

In the case where disc eccentricity exists, it is

$$L_c \approx v(t)T_c' = T_c'R\omega_d(1 + \delta\cos\omega_d t) = T_c R\omega_d \qquad \text{(A3.3–73)}$$

$$l_c \approx v(t)\tau_c' = \tau_c R\omega_d \qquad \text{(A3.3–74)}$$

$$L_m \approx v(t)T_m' = T_m R\omega_d \qquad \text{(A3.3–75)}$$

The time of occurrence of the steps is given by

$$\omega_c\left[\left(t + \frac{\tau_c}{2}\right) + \tau_d\sin\omega_d t\right] + \beta\sin\omega_m(t + \tau_d\sin\omega_m t) = 2n\pi \qquad \text{(A3.3–76)}$$

$$\omega_c\left[\left(t - \frac{\tau_c}{2}\right) + \tau_d\sin\omega_d t\right] + \beta\sin\omega_m(t + \tau_d\sin\omega_m t) = 2n\pi \qquad \text{(A3.3–77)}$$

For this case the output signal is

$$F_r(t) = k_1\phi[d(t)] = S_r\frac{\tau_c}{T_c} + 2S_r\frac{\tau_c}{T_c}\sum_{k=1}^{\infty}\sum_{n,p,q=\infty}^{\infty}\frac{\sin[k(\tau_c/2)T_c]}{k(\tau_c/2)T_c}$$
$$\times J_n(k\beta)J_p(k\beta_{dc})J_q(n\beta_{dm})\cos[k\omega_c + n\omega_m + (p+q)\omega_d]t \qquad \text{(A3.3–78)}$$

or

$$F_r(t) = S_r\frac{\tau_c}{T_c} + 2S_r\frac{\tau_c}{T_c}\sum_{k=1}^{\infty}\sum_{n,p=-\infty}^{\infty}\frac{\sin[k(\tau_c/2)T_c]}{k(\tau_c/2)T_c}$$
$$\times J_n(k\beta)J_p(k_{\beta dc} + n\beta_{dm})\cos(k\omega_c + n\omega_m + p\omega_d)t \qquad \text{(A3.3–79)}$$

Figure A3.3–6 illustrates output spectrum for this case.

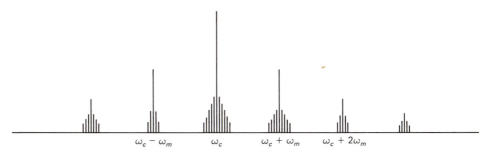

Figure A3.3–6 Eccentricity influence on the pulse FM signal: output spectrum.

A3.3–5 THE CASE WHEN PILOT IS ADDED

Suppose that we have the sum signal, $s(t)$:

$$s(t) = S_c\sin(\omega_c t + \beta\sin\omega_m t) + S_p\sin\omega_p t \qquad \text{(A3.3–80)}$$

where $S_p \ll S_c$.

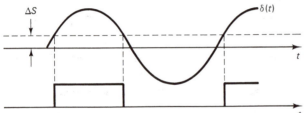

ΔS

$\delta(t)$

t

t

The position of occurrence of zero-crossing points is given by

$$\sin \frac{1}{2}\left(\omega_c t + \beta \sin \omega_m t + \frac{S_p}{S_c} \sin \omega_p t\right)$$

(A3.3–81)

$$\times \cos \frac{1}{2}\left(\omega_c t + \beta \sin \omega_m t - \frac{S_p}{S_c} \sin \omega_p t\right) = 0$$

If a hard limiter has an offset ΔS, the position of occurrence of steps is given by (Fig. A3.3–7)

$$\sin (\omega_c t + \beta \sin \omega_m t) = \frac{\Delta S}{S_c} - \frac{S_p}{S_c} \sin \omega_p t \qquad \text{(A3.3–82)}$$

Taking the approximation

$$\arcsin (x + \Delta x) \approx \arcsin x + \frac{\Delta x}{\sqrt{1 - x^2}} \qquad \text{(A3.3–83)}$$

we have that the edges of the resulting pulse train are given by

$$\omega_c \left(t + \frac{\tau}{2}\right) + \beta \sin \omega_m t + \epsilon \sin \omega_p t = 2n\pi \qquad \text{(A3.3–84)}$$

$$\omega_c \left(t - \frac{t}{2}\right) + \beta \sin \omega_m t - \epsilon \sin \omega_p t = 2n\pi \qquad \text{(A3.3–85)}$$

where

$$\epsilon = \frac{1}{\sqrt{1 - (\Delta S / S_c)^2}} \frac{S_p}{S_c} \qquad \text{(A3.3–86)}$$

Equations (A3.3–84) and (A3.3–85) are useful for the case where the sum signal is composed from a symmetrical triangle wave-frequency modulated and from a sine wave with significantly less amplitude. In that case ϵ is not given by Eq. (A3.3–86). Also, this equation can be used for the case of fixed time delay Δt, by substituting for t in Eq. (A3.3–85) the expression $t - \Delta t$. This case will not be considered here. The expression for the resulting pulse train is in the following form:

$$F(t) = \frac{A}{2\pi j} \sum_{k=-\infty}^{\infty} \frac{1}{k} \{e^{j[k\omega_c(\tau/2) + k\beta \sin \omega_m t + k\epsilon \sin \omega_p t]}$$

$$- e^{-j[k\omega_c(\tau/2) - k\beta \sin \omega_m t + k\epsilon \sin \omega_p t]}\} e^{jk\omega_c t} \quad \text{(A3.3–87)}$$

Using the identity given by Eq. (A3.3–84), the following expressions will be obtained:

$$F(t) = A \frac{\omega_c \tau}{2\pi} + \frac{A\epsilon}{\pi} \sin \omega_p t + \frac{A}{\pi} \sum_{k=1}^{\infty} \sum_{n=-\infty}^{\infty} \sum_{l=-\infty}^{\infty} \frac{1}{k} J_n(k\beta) J_l(k\epsilon)$$

$$\times \{\sin [(k\omega_c + n\omega_m + l\omega_p)t + k\omega_c(\tau/2)]$$
$$- \sin [(k\omega_c + n\omega_m - l\omega_p)t - k\omega_c(\tau/2)]\} \quad \text{(A3.3–88)}$$

or

$$F(t) = \frac{A\omega_c \tau}{2\pi} + \frac{A\epsilon}{\pi} \sin \omega_p t + \frac{2A}{\pi} \sum_{k=1}^{\infty} \sum_{l=-\infty}^{\infty} \frac{1}{k} J_l(k\epsilon) \sin [l\omega_p t$$

$$+ k\omega_c(\tau/2)] \times \{J_0(k\beta) \cos k\omega_c t + \sum_{n=1}^{\infty} J_n(k\beta)[\cos(k\omega_c$$

$$+ n\omega_m)t + (-1)^n \cos (k\omega_c - n\omega_m)t]\} \quad \text{(A3.3–89)}$$

Now we can get components:

$$s_{\omega_p}(t) = \frac{A\epsilon}{\pi} \sin \omega_p t \quad \text{(A3.3–90)}$$

For $k = 1$, $n = -1$, and $x = 0$, from Eq. (A3.3–89),

$$s_{\omega_c \pm \omega_m}(t) = \frac{2A}{\pi} J_1(\beta) J_0(\epsilon) \sin [k\omega_c(\tau/2)]$$

$$\times [\cos (\omega_c + \omega_m)t - \cos (\omega_c - \omega_m)t] \quad \text{(A3.3–91)}$$

For $k = 1$, $n = -1$, and $1 = \pm 1$, from Eq. (A3.3–88),

$$s_{\omega_c - \omega_m \pm \omega_p}(t) = \frac{2A}{\pi} J_1(\beta) J_1(\epsilon) \cos \omega_c(\tau/2)$$

$$\times [\sin (\omega_c - \omega_m + \omega_p)t - \sin (\omega_c - \omega_m - \omega_p)t] \quad \text{(A3.3–92)}$$

For $k = 1$ and $1 = 0$, from Eq. (A3.3–89),

$$s_{\omega_c}(t) = \frac{2A}{\pi} J_0(\epsilon) J_0(\beta) \sin [\omega_c(\tau/2)] \cos \omega_c t \quad \text{(A3.3–93)}$$

For $k = 1$, $n = 0$, and $1 = \pm 1$, from Eq. (A3.3–88),

$$s_{\omega_c \pm \omega_p}(t) = \frac{2A}{\pi} J_0(\beta) J_1(\epsilon) \cos \omega_c(\tau/2)$$

$$[\sin (\omega_c + \omega_p)t - \sin (\omega_c - \omega_p)t] \quad \text{(A3.3–94)}$$

Next, the ratios are

$$\left| \frac{S_{\omega_c - \omega_n}}{S_{\omega_p}} \right| = \frac{2}{\epsilon} J_1(\beta) J_0(\epsilon) \sin \left[\omega_c(\tau/2) \right] \tag{A3.3-95}$$

$$\left| \frac{S_{\omega_c - \omega_m \pm \omega_p}}{S_{\omega_c - \omega_m}} \right| = \frac{J_1(\epsilon)}{J_0(\epsilon)} \, \mathrm{ctg} \left[\omega_c(\tau/2) \right] \tag{A3.3-96}$$

$$\left| \frac{S_{\omega_c \pm \omega_p}}{S_{\omega_c}} \right| = \frac{J_1(\epsilon)}{J_0(\epsilon)} \, \mathrm{ctg} \left[\pi(\tau/T_c) \right] \tag{A3.3-97}$$

If the writing velocity is constant, in the space domain, the signal is given by

$$\phi(d) = D \frac{l_c}{L_c} + \frac{D\epsilon}{\pi} \sin \Omega_p d + \frac{D}{\pi} \sum_{k=1}^{\infty} \sum_{n,l=-\infty}^{\infty} \frac{1}{k} J_n(k\beta) J_l(k\epsilon)$$

$$\times \left\{ \sin \left[(k\Omega_c + n\Omega_m + l\Omega_p)d + k\frac{\Omega_c l_c}{2} \right] \right.$$

$$\left. - \sin \left[(k\Omega_c + n\Omega_m - l\Omega_p)d - k\Omega_c l_c/2 \right] \right\} \tag{A3.3-98}$$

or

$$\phi(d) = D \frac{l_c}{L_c} + \frac{D\epsilon}{k} \sin \Omega_p d$$

$$+ \frac{2D}{\pi} \sum_{k=1}^{\infty} \sum_{l=-\infty}^{\infty} \frac{1}{k} J_l(k\epsilon) \sin (l\Omega_p d + k\Omega_c l_c/2)$$

$$\times \left\{ J_0(k\beta) \cos k\Omega_c d + \sum_{n=1}^{\infty} J_n(k\beta) \left[\cos (k\Omega_c + n\Omega_m)d \right. \right.$$

$$\left. \left. + (-1)^n \cos (k\Omega_c - n\Omega_m)d \right] \right\} \tag{A3.3-99}$$

The position of occurrence of the steps is given by

$$\Omega_c \left(d + \frac{l_c}{2} \right) + \beta \sin \Omega_m d + \epsilon \sin \Omega_p d = 2n\pi \tag{A3.3-100}$$

$$\Omega_c \left(d - \frac{l_c}{2} \right) + \beta \sin \Omega_m d - \epsilon \sin \Omega_p d = 2n\pi \tag{A3.3-101}$$

In the case where disc eccentricity exists, the time of occurrence of the output steps is given by

$$\omega_c \left[\left(t + \frac{\tau_c}{2} \right) + \tau_d \sin \omega_d t \right] + \beta \sin \omega_m (t + \tau_d \sin \omega_d t)$$

$$+ \epsilon \sin \omega_p (t + \tau_d \sin \omega_d t) = 2n\pi \tag{A3.3-102}$$

$$\omega_c \left[\left(t - \frac{\tau_c}{2} \right) + \tau_d \sin \omega_d t \right] + \beta \sin \omega_m (t + \tau_d \sin \omega_d t)$$

$$+ \epsilon \sin \omega_p (t + \tau_d \sin \omega_d t) = 2n\pi \tag{A3.3-103}$$

In this case, the output signal is

$$F_r(t) = S_r \frac{\tau_c}{T_c} + \frac{S_r \epsilon}{\pi} \sin \omega_p(t + \tau_d \sin \omega_d t)$$

$$+ \frac{S_r}{\pi} \sum_{k=1}^{\infty} \sum_{n,l=-\infty}^{\infty} \frac{1}{k} J_n(k\beta) J_l(k\epsilon)$$

$$\times \{ \sin [(k\omega_c + n\omega_m + l\omega_p)(t + \tau_d \sin \omega_d t) + k\omega_c(\tau/2)]$$

$$- \sin [(k\omega_c + n\omega_m - l\omega_p)(t + \tau_d \sin \omega_d t) - k\omega_c(\tau/2)]\} \qquad \text{(A3.3–104)}$$

or

$$F_r(t) = S_r \frac{\omega_c \tau_c}{2\pi} + \frac{S_r \epsilon}{\pi} \sin \omega_p(t + \tau_d \sin \omega_d t)$$

$$+ \frac{2S_r}{\pi} \sum_{k=1}^{\infty} \sum_{l=-\infty}^{\infty} \frac{1}{k} J_l(k\epsilon) \sin [l\omega_p(t + \tau_d \sin \omega_d t) + k\omega_c(\tau_c/2)] \qquad \text{(A3.3–105)}$$

$$\times \{ J_0(k\beta) \cos k\omega_c(t + \tau_d \sin \omega_d t) + \sum_{n=1}^{\infty} J_n(k\beta)$$

$$\times [\cos (k\omega_c + n\omega_n)(t + \tau_d \sin \omega_d t)$$

$$+ (-1)^n \cos (k\omega_c - n\omega_n)(t + \tau_d \sin \omega_d t)]\}$$

After simplification, it is

$$F_r(t) = S_r \frac{\tau_c}{T_c} + \frac{S_r \epsilon}{\pi} \sum_{p_1=-\infty}^{\infty} J_{p_1}(\beta_{dp}) \sin (\omega_p + p_1 \omega_d) t$$

$$+ \frac{S_r}{\pi} \sum_{k=1}^{\infty} \sum_{n,l=-\infty}^{\infty} \frac{1}{k} J_n(k\beta) J_l(k\epsilon)$$

$$\times \Bigg\{ \sum_{p=-\infty}^{\infty} J_p(k\beta_{dc} + n\beta_{dm} + l\beta_{dp})$$

$$\sin [(k\omega_c + n\omega_n + l\omega_p + p\omega_p)t + k\omega_c(\tau/2)]$$

$$- \sum_{p=-\infty}^{\infty} J_p(k\beta_{dc} + n\beta_{dm} - l\omega_p)$$

$$\sin [(k\omega_c + n\omega_m - l\omega_p + p\omega_p)t - k\omega_c(\tau/2)]\} \qquad \text{(A3.3–106)}$$

or

$$F_r(t) = S_r \frac{\tau_c}{T_c} + \frac{S_r \epsilon}{\pi} \sum_{p=-\infty}^{\infty} J_p(\beta_{dp}) \sin (\omega_p + p\omega_d) t$$

$$+ \frac{S_r}{\pi} \sum_{k=1}^{\infty} \sum_{n,l=-\infty}^{\infty} \frac{1}{k} J_n(k\beta) J_l(k\epsilon)$$

$$\Bigg\{ \sum_{q,r,s=-\infty}^{\infty} J_2(k\beta_{dc}) J_r(n\beta_{dm}) J_s(l\beta_{dp})$$

$$\sin [(k\omega_c + n\omega_m + l\omega_p + (q + r + s)\omega_p)t + k\omega_c(\tau/2)]$$

$$- \sin [(k\omega_c + n\omega_m - l\omega_p + (q + r - s)\omega_p)t - k\omega_c(\tau/2)])\} \qquad \text{(A3.3–107)}$$

or

$$F_r(t) = S_r \frac{\omega_c \tau_c}{2\pi} + \frac{S_r \epsilon}{\pi} \sum_{p_1 = -\infty}^{\infty} J_{p_1}(\beta_{dp}) \sin (\omega_p + p_1 \omega_d)t$$

$$+ \frac{2 S_r}{\pi} \sum_{k=1}^{\infty} \sum_{l=-\infty}^{\infty} \frac{1}{k} J_l(k\epsilon)$$

$$\left(\sum_{p=-\infty}^{\infty} J_p(l\beta_{dp}) \sin [(l\omega_p + p\omega_d)t + k\omega_c(\tau_c/2)] \right) \qquad \text{(A3.3–108)}$$

$$\left\{ J_0(k\beta) \sum_{p_2=-\infty}^{\infty} J_{p_2}(k\beta_{dc}) \cos (k\omega_c + p_2 \omega_d)t \right.$$

$$+ \sum_{n=1}^{\infty} J_m(k\beta) \left[\sum_{q=-\infty}^{\infty} J_q(k\beta_{dc} + n\beta_{dm}) \cos (k\omega_c + n\omega_m + q\omega_d)t \right.$$

$$\left. \left. + (-1)^n \sum_{r=-\infty}^{\infty} J_r(k\beta_{dc} - n\beta_{dm}) \cos (k\omega_c - n\omega_m + r\omega_d)t \right] \right\}$$

A3.3–6 DISCUSSION

So far, the ideal case has been considered: no allowance was made for the frequency response of the system. In the real case, however, the influence of the frequency characteristic of the transfer function must be considered. Probably the most critical influence is that of the lens. One possible approximation for MTF is

$$H(f) = \begin{cases} 1 - \left| \dfrac{f}{f_o} \right| & \text{for } |f| \leq f_o \\ 0 & \text{elsewhere} \end{cases} \qquad \text{(A3.3–109)}$$

where

$$f_o \approx 13 \text{ MHz}$$

REFERENCES

1. P. Panter, *Modulation, Noise, and Spectral Analysis*, McGraw-Hill, New York, 1965, Secs. 7.2 and 17.4.

2. J. W. Goodman, *Introduction to Fourier Optics*, McGraw-Hill, New York, 1968.

3. J. Isailović, "The influence of eccentricity on the FM information contained in a videodisc" MCA DiscoVision TN No. 364, Apr. 1977.

APPENDIX 3.4 ANALYSIS OF THE TIME-BASE SERVO ERROR SIGNAL

Here we investigate the performance of the time-base correction servo using either a pilot signal or the chroma burst as the error signal. The primary aim is to determine the amplitude of the pilot signal required to have identical performance to burst correction for the typical videodisc system.

A3.4–1 THE INFLUENCE OF ECCENTRICITY ON THE COLOR SIGNALS

The color picture signal (NTSC), $e_m(t)$, is specified by [1]

$$e_m(t) = E_y(t) + E_Q \sin(\omega t + 33°) + E_I \cos(\omega t + 33°) \qquad (A3.4–1)$$

where E_y is the luminance and E_Q and E_I are color signals. Equation (A3.4–1) can be expressed in the form

$$e_m(t) = E_y(t) + E_c \cos[\omega t + \Theta(t)] \qquad (A3.4–2)$$

where

$$E_c = \sqrt{E_Q^2 + E_I^2}$$

and

$$\Theta(t) = 33° + \arctan\left(\frac{-E_Q}{E_I}\right)$$

For the sake of simplicity we will suppose that the luminance and hue are constant:

$$E_y = \text{const.}$$

and

$$\Theta(t) = \text{const.} = \Theta_R(t)$$

In the case where disc eccentricity exists ($\Delta R \neq 0$), the output modulated video signal, $s_r(t)$, output pilot, $s_{pr}(t)$, and output burst, $s_{br}(t)$, are (Appendix 3.3)

$$s_r(t) = S_r \cos\{\omega_c(t + \tau_d \sin \omega_d t) + \beta \sin[\omega_m(t + \tau_d \sin \omega_d t) + \Theta_R]\} \qquad (A3.4–3)$$

$$s_{pr}(t) = S_{pr} \sin[\omega_p(t + \tau_d \sin \omega_d t)] \qquad (A3.4–4)$$

$$s_{br}(t) = S_{br} \cos\{\omega_c(t + \tau_d \sin \omega_d t) + \beta \sin \omega_m(t + \tau_d \sin \omega_d t)\} \qquad (A3.4–5)$$

where $\omega_d = $ disc velocity ($\omega_d = 2\pi f_d$, $f_d = 30$ Hz)
$\tau_d = (1/\omega_d)(\Delta R/R)$
$R = $ current (riding) radius

ΔR = eccentricity (Fig. A3.3–2)
f_m = color carrier frequency
f_c = carrier frequency
f_p = pilot frequency

A3.4–1a Pilot for Time-Base Error Correction

If a pilot is taken for correction of the influence of eccentricity, the error signal (open loop) is

$$e_p = \omega_p \tau_d \sin \omega_d t \qquad\qquad (A3.4\text{–}6)$$

With a closed loop, the output error signal, g_p, is reduced (ideally $g_p \to 0$), and the output video signal is

$$s_r(t)_{\text{cor}} = S_r \cos\left\{\omega_c t + g_p \frac{\omega_c}{\omega_p} + \beta \sin\left[\omega_m t + \Theta_R + g_p \frac{\omega_m}{\omega_p}\right]\right\} \qquad (A3.4\text{–}7)$$

According to Eq. (A3.4–7), the influence of eccentricity in luminance and color can be found. The color change is

$$(\Delta\Theta)_p = g_p \frac{f_m}{f_p} \qquad\qquad (A3.4\text{–}8)$$

A3.4–1b Burst for Time-Base Error Correction

Now the error signal (open loop) is

$$e_{br} = \omega_m \tau_d \sin \omega_d t \qquad\qquad (A3.4\text{–}9)$$

With a closed loop, the output error signal, g_{br}, is reduced (ideally $g_{br} \to 0$), and the output video signal is

$$s_r(t)_{\text{cor}} = S_r \cos\{\omega_c t + g_{br} + \beta \sin[\omega_m t + \Theta_R + g_{br}]\} \qquad (A3.4\text{–}10)$$

The color change is

$$(\Delta\Theta)_{br} = g_{br} \qquad\qquad (A3.4\text{–}11)$$

From Eqs. (A3.3–8) and (A3.3–11), it follows that

$$\frac{(\Delta\Theta)_p}{(\Delta\Theta)_{br}} = \frac{\omega_m}{\omega_p} \frac{g_p}{g_{br}} = 1 \qquad\qquad (A3.4\text{–}12)$$

because [Eqs. (A3.4–6) and (A3.4–3)]

$$\frac{g_p}{g_{br}} = \frac{e_p}{e_{br}} = \frac{\omega_p}{\omega_m} \qquad\qquad (A3.4\text{–}13)$$

It must be noticed that when an ideal case is supposed, no noise influence is taken into consideration.

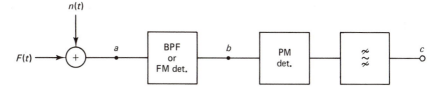

Figure A3.4–1 Block diagram for SNR calculation.

A3.4–2 CALCULATION OF SNR

There are four signal-to-noise ratios. The block diagram (Fig. A3.4–1) and spectra in corresponding points (Fig. A3.4–2) illustrate this.

A3.4–2a SNR for Pilot

The pilot signal is

$$s_p(t) = S_p \sin \omega_p t \qquad (A3.4–14)$$

[From Eq. (A3.3–106), Appendix 3.3, it follows that $S_p = S_r(2/\pi)$.] The noise is

$$s_n(t) = A_n \sin \omega_n t \qquad (A3.4–15)$$

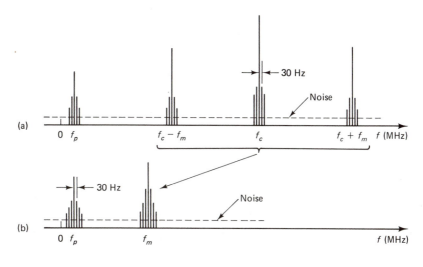

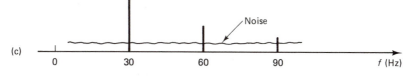

Figure A3.4–2 Spectra in corresponding points in Fig. A3.4–1.

where $A_n = \sqrt{2S'(f_n)\Delta f}$, $S'(f_n)$ being the noise power spectral density in watts per cycle.

The powers are

$$S = \tfrac{1}{2}S_p^2 \qquad (A3.4\text{--}16)$$

and

$$N = 2N_0(f_p)\text{BF}_p \qquad (A3.4\text{--}17)$$

where N_0 is the noise power spectral density (white noise is supposed) and BF_p is the filter's bandwidth.

The signal-to-noise ratio is

$$\left(\frac{S}{N}\right)_p = \frac{S_p^2/2}{2N_0(f_p)\text{BF}_p} \qquad (A3.4\text{--}18)$$

A3.4–2b Time-Base Error for Pilot $\Delta R \neq 0$

The pilot signal is phase modulated:

$$s_p(t) = S_p \sin \omega_p(t + \tau_d \sin \omega_d t) \qquad (A3.4\text{--}19)$$

Table A3.4–1 contains the results for $\Delta R = 5.08$ nm ($= 2 \times 10^{-3}$ in.).

TABLE A3.4–1

R	$\tau_d \times 10^6$ (s)	$\beta_{dp} = \omega_{pd}$
15.24 cm (= 6 in.)	1.769	4.4
7.62 cm (= 3 in.)	3.53857	8.8

The instantaneous phase deviation due to the noise contained within a narrow bandwidth, Δf, is approximated by [2]

$$\Theta(t) \approx \frac{A_n}{S_p} \sin (\omega_n - \omega_p)t \qquad (A3.4\text{--}20)$$

The output mean square value due to the discrete noise component is therefore given by

$$\overline{\Theta^2(t)} = k^2 \frac{A_n^2}{2S_p^2} \qquad (A3.4\text{--}21)$$

The total output noise power $(N)_o$ becomes

$$(N)_o = \left(\frac{k}{S_p}\right)^2 N_0(f_p)\,\text{BF}_p \qquad (A3.4\text{--}22)$$

and $(S/N)_{op}$ is

$$\left(\frac{S}{N}\right)_{op} = \beta_{dp}\ \frac{S_p^2/2}{N_0(f_p)\mathrm{BF}_p} \tag{A3.4–23}$$

From Eqs (A3.4–18) and (A3.4–23), it follows that

$$\left(\frac{S}{N}\right)_{op} = 2\beta_{dp}^2 \left(\frac{S}{N}\right)_p \tag{A3.4–24}$$

A3.4–2c SNR for Burst

Before FM detection, the modulated signal is

$$s_r(t)_{br} = S_{br} \cos\left(\omega_c t + \sin \omega_m t\right) \tag{A3.4–25}$$

The instantaneous angular frequency deviation caused by the incremental noise element is given by

$$\frac{d\Theta(t)}{dt} \approx (\omega_m - \omega_c)\frac{A_n}{S_{br}} \cos\left(\omega_m - \omega_c\right)t \tag{A3.4–26}$$

The output mean-square value due to the discrete noise component is therefore given by

$$\overline{\left[\frac{d\Theta(t)}{dt}\right]^2} = k^2(\omega_n - \omega_c)^2 \frac{A_n^2}{2S_{br}^2} \tag{A3.4–27}$$

The total output noise power $(N)_o$ becomes

$$(N)_o = \frac{1}{3}\left(\frac{k\pi}{S_{br}}\right)^2 \left(\mathrm{BF}_{br}\right)^3 N_0(f_c) \tag{A3.4–28}$$

where BF_{br} is the filter's bandwidth. Thus $(S/N)_{br}$ is

$$\left(\frac{S}{N}\right)_{br} = 12\beta^2 \left(\frac{f_m}{\mathrm{BF}_{br}}\right)^2 \frac{S_{br}^2/2}{N_0(f_c)\mathrm{BF}_{br}} \tag{A3.4–29}$$

A3.4–2d Time-Base Error for Burst, $\Delta R \neq 0$

After FM detection, the burst signal is

$$s_{br}(t) = A_{br} \cos \omega_m (t + \tau_d \sin \omega_d t) \tag{A3.4–30}$$

The powers in output of phase detector are

$$S = \tfrac{1}{2} k^2 (\omega_m \tau_d)^2 = \tfrac{1}{2} k^2 \beta_{dm} \tag{A3.4–31}$$

and

$$(N)_0 = \left(\frac{k}{A_{br}}\right)^2 N_0(f_m)\text{BF}_{br} \qquad \text{(A3.4–32)}$$

Thus $(S/N)_{obr}$ is

$$\left(\frac{S}{N}\right)_{obr} = \beta_{dm}\frac{A_{br}^2/2}{N_0(f_m)\text{BF}_{br}} \qquad \text{(A3.4–33)}$$

A3.4–3 INFLUENCE ON THE COLOR SIGNAL IN THE PRESENCE OF NOISE

Influence of noise can be taken into consideration if a relationship between $(S/N)_{br}$ and $(S/N)_{obr}$ will be established in the same way as the relation between $(S/N)_p$ and $(S/N)_{op}$ in Eq. (A3.4–24). Here we will take another way.

From Eqs. (A3.4–8) and (A3.4–11), it can be seen that in the presence of noise, the influence on color will be the same in both methods (pilot and/or burst) if it is

$$\overline{\left[\frac{\omega_m}{\omega_p}(g_p + n_1)\right]^2} = \overline{(g_{br} + n_2)^2} \qquad \text{(A3.4–34)}$$

(For the sake of generality, it is not supposed that the amount of noise is the same.) If $(S/N)_{(g)} \gg 1$ for both closed loops, it is

$$\left(\frac{\omega_m}{\omega_p}\right)^2\left(\frac{S}{N_1}\right)_{op(g)} \approx \left(\frac{S}{N_2}\right)_{obr(g)}\frac{N_2}{N_1} \qquad \text{(A3.4–35)}$$

where the subscript (g) indicates a closed-loop SNR. After substituting Eq. (A3.4–24) into Eq. (A3.4–35), it follows that

$$\left(\frac{S}{N}\right)_p(\text{dB}) = \left(\frac{S}{N}\right)_{obr}(\text{dB}) - \left(3\text{ dB} + 20\log\frac{f_m}{f_p}\right.$$
$$\left. + 20\log\beta_{dp} + 10\log\frac{N_1}{N_2}\right) \qquad \text{(A3.4–36)}$$

A3.4–4 DISCUSSION

For typical data [3] such as $f_p = 400$ kHz, $f_m \simeq 3.6$ MHz, $A_{(\text{dB})} = 50$ dB, $(S/N)_{obr(g)} = 25$ dB, and for $N_1 = N_2$, the SNR for the pilot is

$$\left(\frac{S}{N}\right)_p \simeq \begin{cases} 40\text{ dB} & \text{for } \beta_{dp} = 4.4 \quad \text{(outside } R) \\ 34\text{ dB} & \text{for } \beta_{dp} = 8.8 \quad \text{(inside } R) \end{cases}$$

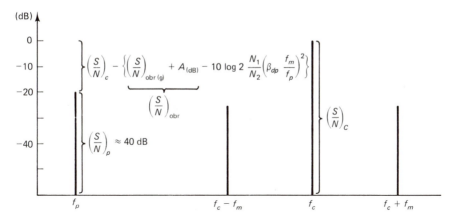

Figure A3.4–3   Practical example.

Figure A3.4–3 illustrates the first case: $(S/N)_c$ is the SNR for carrier frequency, f_c. In the real case, the noise in f_p is 2 dB $\div$ 4 dB larger than in f_c, which means that S_p for the same amount has to be larger.

Without noise, both the burst and pilot techniques give acceptable results. In the presence of noise, the pilot must have a minimum amplitude determined by Eq. (A3.4–36) to be acceptable. For typical conditions the pilot signal would have to be about -20 dB from the main carrier (unmodulated) to have equivalent results with burst correction.

REFERENCES

1. Howard W. Sams Editorial Staff, *Color-TV Training Manual*, 3rd ed., Howard W. Sams, Indianapolis, Ind., 1970, Chap. 3.
2. P. Panter, *Modulation, Noise and Spectral Analysis*, McGraw-Hill, New York, 1965, Sec. 16.1.
3. R. Dakin and G. Yoshida, private communication.

Chapter 4

CAPACITIVE
PICKUP PLAYBACK

4.1 INTRODUCTION

Development of a low-cost capacitive videodisc system is based on three basic design decisions. First is the use of a grooved disc to provide mechanical tracking of a stylus along the signal path. The idea is to eliminate the electronic servos.

Second is the use of a capacitance stylus (which incorporates a metallic electrode deposited on it) to read out the signal from the surface of the disc. The reasons for this are (1) the stylus is easy and inexpensive to fabricate compared to laser pickup and (2) the capacitance pickup is capable of resolving signal elements smaller than the wavelength of visible light.

Third is the choice of a 450-rpm disc rotation speed. The reasons for this decision are (1) lower tolerances are required, at lower rpm rates, for both the disc and the player; and (2) time-base errors can be corrected electromechanically, at a lower rotational speed.

Instead of grooves, tracking signals can be recorded on the side of the information track to provide capacitive pickup orientation on the disc surface. These are flat, or grooveless, capacitive discs. Eliminating the groove allows a narrower pitch, corresponding to 900 rpm, and more versatile control properties, but increases the cost of the pickup.

For the signal to be detected capacitively, the disc itself is conducting or has a conductive layer. Signal information is recorded on the surface of the disc along a spiral groove, beginning at the disc's outer diameter, and extending inward toward the center of the disc. There are 3793 grooves per radial centimeter. In the transverse plane of reference, the width of the slots is constant across the width of the groove

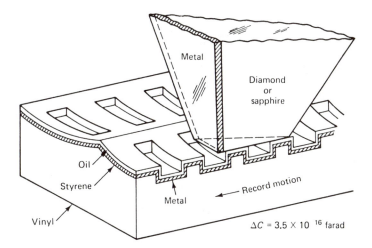

Figure 4–1   Stylus tip and capacitive videodisc surface [1].

itself. There is no signal information contained in the transverse width, or the depth, of the slots. Signal information is contained along the bottom of the groove, in the length of the slot, and in the spacing between the slots (Fig. 4–1). Slot length varies from about 2300 Å up to about 3.5 μm.

The replicated disc is composed of the PVC substrate, a 250-Å layer of metal, a 200-Å layer of styrene dielectric, and a 200-Å layer of oil, which serves as a lubricant. The discs are 30.2 cm in diameter and 1.9 mm thick, about the same dimensions as an optical videodisc or an audio LP record.

The stored information is read by measuring the capacitive variations between the electrode on the stylus and the surface of the disc. Detection of the small capacitive variations is performed by forming a resonant circuit with the stylus–disc capacitance. Pickups made in this manner can resolve signal elements smaller than the wavelength of light with sufficient reliability to reproduce high-quality video and audio signals.

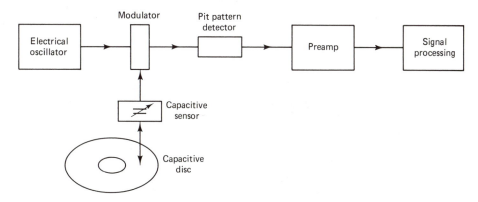

Figure 4–2   Model for the capacitive disc reading process.

A simplified model for the capacitive disc reading process is shown in Fig. 4–2. An electrical oscillator generates a periodical signal with a high carrier frequency f_0. In the modulator this carrier is amplitude modulated by the pit pattern on the disc surface; the modulation is performed through capacitive sensor. A pit-pattern detector demodulates the carrier with frequency f_0 and thus converts the mechanical pattern relief from the disc surface to the electrical signal. This model, basically, is valid for all capacitive systems (pregrooved or grooveless discs).

4.2 PRINCIPLE OF THE CAPACITIVE PICKUP

Two things should be made clear here:

1. The relation of recorded signal elements and capacitance
2. The detection of capacitance variations

There is more than one way to form signal elements as capacitors and to read capacitive variations. For example, a Vidicon-like principle with a V-shaped spiral rather than a raster structure in the reading process could be used.

4.2.1 Recorded Signal Elements

A hypothetical capacitive structure is shown in Fig. 4–3: two capacitors, C_0 and C_1, represent two signal elements. The upper covers are in the same plane, but the lower covers can be either at a distance h_0 or h_1. In a very rough approximation (e.g., edge effects are disregarded) the ratio of those two capacitances is equal to

$$\frac{C_1}{C_0} = \frac{h_0}{h_1}$$

If instead of that surface, an accordian surface is used, a larger elementary capacitance can be obtained for the same number of signal elements on the (disc) surface. This

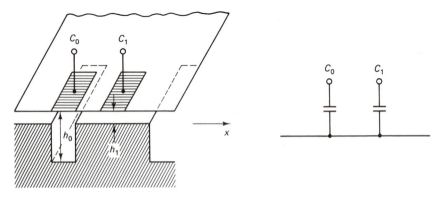

Figure 4–3 Hypothetical capacitive structure.

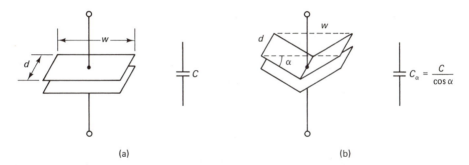

Figure 4–4   Increased capacitance: (a) flat surfaces; (b) "broken surfaces."

increase in capacitance is illustrated in Fig. 4–4. In practice, the upper covers in the hypothetical structure in Fig. 4–3 are replaced (substituted) by one moving cover.

Figure 4–5a shows cutaway views of the stylus and conductive disc surface with typical dimensions indicated [1]. Basically, this is the only approach used for capacitive systems. The stylus body can be made of either sapphire or diamond. Its function is to track the groove, and it supports an electrode that is coated on its trailing surface. The stylus foot is shaped to fit the triangular groove cross section and is sufficiently long to cover several of the longest recorded wavelengths so that the stylus tip rides smoothly on the crests of the recorded signal. A typical size is 2 μm across the groove and 4 μm along the groove. The rest of the stylus shape is determined by mechanical requirements. It must have strength to resist breaking and its contact area should not change greatly with wear. In normal use, the diamond wear is less than 1 nm/h. The electrode thickness is usually 1000 to 1500 Å.

The capacitance between the electrode and the conductive surface decreases as the distance between them increases. The capacity is a nonlinear function of the

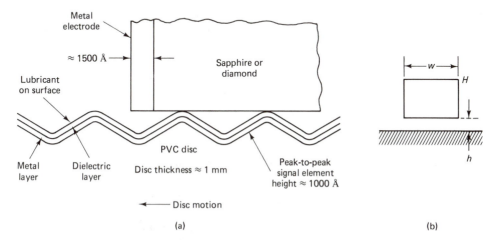

Figure 4–5   (a) Cutaway views of stylus and conductive disc; (b) simplified cutaway views [1].

distance. The degree of nonlinearity depends on the electrode size, the magnitude of the distance change relative to the average distance, and the dielectric constants of the surrounding material, including the stylus body material. For a thin electrode surrounded by material of the same dielectric constant, the capacitance versus distance is related by

$$C_{\text{tot}} = \epsilon_0 \epsilon w \, \frac{4}{\pi} \ln \frac{H}{2h} \qquad (4\text{--}1)$$

where ϵ is the dielectric constant, w the width of the electrode (looking into the paper in Fig. 4–5b), h the elevation of the electrode edge from the conducting surface, and H the height of top edge of the electrode (Fig. 4–5b).

For example, for an elevation change from 250 Å to 1250 Å approximating 1000-Å peak-to-peak signal elements, a stylus width of 2.5 μm, and a dielectric constant of 2.5, the change in capacitance is

$$\Delta C = 1.1 \times 10^{-16} \text{ F}$$

The change in capacitance for small changes in elevation is

$$\frac{dC_{\text{tot}}}{dh} = -\epsilon_0 \epsilon W \, \frac{4}{\pi} \frac{1}{h} \qquad (4\text{--}2)$$

The resolution of the pickup is also dependent on the height of the electrode above the disc surface. Under the same assumptions as those used previously, the equivalent transfer characteristic, for a small signal amplitude of average elevation h for the electrode edge, is [1]

$$A(h, L) = \frac{h_0 e^{-2\pi h/L}}{h}$$

where h_0 is an arbitrary reference height and L is the wavelength of the recorded signal. Thus the idealized transfer function is of the $y = e^{-1/x}$ form. If a spatial frequency, ν, is introduced:

$$\nu = \frac{1}{L}$$

The normalized transfer function, T, is

$$T(\nu) = e^{-K\nu}$$

The curve is illustrated in Fig. 4–6, and it corresponds to the MTF of the optical pickup. Although obtained under simplified assumptions, the transfer function shows some important properties of the capacitive pickup:

- Shorter wavelengths will be detected with smaller amplitudes than equal-amplitude recorded signals of longer wavelengths.
- The amplitude for a given wavelength decreases as h is increased.

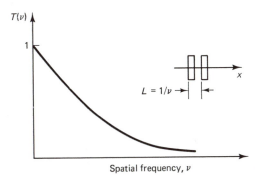

Figure 4-6 Simplified transfer function of the capacitive pickup.

- The shortest resolvable wavelength is a function of the noise in the system and the desired signal-to-noise ratio.

The expressions derived previously are based on a number of simplifications: a thin electrode, small signal element heights compared to the electrode elevation, and an electrode surrounded by material of uniform dielectric constant. None of these simplifications is absolutely true, but all of the modifications toward more realistic conditions are second-order effects. The electrode thickness is actually approximately 1500 Å but is much smaller than the wavelength of the recorded signals. The peak-to-peak signal elements are roughly 1000 Å, which is approximately the electrode elevation. This results in nonlinear response. Also, the electrode is surrounded by various materials of different dielectric constants. On the leading side of the electrode is a sapphire or diamond stylus with approximate dielectric constants of 9 and 4, respectively. On the trailing side of the electrode is air, with a dielectric constant of 1.

Under the stylus are a lubricant and a partly conducting insulator, each with a dielectric constant of approximately 2.5. If the lubricant is fluid, it can also surround the parts of the electrode and disc that are of interest. The most significant of these second-order effects is the presence of the stylus body. Its high dielectric constant can force the electric field to the stylus side of the electrode, causing the aperture response to be unsymmetrical. The asymmetry of this aperture is relatively more significant as the electrode (and stylus) height is elevated. The result is an apparent shift in the position or phase of the detected signal as the stylus is elevated. When the video and audio carriers are recorded, the amplitude of the phase disturbance is proportional to the product of the video carrier frequency and the audio carrier amplitude.

4.2.2 Detection of Capacitive Variations

The change in capacitance experienced by the stylus electrode is very small, perhaps 10^{-16} F, about a ten-thousandth of a picofarad. So small change in capacitance can be detected only at high frequencies. The front (arm) circuitry operates at about

1 GHz and uses transmission-line techniques, but its function is usually described schematically by the lumped-element circuit [1] shown in Fig. 4–7.

The circuit path described above is a resonant circuit, including the stylus–disc capacitance. The resonant peak of this circuit changes as the capacitance changes. The resonant circuit is shown in Fig. 4–7a and its response curve is shown in Fig. 4–7b. This resonant circuit is stimulated by an oscillator with a frequency on the slope of the resonant curve. The response to the oscillation is detected with a pickup loop. The magnitude of the detected response will depend on where the oscillator frequency falls on the resonant curve. As the stylus–disc capacitance changes, the resonant frequency will change such that the detected carrier will be amplitude modu-

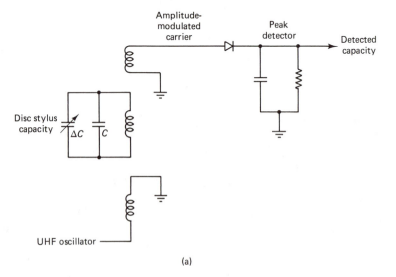

(a)

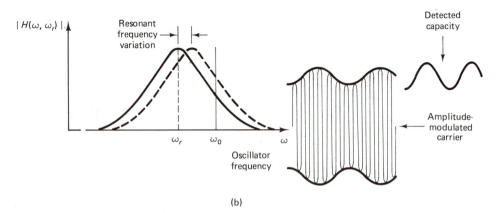

(b)

Figure 4–7   Capacitive variation to AM signal converter [1].

lated, and the voltage obtained by peak detection of this carrier follows the stylus–disc capacitance changes.

Typical values are $f_r = 910$ MHz and $f_0 = 915$ MHz. Let $H(\omega, \omega_r)$ represent the transfer function of the resonant circuit from the oscillator port to the carrier output port, and ω_r represent the resonant frequency of an assumed single-pole resonance. For frequencies close to the resonance, the degree of amplitude modulation is

$$\frac{\Delta H}{H} = -\frac{\omega_r}{4(\omega_o - \omega_r)}\frac{\Delta C}{C} = -\frac{Q}{2}\frac{\Delta C}{C} \tag{4–3}$$

It is assumed that circuit operates at the 3-dB point of the transfer characteristic, where $\Delta H/H$ has a maximum sensitivity with respect to ω_r. Therefore, for small peak-to-peak capacitance variations, ΔC, the degree of peak-to-peak amplitude modulation is proportional to ΔC. If the average peak voltage of the RF carrier across the capacitance is V_o, then the amplitude of the largest peak-to-peak variation in the voltage across the capacitance is $V_o(Q/2)(\Delta C/C)$.

For example, for a $\Delta C = 10^{-14}$ pF, a standing capacitance $C = 0.5$ pF, a bandwidth between 3-dB points of 30 MHz, and a resonant frequency of 900 MHz, $Q = 30$ and $(Q/2)(\Delta C/C) = 3 \times 10^{-3}$. The equivalent shunt resistance across the capacitance is $Q/\omega C - 10^4 \ \Omega$, and if $V_o = 10$ V, the peak-to-peak voltage will vary by 30 mV over this 10-kΩ resistance. This is equivalent to 2.6 mV if the output is stepped down to 75 Ω.

4.3 SYSTEM PARAMETERS

The minimum recording wavelength on the disc, L_{min}, is

$$L_{min} = \frac{V_{min}}{f_{max}}$$

where f_{max} is the highest instantaneous frequency in the recorded signal, and

$$V_{min} = 2\pi R_i n$$

R_i is the inside radius, and n the rotations per second. This can be expressed as [1]

$$L_{min} = \frac{2\pi(R_o - R_i)R_i}{T_p f_{max} b}$$

where R_o is outside radius, T_p is playing time, and b is in grooves per millimeter (defined by groove pitch).

For fixed b, T_p, and f_{max}, L_{min} is a maximum when $R_i = R_o/2$. L_{min} should be maximized because long wavelengths are easier to resolve with the pickup (Fig.

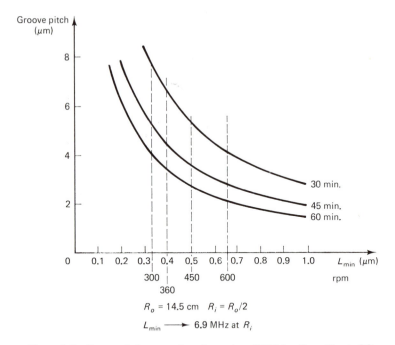

Figure 4–8 Groove pitch versus L_{min} for various RPM for $T_p = 60$ min [1].

4–6), causing less signal loss due to electrode lifting from either debris or severe disc warp. Also, long wavelengths are easier to record and easier to replicate.

Figure 4–8 shows b versus L_{min} for various playing times and for $R_o = 2R_i = 14.5$ cm. In this case, L_{min} was calculated for $f_{max} = 6.9$ MHz. (This frequency is the limit of the video in the CED system.)

Rpm is also indicated for certain specific rpm rates on the L_{min} axis. It is desirable to keep an integral number of TV frames in one revolution. This allows the possibility of searching for a particular spot on the disc with the stylus down, without affecting the periodicity of the sync signals. The rates 600, 450, 360, and 300 rpm correspond to 3, 4, 5, and 6 frames per revolution, respectively.

At the early stage of the technology development, the system goal was a disc with 60 min of playing time per side. The parameters were then set with $T_p = 60$ min. Since there is not a sharp optimum at $R_i = R_o/2$, R_i can be varied to match other constraints. Figure 4–9 shows L_{min}; however, there is an optimum rpm and b at this limit. Due to the constraints mentioned above, the system parameters selected are

$$T_p = 60 \text{ min}$$
$$\text{rpm} = 449.550 \text{ (four fields per revolution); rpm} = 60n$$
$$\text{groove pitch} = 2.66 \text{ } \mu m \text{ (b is 9541 grooves per inch or 3756 grooves/cm)}$$
$$R_o = 14.5 \text{ cm (5.7 in.)}$$
$$R_i = 7.3 \text{ cm (2.9 in.)}$$

$$L_{\min} = 0.50 \ \mu m \ (6.9 \ MHz \ at \ R_i)$$
$$L_{\max} = 1.75 \ \mu m \ (3.9 \ MHz \ at \ R_o)$$

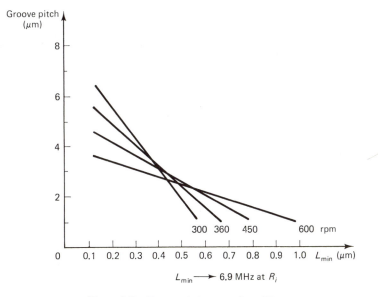

Figure 4-9   Groove pitch versus $L_{\min}$ [1].

4.4 TIME-BASE ERROR CORRECTION

Due to mechanical tolerances in the disc playback system, such as imperfect centering and warp, the signal recovered from the disc may have large time-base errors [2]. For example, a centering error of about 0.18 mm would cause a 50-μs peak-to-peak error varying at a one revolution rate. This causes poor color reproduction and horizontal sync instability, since errors of this size and rate will not normally be corrected by the color or horizontal phasing circuits of a TV receiver due to bandwidth limitations in these circuits. A shift of a few degrees in the 1.53-MHz (color subcarrier of the recorded video signal) chroma can cause objectionable hue shifts. Therefore, these chroma phase errors must be reduced by at least 70 dB. The horizontal sync variations can cause objectionable motion of the picture, but these tend to be tracked to some extent by the receiver, and 20 to 30 dB correction is sufficient.

Time-base error correction can be accomplished by two closed-loop correction systems driven from the same error signal. The first is the arm-stretcher transducer system, which reduces the error by moving the stylus along the groove so that essentially constant relative velocity between the stylus and the recorded information on the disc is maintained. The second provides the final 3.58-MHz color subcarrier with the stability and accuracy required to assure accurate color display. A phase-locked loop is used for this purpose.

Because the color subcarrier for the recorded video signal is 1.53 MHz, a hetero-dyne technique is used to recover a burst with a nominal frequency of 3.58 MHz from the signal obtained from the disc. This 3.58-MHz component is then compared with a reference 3.58-MHz carrier from a crystal oscillator. The difference between the two is applied as an error signal to a variable oscillator with a nominal frequency of 5.11 MHz in such phase as to minimize the difference and attain the required accuracy. The loop is designed with a bandwidth of about 3kHz, which is approximately 20 times as wide as that permitted in color TV receivers, so it can track and correct fairly rapid variations in the carrier frequencies from the disc. The correction factor is 50 dB at the disc revolution rate of 7.5 Hz.

The bandwidth of the time-base correcting arm-stretcher system is 200 Hz. Since the back-and-forth motion of the transducer cannot correct average speed errors, the very low frequency portion of the error signal is extracted with a low-pass filter and added to the 5.11-MHz oscillator control voltage. This causes an average speed error to be tracked by the oscillator rather than the transducer, and limits the response of the mechanical system to rates greater than 2 Hz.

The closed-loop resonance for the time-base correction system, that is the time-base error transfer characteristics of the player, are shown in Figs. 4–10 and 4–11. These curves are the magnitude of the time-base error at the output for a fixed sinusoidal error at that input, as a function of the frequency of the disturbance.

Figure 4–10 shows the response of the electromechanical system. The largest error is due to the disc eccentricity, and results in a 7.5-Hz sinusoidal disturbance (2 frames per revolution). The maximum correction near this frequency is required. The correction peak near 65 Hz is due to the mechanical resonance of the transducer. An accentuation of errors occurs near 2.2 Hz, due to the limited phase margin on the low-frequency side of the system, but since no large time-base errors occur at this rate, it presents no problem. Figure 4–11 shows the transfer characteristics for

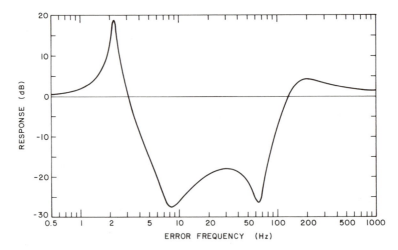

Figure 4–10   Arm-stretcher system, TBE transfer characteristic (2).

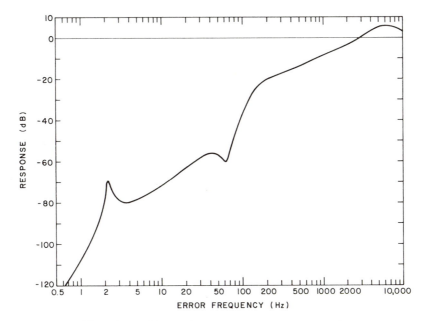

Figure 4–11   Chroma phase error transfer characteristic [2].

chroma phase errors. This response results from the sum of the correction provided by the chroma phase-locked loop and the electromechanical system. Below 2 Hz the effect of adding in the dc error from the electromechanical system can be seen. The general discussion about TBE given in Chapter 3, including Appendix 3.3, is also valid for capacitive systems.

4.5 ELECTRONICS IN THE PLAYER

To illustrate, the signal processing in the player [2, 3], control, servo systems, and dropout compensation—the principal parts of the player electronics and operations—will be discussed briefly. For capacitive systems, as basically for mechanical ones, it is difficult to record full NTSC, PAL, or SECAM bandwidth: a compressed-bandwidth (extended play) approach is used.

In the recording, the combed luminance and combed color signals are added together to form a composite video signal in which the luminance and chrominance spectral components are interleaved. This process is called buried-subcarrier encoding because the color subcarriers are located in the middle of the luminance band. The composite video signal is applied to a frequency modulator such that a peak white input produces an output frequency of 6.3 MHz, the black level produces 5.0 MHz, and the sync tips produce 4.3 MHz, as shown in Fig. 4–12. Because the composite video signal contains components as high as 3 MHz, the first-order sidebands of the video FM signal extend from 2 to 9.3 MHz, as shown. The audio signal is applied

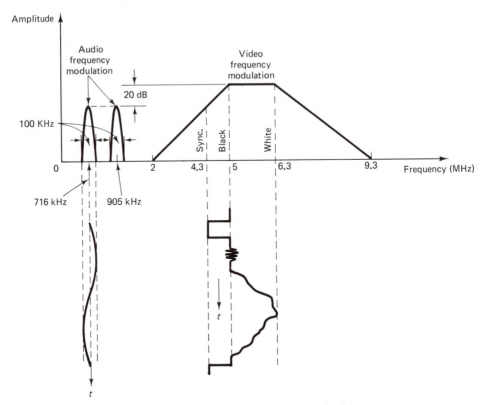

Figure 4–12   Frequency spectrum in the channel.

to a second frequency modulator with center frequency at 716 kHz (and 905 kHz, with the addition of stero) and produces a maximum deviation of ±50 kHz. The audio FM signal is added to the video FM signal to provide a single voltage waveform.

The videodisc player comprises a turntable, driven at 450 rpm synchronously with the 60-Hz power line; a mechanism for inserting and retrieving a disc; a stylus cartridge pickup arm; signal processing circuitry; and control elements. A simplified block diagram of the player is shown in Fig. 4–13. At the left is the 915-MHz oscillator, the output of which is modulated by the stylus–disc capacitance and rectified to provide the FM signal that follows the rise and fall of the disc surface. This signal is fed to both video and audio demodulators. When either the audio or video FM carrier is lost, measures are taken to prevent a noticeable disturbance in the audio or video output. When no audio disturbance is detected, the audio feeds through the track-and-hold circuit to the 4.5-MHz frequency modulator to generate the audio portion of the NTSC signal. When a defect is detected, the audio level is maintained at its last valid value until the disturbance is eliminated.

When no disturbance occurs in the video signal, it is passed directly to a 1-H delay line, the principal component in the comb filter. The input of the delay

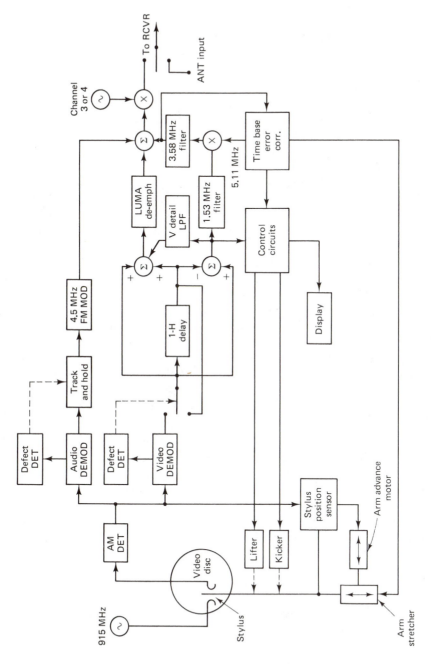

Figure 4–13 Block diagram of a videodisc player.

line and the output are added together to realize the luminance comb filter. After appropriate deemphasis, the luminance signal is added with other signals to make up the composite NTSC signal.

The output of the delay line is subtracted from the input to realize the color comb filter. The output of this filter, containing the 1.53-MHz color subcarrier, is mixed with a 5.11-MHz oscillator signal to convert the color subcarrier to 3.58 MHz, as required for the NTSC signal. The up-converted 3.58-MHz color signal is added to the luminance and audio to provide the composite NTSC signal, which is then translated to either channel 3 or 4, and provided to the output jack on the rear of the player. When the player is turned off, the antenna input signal is fed directly to the output jack. The passage of the luminance signal through its comb filter will remove vertical detail from this signal because the CCD delay line is used, and the luminance signal is combed from dc up. For example, the luminance video signal required to write alternate black-and-white lines is completely removed when the input and output of the delay line are added together. To replace the low-frequency signal required for vertical detail in the picture, the low-frequency portion of the combed chroma signal, which is truly luminance information rather than chrominance information, is added to the combed luminance signal to provide an uncombed signal below 500 kHz.

When a dropout is detected in the video FM carrier, the automatically controlled switch at the delay line input causes the output of the delay line to be fed back to the input so that the signal is reused. Thus, when a defect is detected on any one horizontal line in the picture, the throwing of the reuse switch causes the signal from the preceding line to be substituted for the bad video. In most cases, such substitution is nearly invisible. As a result, disturbances on the disc that would otherwise be objectionable are reduced to near-invisibility.

The arm stretcher shown in the lower left corner of Fig. 4–13 has been included in the player to correct for the effects of off-centering (eccentricity). This device is similar to a moving-coil loudspeaker mechanism which is fastened by a flexible coupling to the rear end of the stylus arm and moves the stylus arm tangentially along the groove. When the groove speed is too low, the stylus arm is moved in reverse to the groove velocity to increase the relative speed. When the groove speed is too high, the stylus arm is moved in the direction of the groove velocity to reduce the relative speed. The signal for driving the arm stretcher is obtained by comparing the 3.58-MHz color burst from the disc with a fixed oscillator at the same frequency. The net result is that off-centered conditions as great as 254 μm (10 mils) do not produce detectable jitter in playback.

During the play of a disc, the carriage for the stylus cartridge is moved in such a way as to keep the stylus arm and flylead centered in the cartridge housing. This is accomplished by sensing the lateral position of the stylus relative to the cartridge housing by capacitive coupling of the stylus flylead to varactor diodes driven out of phase by a 260-kHz oscillator. As the capacitance of one diode increases, the other decreases, and vice versa. These diodes are located one on each side of the stylus flylead, so that a fraction of their capacitance is added to the stylus capacitance as a function of how close the diodes are to the flylead. When the stylus is centered, the capacitance variations of the two varactor diodes, being out of phase, cancel one another. However, in off-centered conditions, the effect of one diode is greater than that of the other. The resultant capacitive variations cause a change in the tuning of the stylus resonant circuit, giving rise to 260-kHz components in the output of the 915-MHz amplitude detector, indicating an off-center condition. The amplitude and phase of the 260-kHz signal indicate the amount and direction of the stylus off-centering. A dc arm-advance motor is driven in to the 260-kHz error signal to center the stylus arm and return the error signal to zero.

The stylus kicker shown in the lower left of Fig. 4–13 is included in the player to provide small, rapid lateral movements of the stylus during play. A small permanent magnet mounted on the stylus arm near the stylus is forced sideways by magnetic fields from small coils mounted in the stylus cartridge housing. When movement of the stylus is desired, an appropriate pulse of current through the coils causes the stylus to jump sideways one or more grooves in either the forward or reverse direction. This operation is activated during the visual-search mode. When the visual-search button is depressed, a pulse is applied to the kicker coil just prior to each vertical blanking interval with an appropriate magnitude to move the stylus two grooves. Since there are eight fields per rotation, the program moves at 16 times normal speed. However, since stylus movement takes place just ahead of the blanking interval, the TV synchronizing pulses are continuous and no picture breakup occurs.

The stylus kicker is also used to correct for locked-groove defects on discs. To facilitate this operation, a unique number is recorded with each TV field on the disc. The numbers increase monotonically from the beginning to the end of the disc. Circuits built into the player decode and keep track of the field numbers. During normal play, these numbers progress regularly. When locked-groove situations arise and the numbers jump backward instead of progressing normally, player recognition of this fact causes the application of pulses to the kicker coils to move the stylus ahead by two grooves. These pulses continue until field numbers read from the disc equal or exceed the numbers predicted by the player circuits. In most cases, locked-groove defects are corrected before an observer is aware of the problem. No corrective action is taken when forward groove skips occur. In general, these forward skips cause little disturbance to the viewer.

Also, the field numbers recorded on the disc convert to time of play in minutes from the beginning of the disc and are shown on the LED displays on the front of the player. The stylus lifter is activated to raise and lower the stylus as required for proper player operation. The stylus is lowered during normal play and visual search, and it is lowered momentarily onto a stylus cleaner each time a disc is removed from the player. In all other conditions, including power-off, the stylus is lifted off the disc.

4.6 STYLUS

The stylus tip or substrate with a thin metal electrode on its flat trailing edge [2–4] is shown in Fig. 4–14. The end of the stylus electrode is about 2 μm wide by 0.2 μm thick. The sides of the stylus tip are cut away as shown, providing the "keel-lapped" shape, to prolong the useful life of the stylus as the tip is abraded during play. The waves pressed into the disc have a peak-to-peak amplitude of about 850 Å.

The desirable characteristics of the stylus are [2]:

- Smooth tracking in the groove, maintaining the bottom of the electrode in contact with the signal elements in the groove. If perturbed by large dust particles or disc defects, the electrode should return quickly to tracking without tending to skip to either side of the proper groove.
- The stylus should last for at least 200 playing hours without noticeable degradation of its pickup performance. Discs played with the styli should exhibit no loss of performance due to wear in at least 50 plays.
- The stylus should be capable of playing through the small quantities of dust without mistracking or lifting of the electrode away from the signal elements.
- The stylus should be manufacturable at low cost and with high yield.

The design parameters that can be manipulated to achieve these characteristics are: stylus material, electrode, tip geometry, arm dynamics, arm-stretcher response, and fabrication and assembly techniques.

The tracking force is low, approximately 70 mg, but the small shoe area produces pressure on the shoe of the stylus in contact with the disc of the order of 10^7 kg/m². Sapphire and diamond, in proper crystallographic orientation, have been found to perform satisfactorily. Diamond has a higher cost and is harder to shape than sapphire, but its wear life and much greater resistance to breakage make it favorable. Crystal flaws are numerous in diamond and care must be taken to use the best orienta-

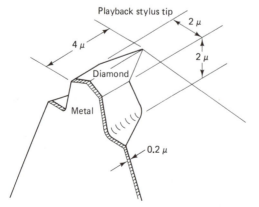

Figure 4–14 Stylus tip: inverted view.

tions. Optimum electrode thickness is determined primarily by the shortest wavelength on the disc and the amount of E-field spreading that occurs in the stylus substrate due to its dielectric constant. The electrode is, for example, sputtered titanium deposited on diamond to a thickness of 1500 Å. The shaping of the tip is critical. In the limit, if the shoe is too sharp, it will tend to cut into the disc. If it is too wide, the life of the electrode will be short, since end of life occurs when the width of the shoe extends beyond that of a single groove (approximately 2.66 μm). The situation can be optimized by varying the angle of the V. As the angle is made more acute, wear life increases but the tendency toward breakage also increases. Broadening the V strengthens the tip, but reduces life.

The length of the keel or shoe also affects both life and breakage. Here the limitation is lift-off of the electrode when playing a warped disc as the shoe becomes too long. The remaining key dimension is the "prow" angle, shown in Fig. 4–14. This is the angle that the leading edge of the stylus makes with the disc as it diverges from the shoe. Its principal effect is on tip strength and/or the separation loss due to debris; if the angle is steep, the tip may become fragile. In addition, whereas some fine debris is pushed out of the way, some tends to wrap around the tip, lifting it off. If the angle is shallow, the tip becomes stronger and critical wraparound is reduced, but the tip will not push as much debris out of the way and will tend to lift up to ride over the debris, causing loss of signal pickup. It has been found that a prow angle of about 30° represents a good compromise.

The diamond stylus is mounted on the end of a 8.4-cm-long stylus arm made from thin-walled aluminum tubing, as shown in Fig. 4–15. A rubber mounting supports the stylus arm with enough flexure at the stylus end so that the stylus will follow

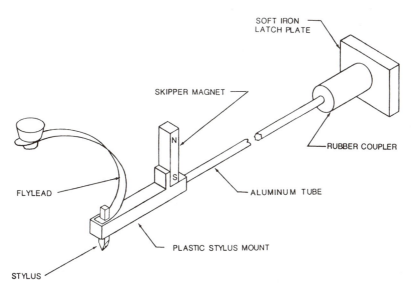

Figure 4–15 Stylus support arm [1].

irregularities in the disc in both the vertical and horizontal directions. A small permanent magnet mounted on the stylus arm near the stylus imparts a small lateral motion to the stylus when acted upon by the magnetic fields of the stylus kicker coils. The rear of the stylus arm is fitted with a soft-iron plate that is attracted to and held by a cup magnet on the arm-stretcher transducer. The electrode on the diamond stylus is connected to the VHF circuitry by a flexible flylead, which also serves as a spring to hold the stylus against the disc with about 65 mg of force. The stylus arm, flylead, and support are mounted in the stylus cartridge, a plastic case that allows easy replacement of the stylus in a player. The replacement of a stylus cartridge requires no tools or adjustments of any kind.

The basic tracking problem in a grooved capacitive videodisc is quite similar to that encountered in audio players. The tracking forces are applied vertically primarily by means of the spring metal flylead that connects to the tip on the stylus support arm (see Figure 15). The force remains vertical as long as the stylus remains in the bottom of the groove. Under ideal conditions, the vertical force required approaches zero. However, if there is a disturbance in the horizontal direction, a restoring force is required. High tracking forces provide high restoring forces but tend to cause increasing disc and stylus wear.

Problems with tracking become more difficult as the speed of the perturbations in both the horizontal and vertical directions increases. Ideally, what is desired is a well-damped, very low mass system in which the tracking force is obtained by means of a flexible element such as a spring. Spring rate represents the amount of force required to move the arm a specified distance. To minimize the effects of side bias caused by either improper set-down or disc runout, the horizontal component should be as low as possible. Similarly, to maintain constant tracking force regardless of disc warp, the vertical spring rate should be low. To eliminate vibrations, the rear end of the support arm is attached to a rear coupler made of a specially formulated butyl rubber that mechanically terminates and damps the arm.

The stylus arm structure is a dominant component in its drive and loop response. The principal problem is to obtain enough phase margin for closed-loop stability, so that delay times must be minimized and resonances eliminated or at least greatly

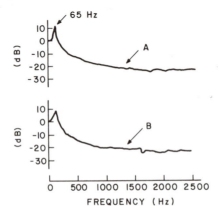

Figure 4–16 Open-loop transfer curves of (a) armstretcher transducer along, and (b) stylus tip on stylus support arm attached to transducer [1].

damped. The rear coupler provides much of the damping. In addition, it is designed to have a very low horizontal and vertical spring rate so that any slight distortions that it might suffer in fabrication or with life will not affect the neutral position of the stylus tip. However, since the coupler must convey the arm-stretcher motion to the stylus tip with minimum delay, it is designed to be very stiff in the longitudinal direction. The wideband open-loop response of the arm-stretcher alone (Fig. 4–16, curve a) can be compared to that of the stylus tip at the end of the arm stretcher (Figure 4–16, curve b). As can be seen, curve b is quite smooth out to 1500 Hz, which assures the maintenance of good phase margin. The low-frequency peak is the resonance of the transducer at 65 Hz. This is in the high control range, but it is adequately damped by the circuit constants.

4.7 CAPACITIVE READOUT FROM A FLAT DISC

This technology (grooveless capacitance systems) represents a compromise between two basic technologies. As a result of this compromise, the two principal characteristics of the system are:

1. Relatively inexpensive readout system
2. Because of the flat disc (grooveless) some special effects are possible, such as slow forward and backward motion, fast forward and backward motion, random access, and freeze frame.

The videodisc itself is a conducting PVC disc. Because the contacting surface between the disc and the reading head is about 10 times smaller than what this surface would be in the case of the disc with grooves, the useful life of the disc, as well as of the readout head itself, is increased.

The characteristic of this system is that these are two kinds of pits: one used for the information, the other for tracking (Fig. 4–17). The information pits resemble those in the pregrooved system but are narrower, corresponding to the smaller pitch. The tracking pits are placed between the information channels, and are used to guide the readout head over the proper channel.

The principal difference in this system from other existing systems is the way in which the readout head is guided over the disc surface [5,6]. The readout head can follow the channel, move directly over the desired channel, or be used for any of the special effects during the readout.

The method of head control, transversal and longitudinal, is shown on Fig. 4–18. The head is attached at one end of a lever, on the other end of which are an electromagnet with its corresponding coils. Coil A serves for transversal motion of the electromagnet and therefore the readout head. Coil B consists of two parts with two sides in (relation) to coil A. The two parts of coil B are energized at opposite phase angles. Coil B enables longitudinal motion of the readout head. The current

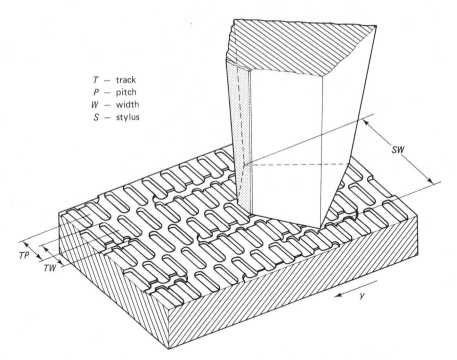

T — track
P — pitch
W — width
S — stylus

TP

TW

SW

y

Figure 4–17 Simplified representation of disc surface and readout head.

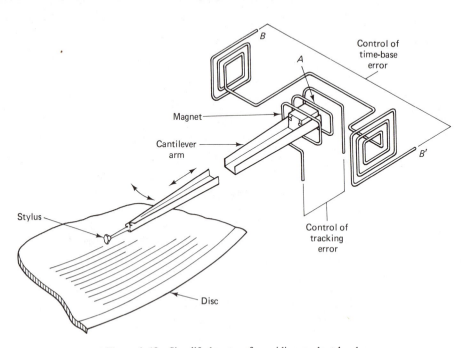

B

A

Control of
time-base
error

Magnet

Cantilever
arm

B′

Stylus

Control of
tracking
error

Disc

Figure 4–18 Simplified system for guiding readout head.

178

through the coils can change due to one of the following signals: signal due to a tracking error, signal due to a base error, or the commanding signal.

A modification of the system for guidance of the readout head [5] is shown in Fig. 4–19. A stylus is mounted on the two codes, forming a V-shape. Two permanent magnets, a V-shaped zone, and two moving coils control the stylus position.

In pregrooved capacitive videodisc systems, the rotational speed of the disc is usually set at a low value, 450 rpm, for example, in order to prolong the life of the stylus tip and to relax other mechanical limitations. However, the wavelength of the information signal recorded in the guide groove becomes short and laser (optical) recording is questionable. In systems with a flat capacitive disc, the rotational speed of the disc is set at the geometrical average (900 rpm) between optical and pregrooved capacitive systems.

On the flat capacitive disc, two pilot signals of mutually different frequencies are recorded on opposite sides of each track carrying the main information signal. When the tracking of the reproduced stylus deviates relative to the main information

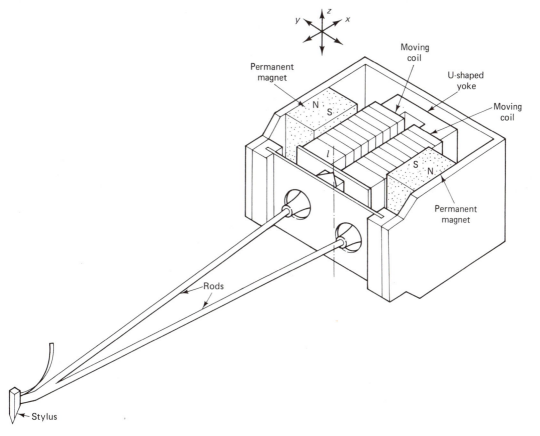

Figure 4–19 Modified pickup from that in Fig. 4–18.

signal track of the recording medium, the reproducing stylus reproduces a pilot signal together with the main information signal. The tracking control circuit operates in response to the pilot signal to produce an output tracking error signal corresponding to the direction and magnitude of the tracking deviation. The reproducing stylus is controlled in response to this error signal so as to track accurately over and along the main information signal track. However, the level of the reproducing signal varies with the radial positional displacement of the stylus on the disc. The reproduced level also depends on variation in the contact between the electrode of the stylus and the disc surface, dust and grime adhering to the disc surface, and undesirable abrasive wear of the electrode part of the stylus.

Figure 4–20 shows principal signal waveforms for the case when two frames (900 rpm) are recorded per revolution [4]. A simplified block diagram of the tracking servo for the videodisc system with flat discs [5] is shown in Fig. 4–21, and typical waveforms are shown in Fig. 4–22. A motor rotates the disc on a turntable with a rotation speed of 900 rpm. The track on the disc is traced by the reproducing stylus, which is fitted into a reproducing transducer. Appropriate resonant circuits are included in the preamplifier. Original information is recovered in the demodulator.

The pilot signals are separated by a low-pass filter. An automatic gain control

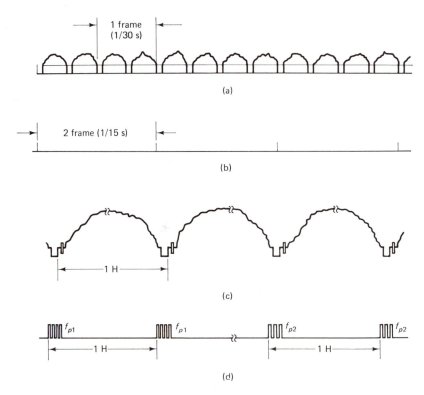

Figure 4–20   Signal waveforms.

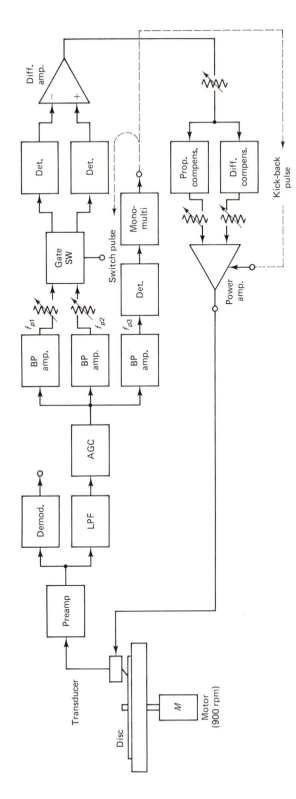

Figure 4–21 Block diagram of the tracking servo.

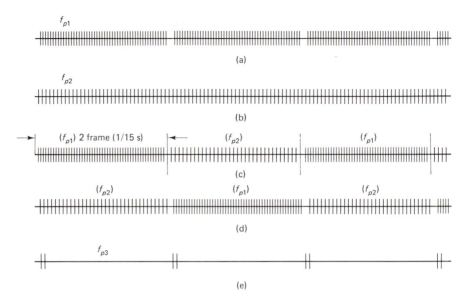

Figure 4–22 Signal waveforms in the tracking servo.

(AGC) circuit monitors the specific levels of the pilot signals. The reproduced levels of the first and second pilot signals f_{p1} and f_{p2} vary in response to the tracking deviation of the stylus. Tracking control is carried out by detecting this variation. Three pilot signals, f_{p1}, f_{p2}, and f_{p3}, are separated by three bandpass filters. The third reference signal, f_{p3}, shows changes between f_{p1} and f_{p2} (Fig. 4–22). A monostable multivibrator is used to reduce noise influence. Based on the f_{p3}, the switch pulses are activated to control a gate switching circuit in normal mode. In a special reproducing mode (e.g., still picture or slow motion), this signal is applied to a kickback pulse-forming circuit.

The gate-switching circuit switches the pilot signals f_{p1} and f_{p2} every revolution period of the disc, in response to the switching pulse supplied. As a result, a switching pulse inverts polarity every two frames ($\frac{1}{15}$ s). The tracking error signal is obtained at the output of the differential amplifier, and this signal indicates the tracking error direction and quantity. After proper compensation and amplification, the tracking error signal is applied to the tracking control mechanism of the transducer. In the still-picture reproducing mode, the transducer stops moving in a radial direction across the disc. This is accomplished by kickback pulse.

REFERENCES

1. J. K. Clemens, "Capacitive pickup and the buried subcarrier encoding system for the RCA videodisc," *RCA Rev.,* Vol. 39, No. 1, Mar. 1978, pp. 33–59.
2. R. N. Rhodes, "The videodisc player," *RCA Rev.,* Vol. 39, No. 1, Mar. 1978, pp. 198–221.
3. H. N. Crooks, "The RCA SelectaVision videodisc system," *RCA Eng.,* Nov./Dec. 1981: special issue devoted to SelectaVision videodisc.
4. J. J. Brandinger, "The RCA CED videodisc system—an overview," *RCA Rev.,* Vol. 42, Sept. 1981, pp. 333–366.
5. K. Goto, O. Tojima, and H. Miyatake, "Signal pickup device with tracking control and jitter compensation for a videodisc," U.S. Patent No. 4,160,269, July 3, 1979.
6. H. Kinjo, "Tracking control apparatus for use in apparatus for reproducing video signals from a rotary recording medium," U.S. Patent No. 4,190,859, Feb. 26, 1980.

Additional References

1. J. J. Gibson, F. B. Lang, and G. D. Pyles, "Nonlinear aperture correction in the RCA videodisc player," *RCA Eng.,* Vol. 26, No. 9 (Nov./Dec. 1981).
2. R. W. Nosker and D. L. Matthies, "Basics of videodisc stylus dynamics and interaction with surface imperfections," *RCA Rev.,* Vol. 43, No. 1.
3. H. Weisberg, "Manufacturing the videodiscs: an overview," *RCA Eng.,* Vol. 27, No. 1. (Jan./Feb. 1982).
4. F. D. Kell, G. John, and J. Stevens, "Videodisc mastering: the software to hardware conversion," *RCA Eng.,* Vol. 27, No. 1 (Jan./Feb. 1982).
5. G. D. Pyles and B. J. Yorkanis, "Videodisc's video and audio demodulation, defect detection, and squelch control," *RCA Eng.,* Vol. 26, No. 9 (Nov./Dec. 1981).
6. D. H. Pritchard, J. K. Clemens, and M. D. Ross, "The principles and quality of the buried subcarrier encoding and decoding system," *IEEE Trans. Consum. Electron.* Vol. CE-27, No. 3, Aug. 1981, pp. 352–360.
7. T. J. Christopher, F. R. Stave, and W. M. Workman, "The SelectaVision player," *IEEE Trans. Consum. Electron.* Vol. CE-27, No. 3, Aug. 1981, pp. 340–351.
8. T. Inoue, T. Hidako, and V. Roberts, "The VHD videodisc system," *SMPTE J.,* Nov. 1982, pp. 1071–1076.
9. E. J. Freeman, G. N. Mehrotra, and C. P. Repko, "Stereo audio on CED videodiscs," *IEEE Trans. Consum. Electron.,* Vol. CE-29, No. 3, Aug. 1983, pp. 153–162.
10. Numerous number of patents.

Chapter 5

FUNDAMENTALS OF OPTICS

5.1 INTRODUCTION

A diversity of phenomena are encompassed in the narrow band of wavelength to which the human eye is sensitive. This band is not even a full octave wide. These are the wavelengths of the electromagnetic spectrum where interactions of waves with electrons first become important. Waves of lower energy mainly stimulate the motion of atoms and molecules, so they are usually sensed as heat. Radiation of higher energy can ionize atoms and permanently damage molecules, so that its effects seem largely destructive. Only in the narrow transition zone between these extremes is the energy of light well tuned to the electronic structure of matter [1–9].

Lately, electronic and communication engineers have been dealing more and more with optics. In capacitive videodisc systems, optics can be involved in the recording process and testing. In the optical videodisc systems, optics is part of both the recording and the reading processes. Some important questions are:

- What is the smallest recording and/or reading information element?
- How does the optical part affect the overall frequency response of the system?
- What is the depth of focus of the readout system?

We introduce here briefly, for the sake of completeness, some of the concepts of optics.

The field of optics has often been divided into three parts: geometrical optics, physical optics, and spectroscopy. For videodisc systems needs, optics could better be divided into more major categories. In geometrical optics, basic elements are mirrors, lenses, and prisms, and in the elementary treatment of image formation, we deal with the laws of reflection and refraction. Rigorous understanding of image formation is possible only by taking into account the wave nature of light. In physical optics, interactions between light and matter, polarization for example, can be discussed. Because of its role in advanced optics, the concept of transformations must be considered. Although holography may be combined with videodiscs, it may be omitted from the basic discussion. Also, quantum optics and relativistic optics are not essential for videodisc system analysis.

We present the basic notations of geometrical optics because valuable information about the size, location, and aberrations of image can be obtained from geometric considerations. Once the fundamentals of geometric optics are understood, the study of diffraction and its effects are much more easily understood.

An understanding of diffraction is necessary for a study of the concept of transformation. In diffraction, the wave nature of light is involved in an essential way. One of the important properties of waves is the superposition principle: The algebraic sum of the amplitude that represents several disturbances also represents a disturbance. This topic will be discussed under the heading of wave optics.

In general, engineers are familiar with linear system theory and Fourier analysis. Because of that, these two concepts are combined here with the theory of diffraction to describe the image-forming process in terms of linear filtering operation for both coherent and incoherent imaging. In doing so, we introduce the concepts of the subject commonly referred to as Fourier optics.

Light is made up of continually varying electric and magnetic fields which may be represented by vectors. In order to provide an understanding of the properties of light and matter, physical optics are discussed briefly. In addition, optic modulators and some characteristics important for optical rideout are discussed.

5.2 GEOMETRICAL OPTICS

The simplest approach to optics is probably to examine how optical systems affect the light rays passing through them; this is the geometrical optics approach. Lines drawn in the direction of propagation of light are called rays. Light rays exist only in theory; they cannot be isolated experimentally. If a medium is transparent, the light will pass through undisturbed. At a boundary between two media, one or more of three things can happen: the light may be transmitted, reflected, or refracted (which means that the light deviates from its initial direction). These are usually referred to as the (three) laws of geometrical optics. Accordingly, geometrical optics can be considered as that part of optics involving image formation and related phenomena that can be discussed within the framework of the three laws of geometrical optics.

The branch of geometrical optics called Gaussian optics is based on an extremely useful approximation to the correct theory [8]. The equations of Gaussian optics are derived by taking first-order approximations, $\sin x \simeq x$, for example, to the equations of the exact theory. Typically, it is assumed that an optical axis can be defined for the optical system and that all light rays and all normals to refracting or reflecting surfaces make small angles with the axis. Such light rays are called paraxial rays (in this case sin $x \approx x$). The small-angle approximations neglect all terms after the first in the Taylor series, of the sine, for example:

$$\sin x = x - \frac{x^3}{3!} + \frac{x^5}{5!} - \cdots \tag{5-1}$$

and this leads to gross inaccuracies in actual optical systems.

The light rays are straight lines in a medium with constant light velocity v. The index of refraction n of a medium is $n = c/v$, where c is the velocity of light in a vacuum. Consider the ray of light in Fig. 5–1, passing from a medium of index n_1 to a medium of index n_2; Fermat's principle states that light originating at some arbitrary point A in one medium and terminating at another point, B, in a second medium, will follow the path for which the travel time is a minimum. It follows, then, that the next relation is satisfied (Fig. 5–1):

$$n_1 \sin \theta_1 = n_2 \sin \theta_2 \tag{5-2}$$

which is also known as Snell's law. Angles θ_1 and θ_2 are the angles of incidence and reflection, respectively.

The law of reflection can be derived from Fermat's principle.

- *First law of reflection:* The angle of incidence is equal to the angle of reflection (this law was known to Euclid of Alexandria in about 300 B.C.).
- *Second law of reflection:* The incident ray, the reflected ray, and the normal created on the surface at the point of incidence all lie in the same plane. This plane, called the plane of incidence, is normal to the reflecting boundary.

If Snell's law is considered as the first law of refraction, then the statement that the incident ray, the refracted ray, and the normal erected on the boundary of the point of incidence all lie in the same plane can be considered as the second law of refraction.

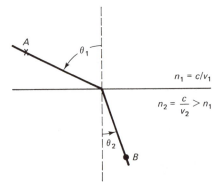

Figure 5–1 Fermat's principle.

The distance between two points, regardless of the medium between them, is called the geometric path length, L. An optical path length, S, is defined as the product of the geometric path length and refractive index, $S = Ln$. In a medium of index n, it takes the light n times as long as in a vacuum ($n = 1$) to traverse a given distance.

The rays bend toward the region of higher refractive index. Consider, for example, air ($n_1 \approx 1$) and glass ($n_2 = 1.5$) boundaries. There are areas in the glass that cannot be reached by the light incident from the air. Also, if the light travels from glass into the air, and if the angle of incidence is made larger than the critical angle, α:

$$\sin \alpha = \frac{n_2}{n_1} \tag{5-3}$$

the light is reflected back into the dense medium (glass). The angle α is called the critical or minimum angle of total internal reflection. The relationships that are postulated or developed for ideal optical systems are based on small-angle approximations and on the assumption that the refractive index is constant with variations of wavelength.

5.2.1 Simple Lenses

Although mirrors and prisms play important roles in many optical systems, we shall restrict our attention here to lenses; the behavior of mirrors is readily obtained by extending the study of lenses.

The simple lens consists of two smooth surfaces bounding a reflecting material and has mainly axial symmetry. Normally, one or both surfaces are paths of a sphere. An ideal lens converts an incident spherical wave into another spherical wave. There are several configurations of single-element lenses that exhibit this ideal behavior to some extent. Cross sections of the major types of simple lens are shown in Fig. 5-2: biconvex, planoconvex, meniscus convex (positive meniscus), biconcave, planoconcave, and meniscus concave (negative meniscus). A lens that tends to move a wavefront more convex or less concave toward the source is called positive. A lens with the opposite effect is said to be negative. A lens is called a thin lens if there is negligible translation within the lens: two refracting surfaces are close enough for their separation

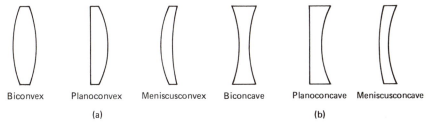

Biconvex   Planoconvex   Meniscusconvex   Biconcave   Planoconcave   Meniscusconcave

(a)   (b)

Figure 5–2   Basic types of simple lenses: (a) positive; (b) negative.

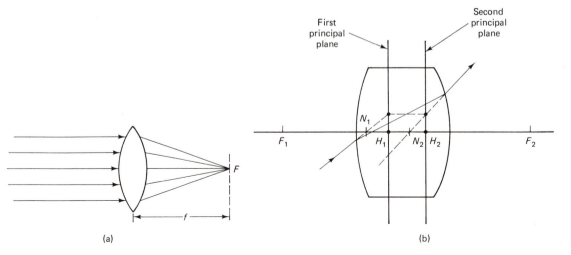

Figure 5–3 Characteristic points: (a) focal; (b) cardinal.

to be negligible with respect to the object and image distances and the focal length. There is no numerical limit at which a thin lens ends and a thick lens begins.

If the rays parallel to the axis at the lens (sunlight) pass through the double convex lens (Fig. 5–3a) at the focal point or focus F, a small intense spot of light is obtained. There are two focal points, F_1 and F_2, for each lens (Fig. 5–3b).

Any lens has six cardinal points; two focal points, F_1 and F_2; two principal points, H_1 and H_2; and two nodal points, N_1 and N_2. The planes constructed of the principal points normal to the optic axis are called principal planes. The principal planes define the planes in a lens where, hypothetically, refraction is assumed to take place without reference to where it actually takes place. If the locations of the cardinal points are known, the behavior of the lens can be determined without going into the details of the refraction process at each surface. Thus the lens can be effectively

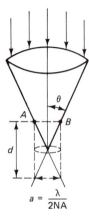

Figure 5–4 Focal depth.

modeled by its two focal points and its two principal planes. All rays passing through the lens behave as if they travel directly to the first principal plane without refraction at the first glass surface, then parallel to the optical axis until they reach the second principal plane, and finally exit directly from that point without refraction at the second glass surface.

The numerical operture (NA) of a lens is

$$NA = n \sin \theta \tag{5-4}$$

where θ is the maximum possible value of the angle between the optical axis and a ray originating on the axial object (or image) point and passing through the lens (Fig. 5-4).

The focal depth, d, is shown in Fig. 5-4, and it can be expressed by geometrical optics. By setting the condition that distance $\overline{AB} = \lambda/2NA = a$, it follows:

$$d = \frac{\lambda}{(NA)^2} \tag{5-5}$$

In Fig. 5-5 it is shown how for given object A, a corresponding image B can be obtained. This is an elementary imaging system. For this figure the following equations can be obtained:

Newton's form of the lens equation:

$$f_1 f_2 = x_1 x_2 \tag{5-6}$$

Gauss's lens equation:

$$\frac{f_1}{s_1} + \frac{f_2}{s_2} = 1 \tag{5-7}$$

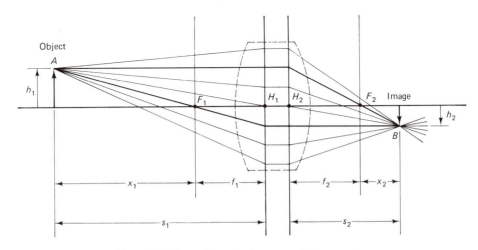

Figure 5-5 Image formation in symmetrical convex lens.

If the media on both sides of the lens have the same index of refraction (e.g., a lens in air), then $f_1 = f_2$.

The lens magnification β is defined as

$$\beta = \frac{h_2}{h_1} = \frac{s_2}{s_1} \tag{5-8}$$

In the optical videodisc systems, the elements that physically limit the angular size of the cone of light accepted by the optical system, and therefore govern the total radiant flux reaching the image plane, are used. A common name for such an element is "aperture stop." The field stop is the element that physically restricts the size of the field of view (or image). The entrance pupil is the image of the aperture stop, as viewed from object space, formed by all of the optic elements preceding it. The exit pupil is the image of the aperture stop, as seen looking back from image space, formed by all of the optical elements that follow it (Fig. 5-6). The aberrations of a system, as well as its resolution, are often associated with the exit pupil.

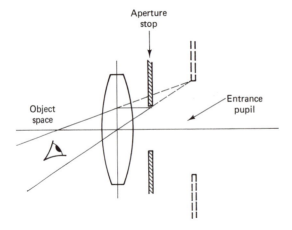

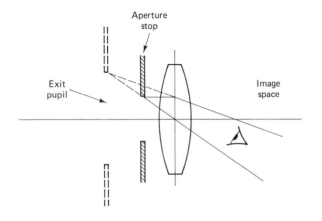

Figure 5-6 Entrance and exit pupils.

5.2.2 Aberrations

In an ideal imaging system, a point object is mapped—projected into a point image at the proper location. Such a system is called "diffraction limited." But there are other limitations besides diffraction that prevent an object from being a true likeness of the object. These are called aberrations.

The seven conventional classes of image defects—aberrations—are:

Monochromatic aberrations (defined by L. Seidel in 1856):

1. Spherical aberrations
2. Coma
3. Astigmatism
4. Curvature of field
5. Distortion

Chromatic aberrations:

6. Longitudinal chromatic aberration
7. Lateral chromatic aberration

Aberrations become more serious if the lens, or mirror, has a large aperture. Also, their influences on the videodisc systems are not equal. The most important are the spherical aberrations; numbers 2, 3, and 4 introduce crosstalk and number 5 interferes with tracking.

Spherical aberration can be defined as the variation of focus with aperture [4]. Figure 5–7 shows spherical aberrations of a lens and a spherical mirror. Those rays passing through the lens near the optical axis are brought to a focus in a different location from those passing through the lens near its edge. The point at which the ideal image is formed is often referred to as the paraxial focus; the transverse deviation

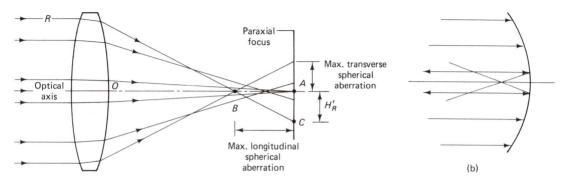

Figure 5–7   Spherical aberation of (a) lens, (b) spherical mirror.

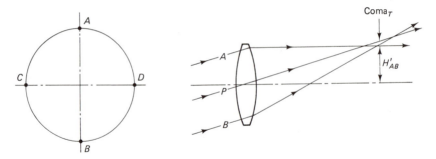

Figure 5–8 Coma.

of a ray from this point is called a transverse (or lateral) aberration, whereas the longitudinal deviation of a ray from this point is called longitudinal (or axial) aberration. Both are known as ray aberrations, as opposed to wavefront aberrations.

For a given aperture and focal length, the amount of spherical aberration in a simple lens is a function of object position and the shape, or bending, of the lens. The image of a point formed by a lens with spherical aberration is usually a bright dot surrounded by a halo of light [4]. The spherical aberration gives a soft contrast of the image and blurs its details.

Coma can be defined as the variation of magnification with aperture, and the aberration is named after the comet shape of the figure obtained for a point image. The distortion is illustrated in Fig. 5–8. Coma is a particularly disturbing aberration because it is nonsymmetrical. Coma is minimized if the optical system is made symmetrical about the aperture stop and if the lateral magnification is unity. In the absence of spherical aberration, coma does not depend on the aperture position; but if spherical aberration is present, a stop position can be found for which coma vanishes, given complete freedom of movement [3].

Astigmatism is another off-axis aberration of the image-blurring type (Fig. 5–9). The image of a point source formed by a fan of rays in the tangential plane will be a line image. This line, called tangential image, is perpendicular to the tangential plane; that is, it lies in the sagittal plane. We define the tangential plane to be the plane containing both the chief ray and the optical axis, and the sagittal plane to be the plane that is both perpendicular to the tangential plane and contains the chief ray. Conversely, the image formed by the rays of the sagittal fan is a line that lies in the tangential plane. Astigmatism occurs when the tangential and sagittal (radial) images do not coincide. In the presence of astigmatism, the image of a point source is not a point, but takes the form of two separate lines, as shown in Fig. 5–9. Between the astigmatic focii the image is an elliptical or circular blur. The amount of astigmatism in a lens is a function of the shape of the lens and its distance from the aperture (diaphragm), which limits the size of the bundle of rays passing through the lens.

The basic field curvature, called Petzval curvature, is associated with every optical system (Fig. 5–10). This curvature is a function of the index of refraction of

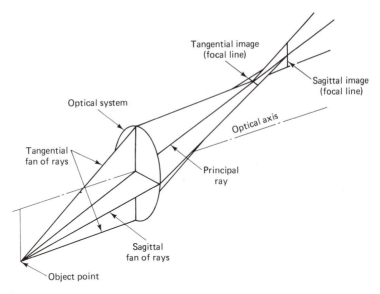

Figure 5–9   Astigmatism.

the lens elements and their surface curvature. With no astigmatism, the sagittal, tangential, and Petzval surfaces coincide with each other. In the presence of astigmatism, the tangential image surface lies three times as far from the Petzval surface as the sagittal image, and both surfaces are on the same side of the Petzval surface (Fig. 5–10).

If all other aberrations have been eliminated, the image is well defined, but images of points are displaced from the paraxial image by an amount proportional to the cube of the object distance off-axis. This type of aberration is called distortion

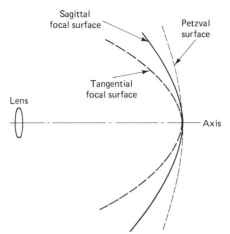

Figure 5–10   Petzval, sagittal focal, and tangential focal surfaces.

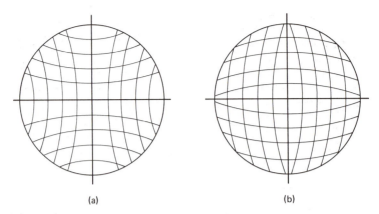

(a)   (b)

Figure 5–11 Distortion of a rectangular object pattern: (a) pincushion; (b) barrel.

and it causes the image of any straight line in the object plane, that does not meet the axis, to be curved (Fig. 5–11).

Distortion is an aberration produced when the lateral magnification varies with off-axis distance. If the magnification increases with off-axis distance, pincushion distortion results. On the other hand, a decrease of magnification with off-axis distance produces barrel distortion. The type of distortion, as well as its severity, may be associated with the location of the aperture stop. For example, a stop located in front of a thin positive lens will produce barrel distortion, whereas a stop placed behind this type of lens will cause pincushion distortion. If the stop is located at the lens, the distortion will vanish.

Spherical aberration, coma, and astigmatism reduce sharpness in the image. Field curvature and distortion cause geometrical distortion of the image.

Chromatic aberrations are not necessary consequences of material inhomogeneity or fabrication errors, and can occur when the laws of reflection and refraction are applied to mathematically correct surfaces.

The focal length of the lens is different for different colors because it depends on the refractive index of the lens material. This index varies with wavelength. Longitudinal chromatic aberration is the variation of focus, or image position, with wavelength

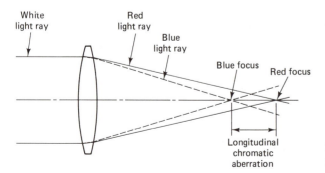

Figure 5–12 Longitudinal chromatic aberration.

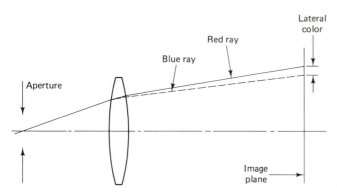

Figure 5–13 Lateral color.

(Fig. 5–12). In general, the index of refraction of optical materials is higher for short wavelengths than for long wavelengths. The image of an axial point in the presence of chromatic aberration is a central bright dot surrounded by a halo. If the screen on which the image is formed is moved toward the lens, the central dot will become blue; if it is moved away, the central dot will become red.

When a lens system forms images of different sizes for different wavelengths, or spreads the image of an off-axis point into a rainbow, the difference between the image heights for different colors is called lateral color or chromatic difference of magnification (Fig. 5–13).

5.3 WAVE OPTICS

From geometric optics it follows that the boundaries of a shadow are sharp and distinct. Experiments show that the boundaries of a shadow are diffused and consist of fringes. This can be explained by assuming a wave nature of light.

5.3.1 Waves

Waves can be treated in a very elementary way in order to review some of their important characteristics. The simplest representation of a wave is

$$s = A \cos kz \tag{5-9}$$

where A is the amplitude and k is called the propagation constant or wave number. The distance between two maxima is the wavelength λ, so

$$\cos k(z + \lambda) = \cos kz \tag{5-10}$$

and therefore

$$k = \frac{2\pi}{\lambda} \tag{5-11}$$

and k is a measure of the number of waves per unit length.

The ordinary units of wavelength measure in the optical region are the angstrom (Å), the micrometer or micron (μm), and the nanometer (nm). One micrometer is a millionth of a meter, and an angstrom is one ten-thousanth of a micrometer:

$$1 \ \mu m = 10^{-6} \ m \qquad 1 \ nm = 10^{-9} \ m \qquad 1 \ \text{Å} = 10^{-4} \ \mu m$$

If there is a time movement with velocity, v, then

$$s = A \cos \left[2\frac{\pi}{\lambda} (z - vt) \right] \tag{5-12}$$

or

$$s = A \cos (kz - \omega t) \tag{5-13}$$

where

$$\omega = vk = 2\pi f = 2\frac{\pi}{T} \tag{5-14}$$

is the angular frequency of the wave, and f is the time frequency of the wave, $f = 1/T$. A wave moving with velocity v is called a traveling wave.

5.3.1.1 Complex amplitude and the quadratic-phase signal.

It is often advantageous to represent physical quantities by phasors, which are complex-valued functions. This representation should be properly formulated to avoid possible confusion. For example, a time-varying voltage is often represented by the phasor $V(t) \exp [j\phi(t)]$, which implies that the voltage consists of a real part and an imaginary part, and this does not make sense physically. In most cases where a physical quantity is represented by a phasor, there is an implicit understanding that it is the real or the imaginary part of the phasor that is of interest. Care must be taken not to violate any of the rules of complex algebra. For example, the real part of the product of two complex quantities, in general, is not equal to the product of the real parts of quantities; the equivalent is true for the imaginary parts.

Monochromatic, linearly polarized wave fields can be adequately described by a real-valued function of position and time, $u(x, y, z; t)$. We associate this scalar function with the magnitude of either the electric field vector or the magnetic field vector:

$$u(x, y, z; t) = A(x, y, z) \cos [2\pi ft - \phi(x, y, z)] \tag{5-15}$$

where f is the temporary frequency of a wave, in vacuum. Many problems of interest can be realized using the quantity called a complex amplitude of the wave field, rather than the real physical wave field. The complex amplitude of the wave field $u(x, y, z; t)$ is defined as

$$u(x, y, z) = A(x, y, z)e^{j\phi(x, y, z)} \tag{5-16}$$

When necessary, $u(x, y, z; t)$ can be obtained from the complex amplitude by

$$u(x, y, z; t) = \text{Re}\,\{u*(x, y, x)e^{j2\pi ft}\} \tag{5-17}$$

where Re $\{\cdot\}$ means "real part of."

Instead of the three independent variables (x, y, z), the vector r, $\mathbf{r}$, can be used. Thus the complex amplitude $u(\mathbf{r})$ of the wave $u(\mathbf{r}, t)$ is defined by the phasor

$$u(\mathbf{r}) = A(r)e^{j\phi(\mathbf{r})} \tag{5-18}$$

so that this wave may be expressed as

$$u(\mathbf{r}, t) = \text{Re}\,\{u*(\mathbf{r})e^{j2\pi ft}\} \tag{5-19}$$

The time average of a periodic function $f(t)$ is

$$\langle f(t) \rangle = \frac{1}{T} \int_0^T f(t)\,dt \tag{5-20}$$

where $T = 2\pi\omega$ is the period.

Using the identity

$$\text{Re}\,\{AB*\} = \text{Re}\,\{A*B\} \tag{5-21}$$

it can be shown that

$$\langle \text{Re}\,(A_{max}e^{-j\omega t})\,\text{Re}(B_{max}e^{-j\omega t}) \rangle = \tfrac{1}{2}\,\text{Re}\{A_{max}B_{max}\} \tag{5-22}$$

For a time-varying electrical field E, the average value of the energy is

$$I = \langle \text{Re}\,(E)\,\text{Re}\,(E*) \rangle = \tfrac{1}{2}\,\text{Re}(E_{max}E^*_{max}) = \tfrac{1}{2}\,E^2_{max} \tag{5-23}$$

The time average of the square of the quantity $u(x, y, z; t)$ is called the irradiance distribution of the light. The irradiance (W/m²) is a measure of radiant-flux density (power per unit area incident on a surface). In the optical region of the electromagnetic spectrum, virtually all detectors respond to this quantity rather than to the field strength.

For monochromatic light, the irradiance distribution may be obtained directly from the complex amplitude:

$$I(x, y, z) = |u(x, y, z)|^2 \tag{5-24}$$

This eliminates the necessity of finding the time average of $u^2(x, y, z; t)$.

A complex-valued function (phasor) whose real part is an even function and whose imaginary part is odd is said to be Hermitian, while a phasor whose real part is odd and whose imaginary part is even is called anti-Hermitian. The Fourier transforms of such functions possess certain special properties.

When the wave fields of interest are described in planes for which $z = $ constant, it is convenient to regard the quantity $u(x, y, z)$ as a two-dimensional function of

x and y with parametric dependence on z, and the complex amplitude in the plane $z = z_i$ is

$$u(x, y, z_i) = u_i(x, y) \tag{5-25}$$

As an example, let's consider a two-dimensional complex amplitude transmittance function t of the various apertures and semitransparent objects. If the object of interest is placed in the $z = z_i$ plane, then

$$t_i(x, y) = \frac{u_i^+(x, y)}{u_i^-(x, y)} \tag{5-26}$$

where the incident wave field is $u_i^-(x, y) = u(x, y, z - e)$, the wave field leaving is $u_i^+(x, y) = u(x, y, z + e)$, and e is a very small distance.

For known amplitude transmittance for a particular object and the complex amplitude of the incident wave field, the transmitted wave field may be described by

$$u_i^+(x, y) = u_i^-(x, y) t_i(x, y) \tag{5-27}$$

For example, if a clear circular aperture of diameter d is placed in the plane $z = z_i$, and the plane wave field

$$u_i^-(x, y) = A e^{j\omega x} \tag{5-28}$$

impinges on it, the transmitted wave field is

$$u_i(x, y) = A e^{j\omega x} \qquad \text{cyl}\left(\frac{\sqrt{x^2 + y^2}}{d}\right) \tag{5-29}$$

It is convenient to define the quadratic-phase signal as

$$q(x, y; a) = e^{j\pi a(x^2 + y^2)} \tag{5-30}$$

or

$$q(r, a) = e^{j\pi ar^2} \qquad r^2 = x^2 + y^2 \tag{5-31}$$

With a, a_1, a_2, and b real constants, this function has the following properties:

$$q(\pm x, \pm y; a) = q(x, y; a) \qquad q(x, y; a_1)q(x, y; a_2) = q(x, y; a_1 + a_2) \tag{5-32}$$

$$q(x, y; -a) = q^*(x, y; a) \qquad q(x, y; a_1)q^*(x, y; a_2) = q(x, y; a_1 - a_2) \tag{5-33}$$

$$q(bx, by; a) = q(x, y; b^2 a) \qquad q(x, y; a_1) ** q(x, y; a_2) = \frac{j}{a_1 + a_2} q$$

$$\left(x, y; \frac{a_1 a_2}{a_1 + a_2}\right) \tag{5-34}$$

The Fourier transform of this function is

$$Q(\xi, \eta; a) = \mathscr{F}\mathscr{F}\{q(x, y; a)\} = \frac{j}{a} q^* \left(\xi, \eta; \frac{1}{a} \right) \tag{5-35}$$

With $s(x, y)$ an arbitrary function, we have

$$s(x, y)**q(x\ y; a) = jaq(x, y; a)[S(ax, ay)**q^*(x, y; a)] \tag{5-36}$$

5.3.1.2 Miscellaneous.   Radiant energy Q_e is energy transported in the form of electromagnetic waves. The known electromagnetic spectrum is diagrammed in Fig. 5–14 and ranges from cosmic rays to radio waves. Each kind of electromagnetic radiation transports energy with a velocity in vacuum of $c = 2.997925 \pm 0.000003 \times 10^8$ m/s. Light is an aspect of radiant energy which a human observer is aware of through the stimulation of the retina of the eye. Thus light is psychophysical rather than purely physical or purely psychological. The visible portion of the electromagnetic spectrum takes up less than one octave, ranging from violent light with a wavelength of 0.4 μm to red light with a wavelength of 0.76 μm.

Radiometry is the measurement of quantities associated with radiant energy. Photometry is the measurement of quantities associated with visually evaluated radiant energy [1]. Radiant flux, $\Phi_e = \partial Q_e / \partial t$ is the rate of change of radiant energy.

Luminous energy Q_v is that aspect of radiant energy which is light. Luminous flux, $\Phi_v = \partial Q_v / \partial t$, is the rate of change of luminous energy. It is visually evaluated radiant flux. The spectral flux Φ_λ denotes the flux per unit wavelength of wavelength λ and is defined as

$$\Phi_\lambda = \frac{\partial \phi}{\partial \lambda} \tag{5-37}$$

The flux within the band from λ_1 and λ_2, $\Phi(\lambda_1, \lambda_2)$, is

$$\Phi(\lambda_1, \lambda_2) = \int_{\lambda_1}^{\lambda_2} \Phi_\lambda(\lambda)\, d\lambda \tag{5-38}$$

Such nonspectral quantities are sometimes called integrated or broadband quantities. Total flux Φ is defined in terms of spectral flux by

$$\Phi = \int_0^\infty \Phi_\lambda(\lambda) d\lambda \tag{5-39}$$

Table 5–1 presents eight quantities that are basic to radiometry and photometry [1].

The velocity of propagation of light waves in vacuum is larger than the velocity in any other medium. In ordinary glass for example, the velocity is about two-thirds of the velocity in free space. Thus the index of refraction for the glass is $n = c/v = \frac{3}{2} = 1.5$.

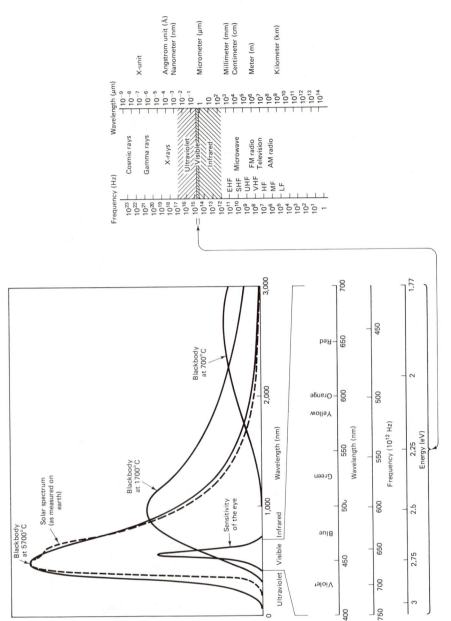

Figure 5–14 Electromagnetic spectrum.

TABLE 5–1 RADIOMETRIC AND PHOTOMETRIC SPECTRAL AND
NONSPECTRAL ENERGY AND FLUX TERMS, SYMBOLS, AND UNITS

	Radiometric, subscript c		Photometric, subscript v	
	Nonspectral	Spectral, subscript λ	Nonspectral	Spectral, subscript λ
Energy, Q	Radiant energy, Q_c joules	Spectral radiant energy, $Q_{c\lambda}$ joules per unit wavelength	Luminous energy, Q_v lumen-seconds	Spectral luminous energy, $Q_{v\lambda}$ lumen-seconds per unit wavelength
Flux, Φ	Radiant flux, Φ_c watts	Spectral radiant fluz, $\Phi_{c\lambda}$ watts per unit wavelength	Luminous flux, Φ_v lumens	Spectral luminous flux, $\Phi_{v\lambda}$ lumens per unit wavelength

5.3.2 Interference

Interference could be studied as a part of physical optics, where the wave nature of light becomes apparent. Interference merely causes a redistribution of the light; in interference, energy is neither changed nor destroyed. If light waves from two different sources are radiating through a common medium, and if the medium is linear, they will be linearly summed. This is according to the principle of superposition, which implies that individual waves do not affect one another; that is, the waves are uncoupled.

In the process of superposition of light waves, it is important how they are added. With a high degree of coherence among the N light patches, amplitudes of the waves would have to be superposed to form an interference pattern. The total disturbance E_T due to N coherent contributions with the same phase is

$$E_T^2 = N^2 E_m^2 \qquad (5\text{--}40)$$

or

$$E_T = N E_m \qquad (5\text{--}41)$$

where a collection of waves of equal amplitude E_m is considered.

The superposition of N waves of equal amplitudes E_m and angle frequencies ω but with random phases produces a wave whose amplitude is $\sqrt{N}$ times the amplitude of single wave:

$$\langle E_T^2 \rangle = E_m^2 N \qquad (5\text{--}42)$$

and

$$E_T = \sqrt{N}\, E_M \qquad (5\text{--}43)$$

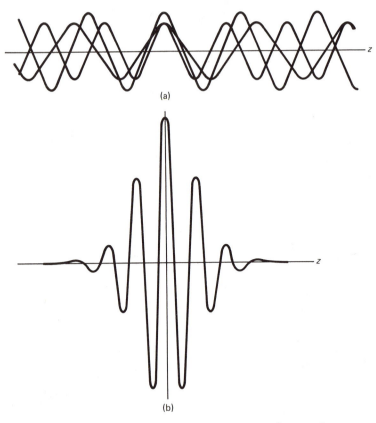

Figure 5–15 Equivalent ways to represent a group of wave trains.

The energy, which is proportional to E^2, is equal to N times the energy of one wave. Two equivalent ways to represent a group of wave trains are shown in Fig. 5–15.

Waves that do not travel but remain stationary are standing waves. They can be produced by summing the incident and reflects waves. For example, if the incident wave is

$$s_1(t) = S_m \sin (kz - \omega t) \qquad (5\text{--}44)$$

and assuming that there are no losses during reflection, the reflected wave is

$$s_2(t) = S_m \sin (-kz - \omega t) \qquad (5\text{--}45)$$

Their sum is

$$s = s_1 + s_2 = A(z) \sin \omega t \qquad (5\text{--}46)$$

where the envelope $A(z)$ depends on position:

$$A(z) = -2S_m \cos kz \qquad (5\text{--}47)$$

Consider two linearly polarized monochromatic waves with identical frequencies and directions of propagation. The resulting intensity of a given point is

$$I = |A_1 e^{j(\phi_1 - \omega t)} + A_2 e^{j(\phi_2 - \omega t)}|^2$$

$$= A_1^2 + A_2^2 + 2A_1 A_2 \cos \delta \tag{5-48}$$

where $\delta = \phi_2 - \phi_1$, phase difference (shift).

A wave field is called fully coherent if all the individual spectral and polarization components have the same phase and amplitude relationship. The visibility of interference (fringes) is

$$V = \frac{I_{max} - I_{min}}{I_{max} + I_{min}} \tag{5-49}$$

For fully incoherent light, the phase shift is random and uniformly distributed between 0 and 2π; thus

$$V = 0 \tag{5-50}$$

In fully coherent light, the resultant amplitude is the sum of the component amplitudes; in fully incoherent light, the resultant intensity is the sum of the component intensities. There is no observable interference with incoherent light. For partially coherent light,

$$0 \le |V| \le 1 \tag{5-51}$$

5.3.3 Diffraction Theory

Electronic engineers are familiar with linear systems. The concept of linearity is based on the superposition theorem, which may be stated as follows: If cause and effect are linearly related, the total effect of several causes acting simultaneously is equal to the sum of the effects of the individual causes acting one at a time. These two elements, the concept and the theorem, are inseparable; the first is essential to the second and the second defines the first.

A linear system h is defined by the relation

$$h[af_1(t) + bf_2(t)] = ah[f_1(t)] + bh[f_2(t)] \tag{5-52}$$

where $f_1(t)$ and $f_2(t)$ are arbitrary inputs, and a and b are constants.

In optics, the superposition theorem is basis for Huygens's principle. Huygens's principle states that each point on a wavefront may be considered as instantaneously and continuously the origin of a new spherical wavefront moving outward from the point. When the original wavefront is infinite in extent, the new wave form is simply the envelope of the wavefronts in the direction of propagation.

If the wave is partially blocked by an opaque barrier or an aperture in a metal plate, the apparent direction of propagation will be altered. The bending of the path is called diffraction. Diffraction is actually a kind of interference. Exact solution of diffraction problems is extremely intricate; in fact, the most difficult calculations in optics are probably diffraction problems [3].

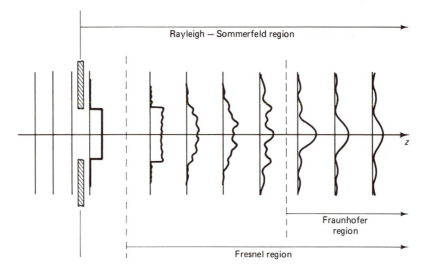

Figure 5–16   Various regions of diffraction associated with an aperture, and nature of the corresponding irradiance profiles.

An unrestricted wave does not produce a diffraction pattern; actually, the finite size of an aperture generates diffraction effects because waves from one part of the opening interfere with waves from the other part. Points on the screen (observation plane), where wavefronts arrive simultaneously, or in phase—reinforcing each other—will be illuminated. However, whenever the waves arrive exactly out of phase, destructive interference will occur and the corresponding points on the screen will be dark.

There are two distinguishable classes of diffraction (Fig. 5–16) that are normally of interest, corresponding to different approximations used in the evolution of the diffraction integral. The field distribution far from the object (at infinity) is called the Fraunhofer diffraction pattern and is mathematically simpler. The field distribution near the object is called the Fresnel diffraction pattern, and the corresponding calculation is performed using Fresnel integrals. The entire space to the right of the diffracting aperture in Fig. 5–16 is called the Rayleigh–Sommerfield region.

Figure 5–17a shows a general case of geometry of diffraction (or radiant energy through an aperture), while Fig. 5–17b shows a simplified case for parallel planes. The phase change δ at P with respect to Q is

$$\delta = kr = \frac{2\pi}{\lambda} r \qquad (5\text{–}53)$$

If the amplitude $A(Q)$ of a wave is given, the amplitude $A(P)$ of point P can be obtained using Huygens's principle: The total amplitude on the observation plane, for any point, is the sum of all the individual contributions from the source. The source is considered to be sum of point radiators, so it is

$$A(P) = C_0 \iint_S \frac{1}{r} A(Q)\phi(P, Q)e^{jkr}\, dS \tag{5-54}$$

where $dS = d\xi d\eta$ and C_0 is the scale constant; S is a source surface, and $\phi(P, Q)$ is an obliquity factor. Only the simplified case (Fig. 5–17), where every point (ξ, η) in the source is approximately equidistant from point (x, y), will be discussed.

The obliquity factor is 1, and $A(Q) = A(\xi, \eta)$, $A(P) = A(x, y)$. The distance $r = \overline{QP}$ is

$$r = \sqrt{r_0{}^2 + (x - \xi)^2 + (y - \eta)^2} \tag{5-55}$$

The first-order approximation (the Fresnel approximation),

$$\sqrt{1 + x} \approx 1 + \tfrac{1}{2} x \qquad \text{for } x \ll 1 \tag{5-56}$$

gives

$$r \approx r_0 + \frac{x^2 + y^2}{2r_0} + \frac{\xi^2 + \eta^2}{2r_0} - \frac{x\xi + y\eta}{r_0} \tag{5-57}$$

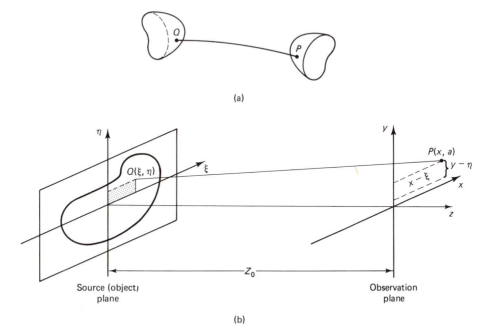

(a)

(b)

Figure 5–17 Geometry of diffraction: (a) general case; (b) simplified case for parallel planes.

If the zero-order approximation for r is introduced in the amplitude part of Eq. (5–54), $1/r \approx 1/r_0$, then

$$A(x, y) = C_1 \iint_S A(\xi, \eta) e^{jkr} \, dS \tag{5–58}$$

where C_1 is a new constant.

5.3.3.1 Fresnel diffraction.

When a wave is generated by a point source and is observed close to the source, it is a spherical wave rather than a plane wave and the spatial dependence of its amplitude varies as $[\exp{(ikr)}]/r$. That means that the intensity falls off as $1/r^2$, where r is the distance from the source. Fresnel diffraction occurs whenever the source or the observation point (screen) is close to the diffracting body.

After substituting a first-order approximation for distance r in Eq. (5–58), a Fresnel transform is obtained:

$$A(x, y) = C_2 \iint_S A(\xi, \eta) e^{jk(\xi^2+\eta^2)/2r_0} e^{-jk(x\xi+y\eta)/r_0} \, d\xi \, d\eta \tag{5–59}$$

Fresnel transform of arbitrary function f, FRT $\{f\}$, can be obtained using the fast Fourier transform, FFT:

$$\text{FRT}\{f\} = \{\text{quadratic phase}\} \times \text{FFT}\{[\text{quadratic phase}] \times [f]\} \tag{5–60}$$

Figure 5–18 shows normalized illumination for the case of a half-plane, or knife edge. The diffraction causes a small but finite amount of light to appear in the shadow region. Also, it can be seen that the system exhibits rather pronounced "ringing." This is analogous to the ringing that occurs in the low-pass filter with

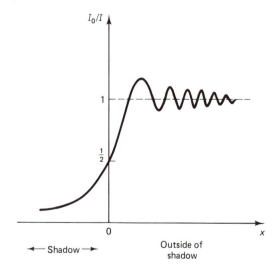

Figure 5–18 Fresnel diffraction: normalized illumination for a long slit.

transfer function that falls too abruptly. With the incoherent illumination, there is no ringing.

5.3.3.2 Fraunhofer diffraction.   The following conditions are assumed:

- The observation plane is located at a long distance from the aperture, so that $r = \overline{QP}$ (Fig. 5–17) is much larger than the dimensions of the source.
- Because $r \gg d$, where d is the largest diameter in the source, $A(\xi, \eta)$ is approximately constant over the source.

Equation (5–61) can be further simplified:

$$A(x, y) = C_1 \iint_S A(\xi, \eta)e^{jk(x\xi \,+\, j\eta)/r}\,dS \tag{5–61}$$

This is the Fraunhofer diffraction equation.

It should be noticed that Huygens's principle deals with the spherical wavefront (exact treatment). In the quadratic Fresnel approximation a parabolic surface is an approximation to a sphere. In the Fraunhofer (linear) approximation a plane (wavefront) is a linear approximation to a parabolic surface; it is not the best linear approximation to a sphere.

Diffraction by a Single Slit.   For a narrow slit, $w \gg l$, where l is length and w is width, consider two parallel rays in the XOZ plane with an angle θ to the Z axis (Fig. 5–19). If r_0 is the distance in the YOZ plane ($x = 0$), the distance r for any x is

$$r = r_0 + \xi \sin \theta \qquad \left(\frac{\xi}{r}\right)^2 \ll \frac{\xi}{r} \tag{5–62}$$

The elementary area in the slit is

$$dS = l\,d\xi \tag{5–63}$$

The Fraunhofer diffraction equation is

$$A(\ ) = Ce^{jkrl}\int_{-w/2}^{w/2} e^{jk\,\xi \sin \theta}\,d\xi \tag{5–64}$$

which gives

$$A = C'\frac{\sin \alpha}{\alpha} \tag{5–65}$$

where

$$\alpha = \tfrac{1}{2}\,kw \sin \theta \tag{5–66}$$

and

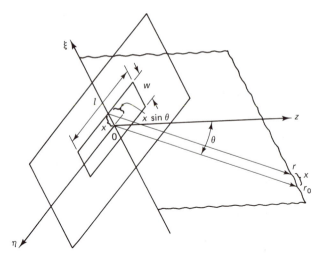

Figure 5–19 Diffraction geometry for a single slit.

$$C' = Ce^{jkr_0} \frac{\omega l}{2} \tag{5-67}$$

The energy carried by the wave (on the surface of the observation plane) is proportional to the intensity I:

$$I = \tfrac{1}{2} AA^* = I_0 \left(\frac{\sin \alpha}{\alpha} \right)^2 \tag{5-68}$$

where $I_0 = \tfrac{1}{2} C'C'^*$ is $I(\theta = 0)$. The normalized intensity $I/I_0 = (\sin \alpha/\alpha)^2$ is plotted in Fig. 5–20.

Rectangular or Square Source.   Integrating over both ξ and η, the normalized illumination on the observation plane for the two-dimensional diffraction pattern is given by

$$\frac{I}{I_0} = \left(\frac{\sin \alpha}{\alpha} \right)^2 \left(\frac{\sin \beta}{\beta} \right)^2 \tag{5-69}$$

where the definition of β is analogous to that of α.

Diffraction by a Circular Source.   When the source is circular with radius R, the amplitude on the observation plane can be obtained, for example, by using for the elementary area a slit parallel to the ξ axis of width $d\eta$ and length $2\sqrt{R - \eta^2}$:

$$A = 2Ce^{jkr_0} \int_{-R}^{R} e^{jk\xi \sin \theta} \sqrt{R^2 - \eta^2} d\eta \tag{5-70}$$

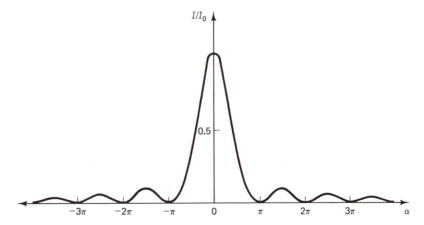

Figure 5–20   Radiant flux density distributed in the diffraction pattern of a slit.

The illumination is given by

$$I = I_0 \left[\frac{2J_1(\rho)}{\rho} \right]^2 \tag{5–71}$$

where

$$I_0 = \tfrac{1}{2} (Ce^{jkr_0}S)^2 \tag{5–72}$$

$J_1(\rho)$ is the Bessel function of the first order, and S is the area of the source.

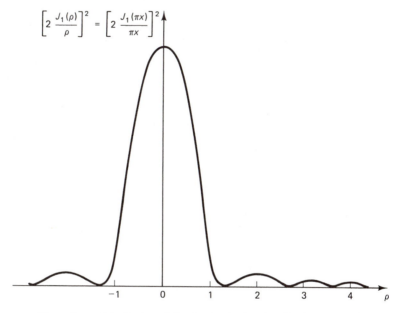

Figure 5–21   Distribution of illumination in the Airy disc (cross section).

The illumination pattern consists of a bright central spot of light surrounded by rings of rapidly decreasing intensity. This pattern is called the Airy disc. The corresponding curve is shown in Fig. 5–21.

The first dark ring corresponds with the first zero of $J_1(\rho)$, for which $\rho = 3.238$. Since

$$\sin \theta = \frac{\rho}{k\rho} \tag{5-73}$$

and $\sin \theta \approx \theta$, the next approximation can be obtained (the Rayleigh criterion):

$$\theta \simeq 3.83 \frac{\lambda}{2\pi R} = \frac{1.22\lambda}{D} \tag{5-74}$$

where D is the diameter of the aperture. This is just an arbitrary criterion and meaningless if not related to the signal-to-noise ratio. Approximately 84% of the energy in the Airy disc is contained in the central spot, and the illumination in the central spot is almost 60 times that of the first bright ring.

5.3.3.3 Diffraction gratings. An equal array of a large number of slits forms a diffraction grating. The diffraction gratings may be realized in the form of one, two, and three-dimensional grids. Gratings deflect some of the energy out of the central diffraction pattern into additional side patterns. The central diffraction pattern is called the zero-order component, while the side patterns are of higher order (first, second, etc.).

Next, we distinguish between transmission gratings and reflection gratings. The reflection type is used more frequently than the transmission type [8].

Gratings can also be classified as amplitude or phase gratings. A grating with opaque, or absorbing, bars and free intervals is called an amplitude grating. In a phase grating alternate strips merely retard the phase of light passing through. The interference maximum produced by amplitude gratings all have the same phase. If the object is a phase grating, a phase difference of 90° is introduced between the light in the zeroth order and in the higher-order maxima. In addition, the zeroth-order maximum is much brighter than with an amplitude object [7]. In other words, the interference patterns caused by amplitude and phase gratings are distinctly different. In a so-called normal grating, bars and intervals have the same width.

When two grids of comparable spacing are superimposed on one another, the points of intersection of their lines determine another repetitive sequence of lines, called moiré fringes.

5.4 FOURIER OPTICS

The concept of frequency analysis is familiar to electronic engineers. It is convenient to describe an electronic circuit in terms of its temporal frequency response. In the same manner it is convenient to describe an ongoing system in terms of its spatial frequency response [6].

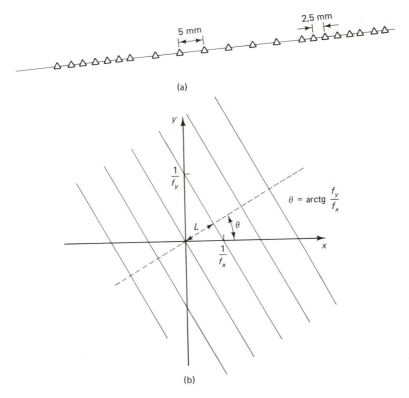

Figure 5–22 Spatial frequency: (a) pattern train; (b) simple two-dimensional periodical structure.

We will now illustrate the concept of spatial frequency. Figure 5–22a shows many identical characters at intervals of 2.5 mm at the ends and 5 mm in the middle. We can describe this train by saying that characters occur at the rate of 400 characters per meter in the middle of line, and 200 characters per meter at the ends. This rate is also called the spatial frequency of the carrier train. High-quality photographic lenses can form images containing structures with spatial frequencies up to 80 or 100 cycles per millimeter; the very finest structures that can be seen through microscopes using visible light have spatial frequencies below 5000 cycles per millimeter.

Figure 5–22b shows a simple two-dimensional periodical structure, for example, a two-dimensional grating, or lines of zero phase for the function [6] exp $[j2\pi(f_x x + f_y y)]$. This elementary function is directed in the xy plane on an angle θ and a spatial frequency $1/L$ in the direction θ, is complex, and is given by the frequency pair (f_x, f_y). The spatial period is

$$L = \frac{1}{\sqrt{f_x^2 + f_y^2}} \qquad\qquad (5\text{–}75)$$

Frequently, the symbol ν is used for the spatial frequency and f is kept for temporal frequency.

Because of an unavoidable blur, the elements surrounding a bright image element (point source) will also receive some of its illumination, so that the original reflectance value will be masked. This effect is expressed quantitatively in terms of the point-spread function, h, of the imaging system, which is analogous to the impulse response of electrical networks. Strictly speaking, h should be defined as a three-dimensional function. But as a rule, only two-dimensional point-spread functions are considered for videodisc systems (and in general). If we use Cartesian coordinates, we may write $h = h(x, y)$. The image illumination is obtained from the object luminance by (two-dimensional) convolution with the point-spread function:

$$s'(x, y) = \iint_{-\infty}^{\infty} s_0(x', y')h(x - x', y - y') \, dx' \, dy' \qquad (5\text{--}76)$$

where $s_0(x, y)$ is the equivalent input and $s'(x, y)$ is the illumination in the image.

5.4.1 Basic Principles of Fourier Optics

It is shown that diffraction phenomena provide a method for computing optical Fourier transforms. More precisely, it is shown that the intensity of the Fraunhofer diffraction pattern is proportional to the Fourier transform of the aperture function with the exactness of a constant.

The Fraunhofer pattern can be observed in two ways. First, the observation plane can be placed at a long distance from the object, as an approximation to infinity. Second, the Fraunhofer pattern can be seen by placing a double convex lens behind the object; the pattern then appears at the focal plane of the lens. The diffraction pattern arises from the aperture, not from the lens.

The Fourier transform property of a lens follows from the fact that the lens's effect on a wave passing through it is in introducing a phase shift which depends on the separation of the two glass surfaces. Suppose that a lens is in the opening, and that a ray of light leaves an object plane at a point $B(x, y)$ and strikes the focal plane at a point $B'(x', y')$ (Fig. 5–23a). The next step would be to examine the behavior of a wave as it passes from the object to the lens and through the lens to the image plane. Because there is no simple way of obtaining the extremely important conclusions, we will give directly the expression for the amplitude $A(x', y')$ of a wave at a point (x', y') on the image plane:

$$A(x', y') = C \iint_{-\infty}^{\infty} A(x, y)e^{-jk(xx' + yy')/f'} \, dx \, dy \qquad (5\text{--}77)$$

where $A(x, y)$ is the amplitude distribution of an object, and C incorporates all the constant factors outside the integral. Introducing the spatial frequencies ν_x and ν_y again, this may be written as

$$A(\nu_x, \nu_y) = C \iint_{-\infty}^{\infty} A(x, y)e^{-jk(\nu_x x + \nu_y y)/f} dx \, dy \qquad (5\text{--}78)$$

Thus, if the lens does not have an apodization or absorption screen, the field at the focal plane F is converted into its Fourier transform [2] in the focal plane F'.

The relation between the fields at other locations is more complicated, and leads to much more complicated transformations and to Fresnel-type diffraction. Properties of the Fourier transform are given in Table A5.1–1 of Appendix 5.1.

Production of a diffraction pattern in the back focal plane of a lens when the object is taken to be a diffraction grating is shown in Fig. 5–23b. The condition for an image point to exist is that all the waves should arrive at the point in the same phase; that is, that all the optical paths between the object point and image point should be equal. In the back focal plane F of the lens, all the waves that are parallel to each other before the lens will come to a focus; $|A(x', y')|^2$ will be observed.

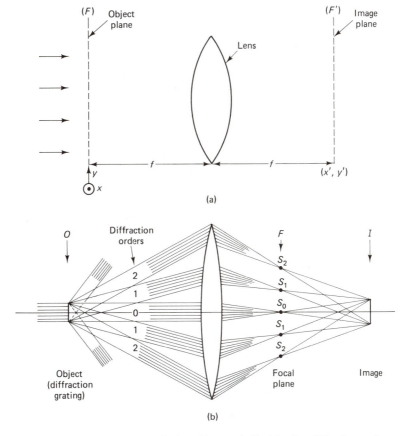

Figure 5–23   Image property of a lens: (a) general; (b) object is a diffraction grating.

Now since the relative phases between object points O and image points I' are the same and may be made zero, it is clear that the phase change between O and F is equal and opposite to the phase change between F and I'. Thus the relation between the wave function in the focal plane of the lens and that in the image is the inverse of that between the wave function in the object and that in the focal plane of the lens. In other words, the image is the inverse Fourier transform of $A(x', y')$.

5.4.2 The Optical Transfer Function

Based on the Fourier transform properties of lenses, an optical image processing system can be arranged as shown in Fig. 5–24. A microscope objective combined with lens 1 serves as a beam spreader. The object (slide) is at the focal point of lens 2, and lens 3 is at a distance equal to the sum of the focal lengths. The negative coordinates in the image plane are a reminder that the output image has inverted coordinates because

$$\text{FT}\{\text{FT}\{f(x, y)\}\} = f(-x, -y) \qquad (5\text{–}79)$$

Signal processing, like filtering for example, can be performed in the Fourier plane. For example, a low-pass filter can be simply realized by one stop with a circular opening in the middle. In general, if $H(\nu_x, \nu_y)$ is the Fourier transform of a stop in the Fourier plane, the Fourier transform of the output picture, $F(\nu_x, \nu_y)$, is

$$F(\nu_x, \nu_y) = H(\nu_x, \nu_y)G(\nu_x, \nu_y) \qquad (5\text{–}80)$$

where $G(\nu_x, \nu_y)$ is the Fourier transition of the object.

In the process of forming an image, a lens cuts out high-spatial-frequency components of an object. These components correspond to fine details. The finite extent of the lens aperture can be accounted for by associating with the lens a pupil function $P(x, y)$ defined by [6]

$$P(x, y) = \begin{cases} 1 & \text{inside the lens aperture} \\ 0 & \text{otherwise} \end{cases} \qquad (5\text{–}81)$$

The Fourier transform of the pupil function is the point-spread function.

A real lens not only cuts out certain spatial frequencies because of diffraction, but also attenuates others. This means that the image-forming quality of a lens could be specified by a function that describes how much each spatial frequency component in an object is attenuated by the lens as it forms an image. In general, the complex frequency response of an optical system which responds to light intensity or power is called its optical transfer function (OTF). The OTF of a system gives a complex coefficient describing magnitude and phase transmission by the system of a sine-wave object of arbitrary spatial frequency. Thus the spatial spectrum of the image, $F(\nu_x, \nu_y)$, is obtained by multiplying the spatial frequency spectrum of the object, $G(\nu_x, \nu_y)$, by the system OTF, $H(\nu_x, \nu_y)$.

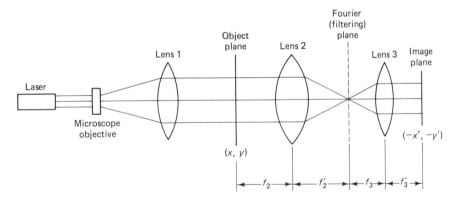

Figure 5–24 Typical arrangement for optical image processing.

The output of the optical system can be predicted, for a known object (input), by OTF, which also measures system quality. The optical system is characterized as a linear system. Instead of temporal frequency, the spatial frequency is used to lay out spectra and responses. Or, in a two-dimensional space domain,

$$f(x, y) = g(x, y) * h(x,y) \qquad (5\text{–}82)$$

where * means convolution, and (f, F), (g, G), and (h, H) are corresponding Fourier transform pairs. The variable $h(x, y)$ is a point-spread function. Equation (5–82) shows that image formation can also be interpreted as the convolution of the object with the transmission properties of the lens or the optical system in general.

The OTF is perhaps employed most often to characterize a lens so that its performance can be predicted under a variety of conditions [8]. Like other frequency response functions widely used in science and engineering, the OTF gives the response of the system to a sinusoidally varying input. For a general linear system a sinusoidal input will yield a sinusoidal output, with in general a frequency-dependent amplitude and phase shift. If a few reasonable assumptions are made, an optical system with incoherent illumination may be regarded as linear, and the statement above will then apply.

Analogies between optical systems and linear electronic circuits are useful, but we should be cautioned against carrying them too far. Optical systems connected together in series will not generally yield a system having a response function that is the product of the individual response function, as is true in the electronic case.

The Fourier transform of the point-spread function can be called a coherent optical transfer function:

$$H(\nu_x, \nu_y) = \text{FTF}\{h(x', y')\} = \text{FTF}\{\text{FTF}\{P(\cdot)\}\} = P\{\nu_x, \nu_y\} \qquad (5\text{–}83)$$

This transfer function is sometimes also called an optical admittance because of its analogy with admittance, $Y(\cdot)$, in electrical circuits.

The normalized Fourier transform of the intensity of the point-spread function, $|h|^2 = h \cdot h^*$, or the optical intensity transfer function, $F(\nu_x, \nu_y)$, is defined as

$$F(\nu_x, \nu_y) = \frac{\displaystyle\iint_{-\infty}^{\infty} h(x', y')h^*(x', y')e^{-j(\nu_x x + \nu_y y)}dx' \, dy'}{\displaystyle\iint_{-\infty}^{\infty} h(x',y')h^*(x', y') \, dx' \, dy'}$$

$$= \frac{\mathscr{F}\{|h|^2\}}{\mathscr{F}\{|h|^2\}}\bigg|_{\nu_x = \nu_y = 0} \quad (5\text{–}84)$$

also known as the optical transfer function (OTF) or incoherent optical transfer function. This function can be expressed as

$$F(\nu_x, \nu_y) = \frac{\displaystyle\iint H(\xi,\eta)H^*(\xi + \nu_x, \eta + \nu_y) \, d\xi \, d\eta}{\displaystyle\iint |H(\xi, \eta)|^2 \, d\xi \, d\eta} \quad (5\text{–}85)$$

where

$$\nu_x = \frac{x}{\lambda f} \qquad \nu_y = \frac{y}{\lambda f} \quad (5\text{–}86)$$

The optical transfer function can be related to the pupil function of an idealized optical system:

$$F(\nu_x, \nu_y) = \iint P(\xi + \frac{f\nu_x}{2k}, \eta + \frac{f\nu_y}{2k})P^*\left(\xi - \frac{f\nu_x}{2k}, \eta - \frac{f\nu_y}{2k}\right) d\xi \, d\eta \quad (5\text{–}87)$$

where the denominator has been dropped because the normalization can be done with respect to some arbitrary value of F. This equation indicates that for an ideal (defect-free) lens, the OTF is simply the area of overlap of two displaced pupil functions, one centered about $(f\nu_x/2k, f\nu_y/2k)$ and the other about the diametrically opposite point $(-f\nu_x/2k, -f\nu_y/2k)$. This geometrical interpretation demonstrates that the OTF of a diffraction-limited system is always real and nonnegative.

Figure 5–25 shows two examples: a circular and square aperture. The shaded area is marked as S. The OTF can be obtained from the area S with appropriate normalization:

For a circular aperture:

$$F(\nu_x, 0) = \frac{S}{\pi r^2} = \frac{2}{\pi}\left[\arccos\frac{S}{2r} - \frac{S}{2r}\sqrt{1 - \left(\frac{S}{2r}\right)^2}\right] \quad 0 \leq \frac{S}{2r} < 1 \quad (5\text{–}88)$$

For a square aperture:

$$F(\nu_x, 0) = \frac{S}{4r^2} = 1 - \frac{S}{2r} \quad \text{and} \quad F(\nu_x, \nu_y) = F_x(\nu_x)F_y(\nu_y) \quad (5\text{–}89)$$

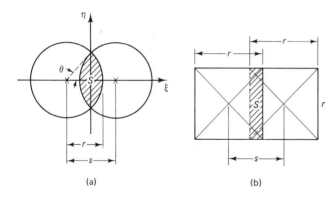

(a)   (b)

Figure 5–25 Overlapping pupils: (a) circular; (b) square.

Figure 5–26 shows corresponding curves for a square and circular aperture. Normalization is done so that $F(0, 0) = 1$.

The transfer function of a diffraction-limited circular lens is sometimes rewritten as

$$\mathcal{H}(\omega) = \begin{cases} \dfrac{2}{\pi}\left[\arccos\dfrac{\omega}{\omega_o} - \dfrac{\omega}{\omega_o}\sqrt{1 - \left(\dfrac{\omega}{\omega_o}\right)^2}\right] & 0 \le \omega \le \omega_o \\ 0, & |\omega| > \omega_o \end{cases} \tag{5-90}$$

The value ν_{x0}, of ν_x when the apertures just touch is called the cutoff frequency, and it represents the maximum spatial frequency passed by the lens and is given by

$$s = 2r \tag{5-91}$$

or

$$\nu_{x0} = 2\frac{rk}{f} = \frac{4\pi r}{\lambda f} \tag{5-92}$$

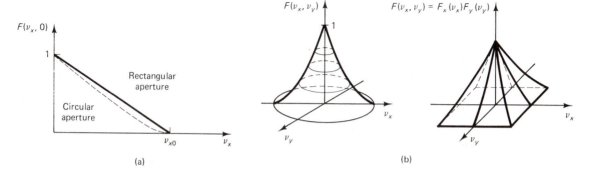

(a)   (b)

Figure 5–26 Normalized OTF for circular and square aperture: (a) one-dimensional, (b) two-dimensional presentation.

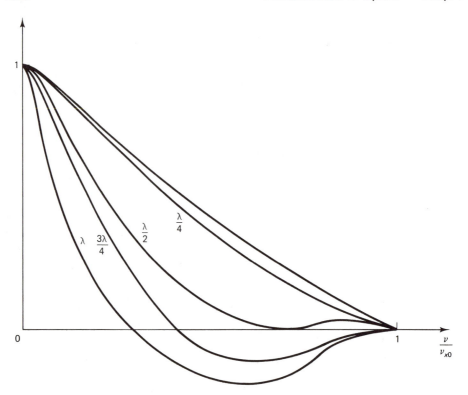

Figure 5–27 OTF of defocused lens.

The aberrations can be included directly in the OTFs [6]. Existing wavefront errors can be obtained if the exit pupil is illuminated by an ideal spherical wave and a phase-shifting plate is introduced within the aperture. If ω is the deviation of the actual image wavefront from an ideal reference sphere (an effective path-length error) the aberration can be incorporated in the OTF at the lens by setting the pupil function $P(x, y)$, which is unity for an ideal lens, to be

$$P(x, y) = e^{ikW(x, y)} \tag{5–93}$$

For an actual system, focusing must also be taken into account. This can be described from a similar viewpoint, although defocusing is not a true aberration. Figure 5–27 shows the OTF functions of a defocused lens with a round exit pupil, for different level of defocusing.

Small defocus produces phase rotation in the OTF; large defocus produces nulls of response of intermediate spatial frequencies. Large apertures show these effects, with waves having shallow phase curvatures, that is, with a smaller absolute amount of defocus. For the case of round, uniform, equal apertures, a defocus of

$$\frac{\lambda}{\pm 2(\text{NA})^2} \tag{5–94}$$

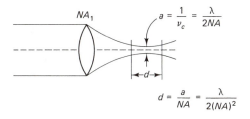

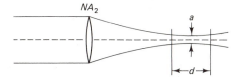

Figure 5–28 Depth of focus d, and the smallest spot diameter a, for different NA.

takes the system halfway to the first null of response; this range is frequently referred to as the depth of focus of the system. If we denote the depth of focus by $d = \lambda/(\text{NA})^2$, and the cutoff spatial frequency for a diffraction-limited system and perfect focus by

$$\nu_c = \frac{2\text{NA}}{\lambda} \tag{5–95}$$

then

$$d = \frac{4}{\lambda(\nu_c)^2} \tag{5–96}$$

This shows how the depth of focus drops rapidly with increasing cutoff frequency; or

$$\nu_c = \frac{2}{\sqrt{\lambda d}} \qquad d^2\nu_c = \frac{2}{\sqrt{\lambda}} \tag{5–97}$$

which shows how the maximum useful cutoff frequency falls as the required depth of focus increases (Fig. 5–28).

5.4.3 The Modulation Transfer Function

The magnitude of the OTF is called the modulation transfer function (MTF) and describes the transfer of contrast from the object to the image as a function of spatial frequency. The phase of the OTF is called the phase transfer function (PTF) and describes image offset versus spatial frequency (Fig. 5–29).

The MTF is sometimes defined[4] as a ratio of the modulation in the image to

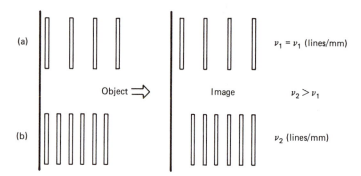

Figure 5–29 Image offset for two spatial frequencies.

that in the object as a function of the frequency (cycles per unit of length) of the sine-wave pattern:

$$\text{OMTF}(\nu) = \frac{M_i}{M_o} \qquad (5\text{–}98)$$

where "modulation" expresses the contrast in the image (for example) and

$$\text{modulation} = M = \frac{I_{\max} - I_{\min}}{I_{\max} + I_{\min}} \qquad (5\text{–}99)$$

where I is the intensity of the radiation. Considering image formation as a convolution process, Fig. 5–30 shows [4] how M changes with spatial frequency.

When the image modulation is plotted as a function of the spatial frequency [4] (Fig. 5–31), the limiting resolution can be presented, although it does not fully describe the performance of the system. Figure 5–31b shows two modulation plots with the same limiting resolution, but plot A will produce crisper, more contrasty images. Figure 5–31c shows two cases with different contrast and resolution and the comparison of these two systems is not obvious.

The MTF of a complex (cascaded) optical system can be obtained as a product of the components MTFs. For example, if the object with contrast (modulation) of 0.2 at 20 cycles per millimeter is photographed by a camera with a lens with an MTF of 0.5 and film with an MTF of 0.7 at this frequency, the image modulation is $0.2 \times 0.5 \times 0.7 = 0.07$.

The transfer function, defined as a ratio of the modulations in the image and in the object, can have any value between -1 and $+1$ and is really the optical modulation transfer function (OMTF); a negative value indicates a reversal of contrast, bright lines of the object appearing as dark lines in the image. The OMTF is frequently referred to as the sine-wave response or contrast transfer.

The OMTF differs from the MTF in the sense that MTF means absolute value, and cannot be negative. On the other hand, the OMTF differs from the OTF because the OTF can have a phase different from 0° and 180°. The OMTF (ν) is an almost

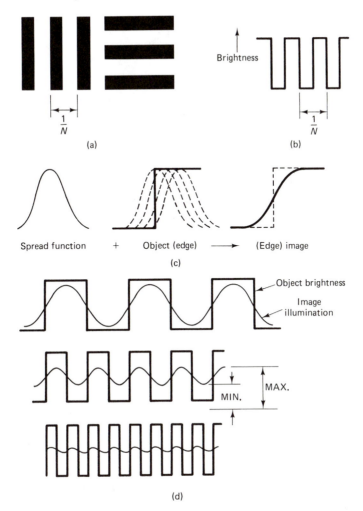

Figure 5–30   Imagery of a bar target: (a) bar train; (b) brightness changes of the object; (c) edge formation ("rounding off the corners"); (d) image contrast for different patterns.

universally applicable measure of the performance of an image-forming system, and can be applied not only to lenses, but also to films, phosphors, image tubes, the eye, and so on.

As an example, consider an object consisting of alternating light and dark bands, the brightness of which varies according to a cosine function:

$$g(x) = b_0 + b_1 \cos 2\pi\nu x \qquad (5\text{--}100)$$

where ν is the spatial frequency of the brightness variation in cycles per unit length. The object modulation is then

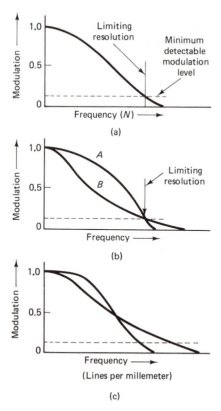

Figure 5–31 Modulation versus spatial frequency.

$$M_0 = \frac{b_1}{b_0} \tag{5-101}$$

The diffraction behavior can be represented by the impulse response h (line-spread function in this example). The image energy distribution can be then expressed as

$$f(x) = \int h(x')g(x - x')\, dx' \tag{5-102}$$

After solving and normalizing with respect to $\int h(x')\, dx'$, it becomes

$$
\begin{aligned}
f(x) &= b_0 + b_1|h(\nu)|\cos(2\pi\nu x - \varphi) \\
&= b_0 + b_1 h_R(\nu)\cos(2\pi\nu x) + b_1 h_I(\nu)\sin(2\pi\nu x)
\end{aligned} \tag{5-103}
$$

where $h_R(\nu)$ and $h_I(\nu)$ are the real and imaginary parts, respectively, of the complex quantity,

$$h(\nu) = \frac{\int h(x)\, e^{j2\pi x}dx}{\int h(x)\, dx} \tag{5-104}$$

and

$$\tan \varphi = \frac{h_I(\nu)}{h_R(\nu)} \tag{5-105}$$

The image energy distribution, $f(x)$, is still modulated by a cosine function of the same frequency, ν. If the line-spread function $h(x)$ is asymmetrical, a phase shift φ is introduced. The image modulation is

$$M_i = \frac{b_1}{b_0} h(\nu) \tag{5-106}$$

which means that

$$\frac{M_i}{M_0} = |h(\nu)| \tag{5-107}$$

Based on the previous results, a practical method for computation of the MTF via a spot diagram has been developed [4].

Technically speaking, there is no single frequency response function. For example, for each of the next cases (and combinations) separate transfer functions can be obtained [3]:

- One for an object area near the axis
- One for each of several areas of varying distances of the axis
- One for each different direction of the spatial frequency lines when asymmetrical aberrations are present
- One for each radiant-energy wavelength region

The effect of defocusing on the transfer function of an aberration-free system has already been shown in Fig. 5–27. Figure 5–32 shows the OMTF of an aberration-free optical system in the presence of a central obscuration of the pupil. This is a form of apodization because the OMTF is reduced at low frequencies and increased at high frequencies.

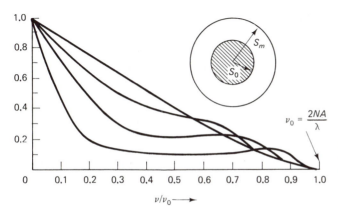

Figure 5–32 Modulation transfer function of an aberration-free optical system with an annual aperture: curve A, $S_0/S_m = 0.0$; curve B, $S_0/S_m = 0.25$; curve C, $S_0/S_m = 0.5$; curve D, $S_0/S_m = 0.75$. (After Ref. 4.)

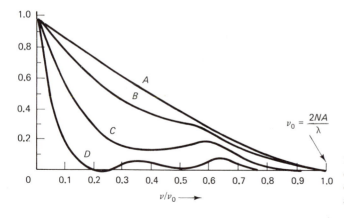

Figure 5–33 Effect of third-order spherical aberration on the MTF when the reference plane is midway between the marginal and paraxial foci. Curve A.

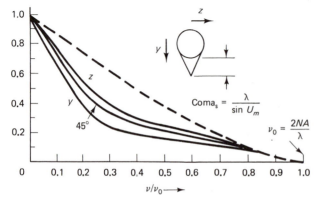

Figure 5–34 Effect of third-order coma on the MTF.

The OMTF of a system with third-order spherical aberration [1] is shown in Fig. 5–33. The OMTF for coma is shown in Fig. 5–34. Because coma is an asymmetrical aberration, the response is different in each meridian.

5.4.4 Resolution of Optical Systems

It is of interest to answer this question: What is the minimum separation of two points which allows them to be distinguished from a single point? If an optical system images two equally bright point sources of light, each point will be imaged as an Airy disc; if the points are close, the diffraction patterns will overlap. The minimum spacing d which can be resolved by an optical system is on the order of the wavelength of light, and is inversely proportional to the numerical aperture (NA) of the optical system.

Figure 5–35 shows the sum of the two diffraction patterns for various amounts of separation [4]. Usually, one of two ad hoc criteria for resolution is used. At a separation of $\lambda/2NA$, Sparrow's criterion, the duplicity of the image points is detectable, although there is no minimum between the maxima from the two patterns. Rayleigh's criterion for resolution is that at a separation of $0.61\ \lambda/NA$, the maximum

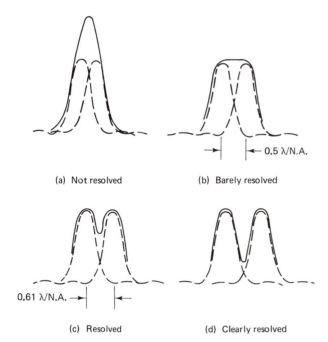

(a) Not resolved

(b) Barely resolved

→| |← 0.5 λ/N.A.

0.61 λ/N.A. →| |←

(c) Resolved

(d) Clearly resolved

Figure 5–35 Resolution of two point objects.

of one pattern is placed on the first dark ring of the other and there is a clear indication of two separate maxima in the combined pattern. Common methods of testing resolution are the use of resolution test charts, bar charts, or spatial equipment for measuring the modulation transfer function.

The criteria mentioned previously can be justified from the point of view of Fourier optics. Fourier optics gives very clear definition of the resolution limit in terms of resolving sinusoidal targets. This is the idea behind the transfer function. When transferring these concepts to resolving two point objects, one also has to include, either explicitly or implicitly, signal-to-noise ratio considerations. For the noise-free case the resolution limit is spatial frequency where MTF goes to zero. In the presence of noise, it is the spatial frequency where MTF drops to some fixed fraction (e.g., 10%) or where SNR possesses a fixed threshold.

An optical system is a passive low-pass filter network; no optical system transmits spatial frequencies higher than a given limit. Experiments have been made to overcome this limit and to design systems which have a higher resolution than that predicted from classical theory. Such systems are said to provide superresolution [7]. For example, color or moiré fringes may be used to improve resolution.

5.4.5 Apodization

If instead of a finite opening, an infinite sheet of transparent material of nonuniform density is taken, the diffraction pattern will be changed considerably. For the Gaussian type of distribution, the light transmission along the x direction is:

$$f(x) = Be^{-x^2/b^2} \qquad (5\text{-}108)$$

where B and b are constants. The diffraction pattern in this case is also Gaussian, and does not show the secondary maxima of the rectangular and circular apertures. This is due to the fact that a Gaussian opening does not terminate abruptly, and there is no sharply defined edge.

For finite openings, most of the energy passing through an opening appears in the central portions of the image rather than in the secondary peaks. This ratio can be changed, and adjusted, if the transmission properties of the opening are changed. The process of redistributing the energy is called apodization. In general, apodization is the use of a variable-transmission filter or coating of the aperture to modify the diffraction pattern. Coating that reduces the transmission at the center of the aperture tends to favor the response at high frequencies, whereas coatings that reduce transmission at the edge of the aperture tend to favor the lower frequencies. The function f (ξ,η), which defines the light transmission through the coating, is referred as the aperture or pupil function. The resolution can never be improved by apodizing; in fact, the resolution is usually appreciably degraded. Figure 5–36 shows the case where the use of an absorption plate has put relatively more energy into the central maximum and thus has reduced the height of the secondary maximum.

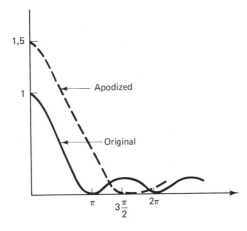

Figure 5–36   Distribution of illumination for normal and apodized case.

5.5 PHYSICAL OPTICS

Physical optics sometimes covers all phenomena associated with light as waves: interference, diffraction, and polarization [2]. Here we deal with the interaction of light and matter.

When light flux is incident on a surface or medium, three processes can occur: absorption, reflection, or transmission. Absorption is the fraction of incident flux that is absorbed. A blackbody is an ideal body of uniform temperature that perfectly absorbs all incident radiation [1]. The radiance of blackbody is the same in all directions. Objects are colored because they absorb some of the components of white light and transmit or reflect others.

The reflectance of a surface or medium for a given flux is the fraction of that flux which is reflected. For regular (specular) or mirror-like surfaces, there is a one-to-one correspondence between incident and reflected rays, and normals to the surface at the point of incidence all lie in the same plane; the angle of reflection equals the angle of incidence.

Transmission is a general term for the process in which incident flux leaves a surface or medium on a side other than the incident side [1]. The transmittance of a medium is the ratio of the flux transmitted to incident flux.

The magnitude of the phase velocity for a monochromatic light traveling through the dielectric is given by

$$v = \frac{1}{\sqrt{\epsilon\mu}} = \frac{c}{n} = \frac{\omega}{k} \tag{5-109}$$

where ϵ is permittivity and μ permeability; these quantities characterize the medium. The phase velocity is also called the wave velocity and it is the velocity of a pure monochromatic, infinitely long wave. The group or envelope velocity is the velocity of energy flow and will not exceed the velocity of light.

If there is no permanent phase relation among the various components of a beam of radiant energy, over a relatively long period of time, and if the orientation of the electric field vectors with respect to each other is random, following no particular pattern, the beam is said to be unpolarized [3]. An unpolarized beam can be polarized by passing the radiation from an ordinary source through a polarizing prism which has a different index of refraction for orthogonal planes of polarization. Since light of one polarization is bent more than the other, it is possible to separate them either by total internal reflection or by deviation in different directions.

Light usually interacts most strongly with materials via its electric field [2]. The magnetic interaction is proportional to v_e/c, where v_e is the velocity of the electrons in the material and c is the velocity of light; this interaction is small.

When the material is unisotropic because of its crystalline structure or its stressed condition, the speed of light depends on the orientation of the electric vector with respect to the transmitting material. Such materials are said to be double refracting or birefringent [3], and are often made up into what is called a wave plate: for example, a quarter-wave plate or a half-wave plate. Such a wave plate can be used to separate the electric field vector of a (linearly) polarized beam into two orthogonal components. The difference in velocity causes one component to be delayed with respect to the other, and the superposition of the two components results in an elliptically polarized beam. Generally, the radiant energy emitted by most sources can be shown experimentally to have at least a small amount of polarization.

The phenomenon of birefringence in which the phase velocity of an optical beam propagating in the crystal depends on the direction of polarization of its E vector is one of the most important consequences of the dielectric anisotrophy of crystals. Birefringence has some interesting consequences. Suppose, for example, a linearly polarized incident field with equal components along x and y. As the wave propagates along the crystal z direction into the crystal, the x and y components get out of phase and the wave becomes elliptically polarized. This phenomenon forms the basis of the electro-optical modulation of light. The phase difference at the output plane (of the crystal) between the two components is called retardation.

The amplitude of a light wave is a vector characterized by its magnitude and direction, and the simplest type of polarization is one in which the amplitude vector points in a fixed direction normal to the direction of propagation. This is also known as plane-polarized light, because the amplitude and the wave vector define a plane. A linear polarizer is anything which, when placed in an incident unpolarized beam, produces a beam of light whose electric vector is vibrating primarily in one plane, with only a small component vibrating in the plane perpendicular to it. If a polarizer is placed in a plane-polarized beam and is rotated about an axis parallel to the beam direction, the transmittance T will vary between a maximum value T_1 and a minimum value T_2 according to the law

$$T = (T_1 - T_2) \cos^2 \theta + T_2 \tag{5-110}$$

If the polarizer is placed in a beam of unpolarized light, its transmittance is

$$T = \tfrac{1}{2}(T_1 + T_2) \tag{5-111}$$

so that a perfect polarizer would transmit only 50% of an incident unpolarized beam [1].

Elliptical polarization is the most general case [3]. The pattern traced out in space by an elliptically polarized wave resembles a flattened helix, and when viewed from behind, the helix looks like an ellipse. The ratio of the major axis to the minor axis of the ellipse can take any value between unity and infinity. When the ratio is unity, light is circularly polarized. In Fig. 5–37a one wavelength of circularly polarized

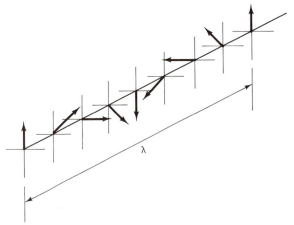

Figure 5–37 Circulary polarized wave.

light is shown. The amplitude of a wave rotates in clockwise manner tracing out a helix in space.

Polarizing beam splitters are a special form of non-normal-incidence interference polarizers in which the beam is incident on a multilayer of dielectric stack at 45°. The transmitted beam is almost entirely plane polarized in the incident plane direction. Generally, the alternating high- and low-index dielectric layers are deposited onto the hypotenuses of two right-angle prisms, which are then cemented together to form a cube. The beam enters a cube face normally and strikes the multilayers on the hypotenuse (the high-index layer is next to the glass) and the reflected and transmitted beams emerge normal to cube face and are separated by 90°. The beam splitter can be made to be insensitive to the polarization of the incident beam.

5.6 REFLECTIVE VIDEODISC PLAYER

The NA of the focusing lens is an important parameter in a videodisc system. Better resolution, that is, smaller pits that can be recorded and read, can be obtained with a higher NA. A high NA is more desirable for both mastering and playback, because the smallest resolvable diameter is $D = \lambda/\text{NA}$. But in a player with an optical pickup, it is better to have a large tolerance for disc vibration. That is, a large depth of focus is desired. But a large depth of focus, d, is obtained for a small NA: $d = \lambda/(\text{NA})^2$. To compromise focal depth and pit size (and thus disc capacity), the NA usually has values of 0.4 to 0.65.

Once the NA is specified, the MTF of the system should be considered. Figure 5–38 shows three cases, three combinations of illuminator and receiver openings. Here we have assumed one-dimensional diffraction-limited systems. Figure 5–38a and b are obviously special cases of Fig. 5–38c. When either NA_i is zero or NA_r is zero, all object spatial frequencies up to NA_i/λ or NA_r/λ, respectively, would be received and all higher object spatial frequencies would be rejected. When the object is illuminated with a broad, uniform spectrum of plane waves whose projections have spatial frequencies up to NA_i/λ and the receiver accepts plane waves whose projections have spatial frequencies up to NA_r/λ, then object spatial frequencies up to $(\text{NA}_i + \text{NA}_r)/\lambda$ send plane waves into the receiving aperture as spatial modulation sidebands of the plane waves in the illuminating spectrum. In videodisc players, the illuminator and the receiver are the same objective lens, and the system response falls gradually throughout the intermediate range of spatial frequency and cuts off the object spatial frequency $2\text{NA}/\lambda$.

For a particular lens, the MTF can be obtained experimentally. Figure 5–39 shows an idealized plot. If the disc has a constant rotation speed (e.g., 30 revolutions per second = 1800 rpm) and recorded signal is periodic with frequency f_o (Hz), the pit size is proportional to its radial position on the disc.

This means that the effective temporal frequency MTF of the recording, which is a corresponding transfer function, is also radius dependent. (Obviously, the lens's

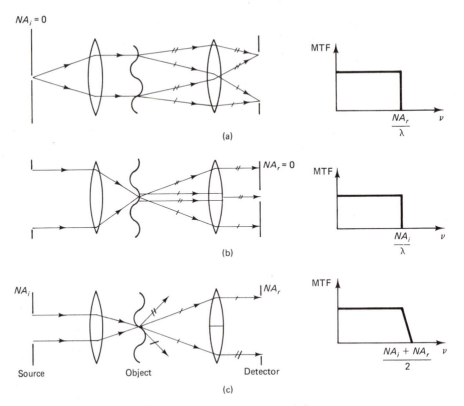

Figure 5–38 Influence of NA_i and NA_r on system frequency response.

MTF does not change unless it has been specified in terms of temporal frequency.) From Fig. 5–39 it can be seen that

$$\frac{\text{MTF}(R_1)}{\text{MTF}(R_2)} = \frac{R_2}{R_1} \tag{5–112}$$

where R_1 and R_2 are the corresponding radii. Figure 5–39b shows the case for the inside radius $R_{\min} = R_1$ and outside radius $R_2 = R_{\max}$. To combine this MTF with the electrical transfer function(s), in order to get a transfer function of the entire system, MTF(f) can be plotted for the corresponding radius as in Fig. 5–39b. The cutoff frequencies are radius dependent and can be expressed as

$$\frac{f_{c1}}{f_{c2}} = \frac{R_1}{R_2} \tag{5–113}$$

The recorded information is transferred from the optical to the electrical "domain" via a photo diode (detector). To find the signal at the detector plane, first, the signal at the videodisc plane should be determined. Figure 5–40 shows a part of the videodisc surface, which can be considered as a two-dimensional grating; this

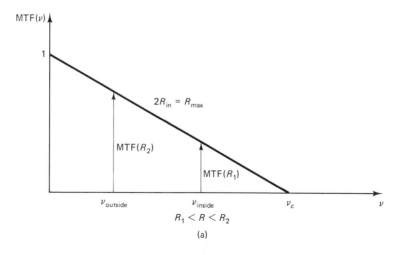

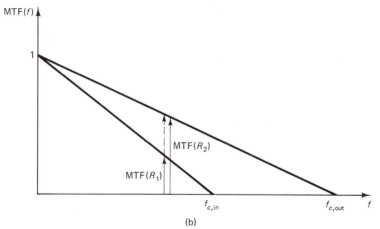

Figure 5–39 Idealized MTF as a function of (a) spatial frequency, ν; (b) output temporal frequency, for different radius.

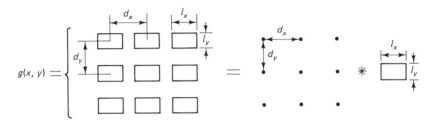

Figure 5–40 Part of the vidiodisc surface: two-dimensional grating.

is a simplified case where a constant spatial frequency is recorded. This function, $f(x, y)$, can be expressed as a convolution of a two-dimensional comb function* ш (x, y) and aperture function $p(x, y)$:

$$g(x, y) = ш\ (x, y)*p(x, y) \tag{5-114}$$

The reading laser beam intensity can be approximated by a Gaussian function:

$$i(x, y) = I_0 e^{-r^2/2\sigma^2} \qquad r^2 = x^2 + y^2 \tag{5-115}$$

If the lens opening is large enough compared with the laser beam diameter defined as $D = 2\sigma$, the reading signal is

$$f(x, y) = g(x, y)*i(x, y) \tag{5-116}$$

In the Fourier transform plane (detector plane) we have

$$F(\nu_x, \nu_y) = G(\nu_x, \nu_y)*I(\nu_x, \nu_y) \tag{5-117}$$

where capital letters denote Fourier transforms of the functions specified by the corresponding lowercase letters. Also

$$G(\nu_x, \nu_y) = ш\ \left(\frac{1}{d_x}, \frac{1}{d_y}\right) P(\nu_x, \nu_y) \tag{5-118}$$

This is illustrated in Fig. 5–41.

Because the laser beam is assumed to be Gaussian,

$$I(\nu_x, \nu_y) = i(-a\nu_x, -b\nu_y) \tag{5-119}$$

where a and b are scaling factors. Finally,

$$F(\nu_x, \nu_y) = \left[ш\ \left(\frac{1}{d_x}, \frac{1}{d_y}\right) P(\nu_x, \nu_y)\right]*i(\nu_x, \nu_y) \tag{5-120}$$

where the minus sign is dropped in the intensity part because of symmetry.*

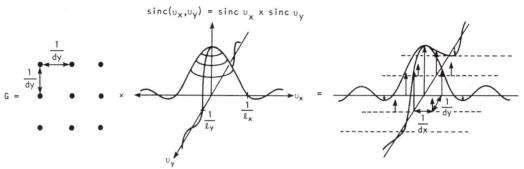

Figure 5–41   Fourier transform of $g(x, y)$.

---

* The symbol III is pronounced "shah" after the cyrillic character ш (∫).
* This discussion was suggested by T. Strand in the summer of 1979.

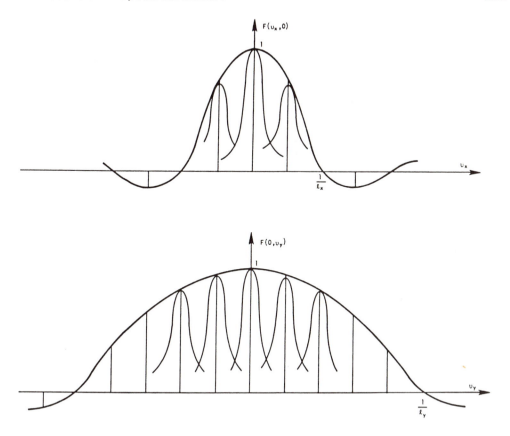

Figure 5–42   Section of the spectrum on the photodetector plane.

Figure 5–42 shows cross sections of $F(\nu_x, \nu_y)$ in the Fourier transform plane (detector plane). It is assumed that $dx = 2l_x = dy = 5l_y$. The Gaussian curves are $\exp(-\nu^2/2\sigma^2)$, which represents the laser beam. If we consider crosstalk from the nearest pits only, it can be seen that crosstalk from track to track is greater than crosstalk in the same track (symbol interference).

5.7 OPTICAL MODULATORS

To convey information on an optical wave, it is necessary to modulate a property of the wave in accordance with the information signal. The wave property may be its intensity, phase, frequency, polarization, or direction; the modulation format may be analog or digital. Electro-optical and acousto-optical modulators are used in the (optical) videodisc systems. Theoretically, magneto-optical effects can be considered based primarily on the Faraday and Kerr effects, although other properties may be more important for the potential recording media.

In the presence of a magnetic field, certain substances become optically active. For example, the plane of polarization of plane-polarized light is rotated through an angle θ when transmitted through glass in a direction parallel to a magnetic field. This is the Faraday effect.

The Kerr magneto-optic effect is characterized by a change in the state of polarization of plane-polarized light when reflected from the polished pole of an electromagnet. The light becomes elliptically polarized. It is most readily observed when the electric vector of the incident light is either parallel or perpendicular to the plane of incidence. This eliminates the ordinary effect of elliptical polarization that occurs by the reflection of plane-polarized light from metals [1].

5.7.1 Electro-optical Modulators

These modulators utilize the phenomenon of induced electrical birefringence in certain materials. An incident plane-polarized wave will, in general, become elliptically polarized upon passage through these materials when a voltage is applied across the material. That is, certain classes of crystals exhibit a linear electro-optical (Pockels) effect whereby an applied electric field E produces a proportional change in refractive index Δn. Since Δn and E are related by a tensor, the field must be applied in special crystal directions for maximum effect and the induced index change depends on the polarization of the optical beam.

Electro-optical modulators utilizing the Pockels effect may be classified as either longitudinal or transverse. In longitudinal modulators, the electric field is parallel to the direction of light propagation; hence semitransparent or ring-electrode structures are required. Transverse modulators operate with the electric field orthogonal to the direction of light propation; thus the field electrodes do not interfere with the optical beam. The induced modulation is converted to a detectable intensity modulation by means of a polarizer, or analyzer. Incorporating this, we get the following expression for the ratio of the output intensity I_o to the input intensity I_i:

$$\frac{I_o}{I_i} = \sin^2\left(\frac{\pi}{2}\frac{V}{V_\pi}\right) \tag{5-121}$$

where V_π is the voltage yielding a retardation π.

If the modulator is biased with a fixed voltage (retardation $\pi/2$) to the 50% transmission point, a small sinusoidal modulation voltage would then cause a nearly sinusoidal modulation of the transmitted intensity, as shown in Fig. 5–43. For the modulation voltage $V_m \cos \omega_m t$, with bias voltage V_π, we obtain

$$\frac{I_o}{I_i} = \sin^2\left(\frac{\pi}{4} + \frac{\Gamma_m}{2}\cos \omega_m t\right)$$
$$= \tfrac{1}{2}[1 + \sin(\Gamma_m \cos \omega_m t)] \tag{5-122}$$

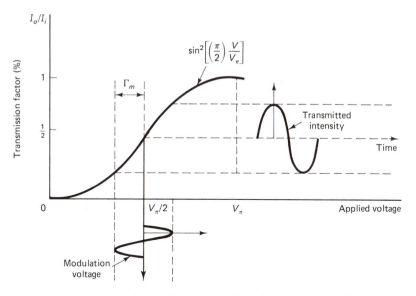

Figure 5–43 Transition factor of a cross-polarized electro-optical modulator as a function of an applied voltage.

where Γ_m is related to the amplitude V_m, $\Gamma_m = \pi(V_m/V_\pi)$.

For relatively small voltages V_m, the intensity modulation is a linear replica of the modulating voltage:

$$\frac{I_o}{I_i} \simeq \tfrac{1}{2}(1 + \Gamma_m \cos \omega_m t) \tag{5–123}$$

A four-crystal configuration of the modulator is shown in Fig. 5–44. Double refraction is compensated by the use of pairs, and two pairs with axes rotated 90° reduced temperature birefringence effects [1].

5.7.2 Acousto-optical Modulators

The principles of acousto-optical devices for the modulation of light are based on the scattering of light by sound (acoustic) waves. Specifically, refraction and diffraction effects are observed when light passes through a transparent material at right angles to a high-frequency sound field propagating in the same medium. The interaction of the light beam with the acoustic beam can result in optical-beam deflection and modulation of the polarization, phase, frequency, or amplitude of the optical energy [1].

The periodic density changes produced in any transparent medium by an acoustic wave give rise to periodic refractive index changes due to the photoelastic effect. The medium then serves as a three-dimensional phase grating. As illustrated in Fig. 5–45a, the incident beam will be partially diffracted from zero order into a multiplicity

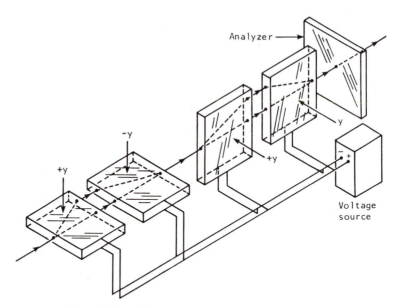

Figure 5–44 Typical electro-optical modulator configuration.

of higher orders when the grating length L is sufficiently small; this is the Raman–Nath regime that applies when $Q = 2\pi\lambda L/n\Lambda^2 < 1$. If L is sufficiently large, the zero-order beam will be partially deflected into only one order; this is the Bragg regime that applies when $Q > 10$. In the definition of Q, L is the width of the acoustic wave of wavelength Λ in a medium of index n at the optical wavelength

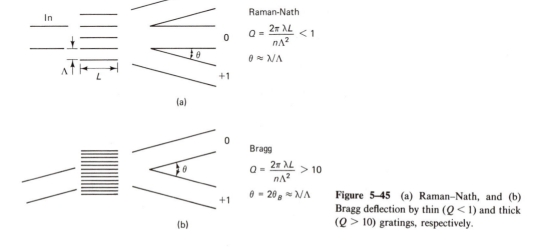

Figure 5–45 (a) Raman–Nath, and (b) Bragg deflection by thin ($Q < 1$) and thick ($Q > 10$) gratings, respectively.

λ. In either of the cases noted above, the device will modulate the intensity of either the zero-order beam or one of the higher-order beams as the intensity of the acoustic subcarrier is modulated.

The Bragg angle modulation type of modulator is designed so that there is a long interaction path between the sound and the incident light whose angle of incidence is placed from the normal (Fig. 5–45b). The necessary condition for the angle of incidence is given by

$$\sin \theta = \frac{1}{2} \frac{\lambda}{n\Lambda} \tag{5–124}$$

where θ is the Bragg angle. In the arrangement shown in Fig. 5–45, the diffracted beam is "minus first order."

The amount of light that is diffracted into the first-order beam for a Bragg cell is given by the equation

$$\frac{I_o}{I_i} = \sin^2 \frac{K}{\lambda} \sqrt{P_{RF}} \tag{5–125}$$

where K is a constant, λ is the laser wavelength, and P_{RF} is the input power. This equation can be rewritten, after normalization, as

$$i_1 = A \sin^2(KE_{RF}) \tag{5–126}$$

where i_1 is the instantaneous intensity in the first-order diffraction beam and E_{RF} is the instantaneous RF envelope voltage across the matched transducer (Fig. 5–46).

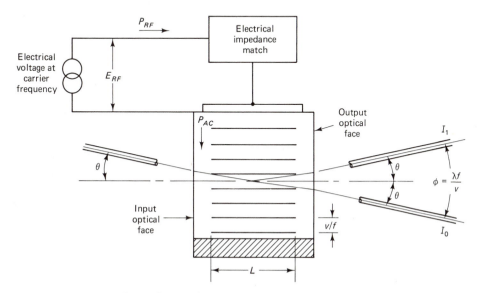

Figure 5–46 Schematic of the acousto-optical modulator.

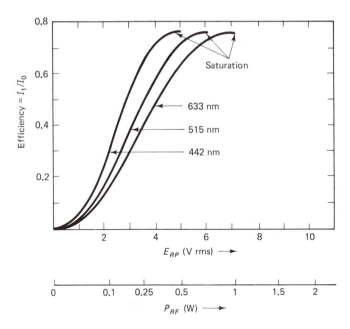

Figure 5–47 Efficiency versus drive voltage curve (typical).

Figure 5–47 shows the intensity versus RF envelope voltage transfer function of the acousto-optical modulator in normalized units.

REFERENCES

1. *Handbook of Optics,* sponsored by the Optical Society of America, ed. W. G. Driscoll and W. Vaughn, McGraw-Hill, New York, 1978.

2. A. Nussbaum and R. A. Phillips, *Comtemporary Optics for Scientists and Engineers,* Prentice-Hall, Englewood Cliffs, N.J., 1976.

3. C. S. Williams and O. A. Becklund, *Optics: A Short Course for Engineers and Scientists,* Wiley-Interscience, New York, 1972.

4. W. J. Smith, *Modern Optical Engineering,* McGraw-Hill, New York, 1966.

5. J. D. Gaskill, *Linear Systems, Fourier Transforms, and Optics,* Wiley, New York, 1978.

6. J. W. Goodman, *Introduction to Fourier Optics,* McGraw-Hill, New York, pp. 1–68.

7. J. R. Meyer-Arendt, *Introduction to Classical and Modern Optics,* Prentice-Hall, Englewood Cliffs, N.J., 1972.

8. M. V. Klein, *Optics,* Wiley, New York, 1970.

9. L. Levi, *"Applied Optics—A Guide to Optical Systems Design,* Vol. 1. Wiley, New York, 1968.

10. L. J. Loub, "Optics of reflective videodisc players," *IEEE Trans. Consum. Electron.,* Aug. 1976.

APPENDIX 5.1
FOURIER TRANSFORM: TRANSFORM PAIRS AND
TRANSFORM PROPERTIES

A5.1–1 TRANSFORM PAIRS

In general, the one-dimensional Fourier transform pair is defined as

$$H(\omega) = a_1 \int_{-\infty}^{\infty} h(t) e^{-j\omega t} dt \qquad \omega = 2\pi f \tag{A5.1–1}$$

$$h(t) = a_2 \int_{-\infty}^{\infty} H(\omega) e^{j\omega t} d\omega \tag{A5.1–2}$$

with imposed requirement that $a_1 a_2 = 1/2\pi$. Different users assume different values for the coefficients a_1 and a_2. Some set $a_1 = 1$, $a_2 = 1/2\pi$; others set $a_1 = a_2 = 1/\sqrt{2\pi}$, or set $a_1 = 1/2\pi$, $a_2 = 1$. We will use the first set of values.

Thus a pair of Fourier transforms, $f(x)$ and $F(\omega)$, are

$$F(\omega) = \int_{-\infty}^{\infty} f(x) e^{-j\omega x} dx \tag{A5.1–3}$$

$$f(x) = \frac{1}{2\pi} \int_{-\infty}^{\infty} F(\omega) e^{j\omega x} d\omega = \int_{-\infty}^{\infty} F(\omega) e^{j2\pi f x} df \tag{A5.1–4}$$

A Fourier transform of Gaussian function,

$$f(x) = \frac{1}{\sqrt{2\pi}\sigma} e^{-x^2/2\sigma^2} \tag{A5.1–5}$$

is also a Gaussian function:

$$F(\omega) = \int_{-\infty}^{\infty} \frac{1}{\sqrt{2\pi}\sigma} e^{-x^2/2\sigma^2} e^{-j\omega x} dx = e^{-2\pi^2\sigma^2\omega^2} \tag{A5.1–6}$$

The Fourier transform of the Dirac delta function;

$$\delta(x) = \begin{cases} 0 & x \neq 0 \\ 1 & x = 0 \end{cases} \qquad \int_{-\infty}^{\infty} \delta(x) \, dx = 1 \tag{A5.1–7}$$

is

$$F(\omega) = 1 \tag{A5.1–8}$$

The delta function has the so-called sampling property:

$$f(x) = \int f(t)\delta(t - x) \, dt \tag{A5.1–9}$$

where $f(x)$ is arbitrary function.

The Fourier transform of the comb function,

$$\text{comb}(t) = \sum_{n=-\infty}^{\infty} \delta(t - nT) = \qquad (t) \qquad \text{(A5.1–10)}$$

is also a comb function with spacing $1/T$:

$$F(\omega) = \text{comb}(\omega) = \sum_{m=-\infty}^{\infty} \delta\left(\omega - m\frac{1}{T}\right) \qquad \text{(A5.1–11)}$$

The Fourier transform of a pulse function:

$$f(t) = \begin{cases} 1 & -\dfrac{\tau}{2} \leqslant t \leqslant \dfrac{\tau}{2} \\ 0 & \text{elsewhere} \end{cases} \qquad \text{(A5.1–12)}$$

is

$$F(\omega) = \tau \frac{\sin x}{x} \qquad x = \pi\tau f = \omega\tau/2 \qquad \text{(A5.1–13)}$$

The two-dimensional Fourier transform pair is defined as

$$F(f_x, f_y) = \iint_{-\infty}^{\infty} f(x,y) \exp\{-j2\bar{u}(f_x x + f_y y)\} \, dx \, dy \qquad \text{(A5.1–14)}$$

$$f(x,y) = \iint_{-\infty}^{\infty} F(f_x, f_y) \exp\{j2\bar{u}(f_x x + f_y y)\} \, df_x \, df_y \qquad \text{(A5.1–15)}$$

The Fourier transform pairs for two-dimensional cases are:

Gaussian function:

$$f(x, y) = e^{-\pi(x^2 + y^2)} \qquad \text{(A5.1–16)}$$

$$F(f_x, f_y) = e^{-\pi(f_x^2 + f_y^2)} \qquad \text{(A5.1–17)}$$

Comb function:

$$f(x,y) = \text{comb}(x, y) = \sum_{n=-\infty}^{\infty} \sum_{m=-\infty}^{\infty} \delta(x - n, y - m) \qquad \text{(A5.1–18)}$$

$$F(u, v) = \text{comb}(u, v) = \sqcup\!\sqcup \ (u, v) \qquad \text{(A5.1–19)}$$

Rectangular function:

$$f(x, y) = \begin{cases} 1 & |x|,|y| \leqslant \frac{1}{2} \\ 0 & \text{otherwise} \end{cases} \qquad \text{(A5.1–20)}$$

$$F(u, v) = \frac{\sin \pi u}{\pi u} \frac{\sin \pi v}{\pi v} \qquad \text{(A5.1–21)}$$

Circular function:

$$f(x, y) = \begin{cases} 1 & \rho \leqslant 1 \quad \rho = \sqrt{x^2 + y^2} \\ 0 & \text{otherwise} \end{cases} \tag{A5.1-22}$$

$$F(u, v) = F(\rho) = \frac{J_1(2\pi\rho)}{\rho} \tag{A5.1-23}$$

where J_1 is a Bessel function of the first kind, order 1.

A5.1-2 TRANSFORM PROPERTIES

TABLE A5.1-1 PROPERTIES OF THE FOURIER TRANSFORM

Spatial Operation	Frequency Operation
1. Linearity $af_1(x, y) + bf_2(x, y)$ where a and b are constants	Linearity $aF_1(u, v) + bF_2(u, v)$
2. Scale change $f(ax, by)$	Inverse scale change $\dfrac{1}{\|ab\|} F\left(\dfrac{u}{a}, \dfrac{v}{b}\right)$
3. Shift of position $f(x - a, y - b)$	Linear phase added $F(u, v) \exp[-j(ua + vb)]$
4. Modulation $\exp[j(u_0 x + v_0 y)]f(x, y)$	Shift of spectrum $F(u - u_0, v - v_0)$
5. Convolution $f(x, y)*h(x, y)$ $= \iint_{-\infty}^{\infty} f(\xi, \eta)h(x - \xi, y - \eta)\, d\xi\, d\eta$	Multiplication $F(u, v)H(u, v)$
6. Multiplication $f(x, y)g(x, y)$	Convolution $F(u, v)*G(u, v)$ $= \iint_{-\infty}^{\infty} F(\xi, \eta)G(u - \xi, v - \eta)\, d\xi\, d\eta$
7. Correlation $\iint_{-\infty}^{\infty} f(\xi, \eta)g(x + \xi, y + \eta)\, d\xi\, d\eta$	Conjugate product $F(u, v)G^*(u, v)$
8. Rotation $f(x', y')$ $x' = x \cos\theta + y \sin\theta$ $y' = -x \sin\theta + y \sin\theta$	Rotation $F(u', v')$ $u' = u \cos\theta + v \sin\theta$ $v' = u \sin\theta + v \cos\theta$
9. Differentiation $\dfrac{d^n}{dx^n} f(x, y)$	High-frequency filter $(ju)^n F(u, v)$
10. Integration $\displaystyle\int \cdots \int_{-\infty}^{\infty} f(x, y)(dx)^n$ n-fold	Low-frequency filter $\dfrac{1}{(ju)^n} F(u, v)$

Source: E. L. Hall, *Computer Image Processing and Recognition*, Academic Press, New York, 1979. pp. 126–127.

A5.1–2a COMMENTS ON THE PROPERTIES OF THE FOURIER TRANSFORM

1. Linearity theorem. Linearity and superposition apply in both domains. The spectrum of a linear sum of images is the linear sum of their spectra. Further, any function may be regarded as a sum of component parts, and the spectrum is the sum of the component spectra.

2. Similarity theorem. Space–bandwidth invariance. Compressing a spatial function expands its spectrum in frequency and reduces its amplitude by the same factor. The amplitude reduces because the same energy is spread over a greater bandwidth. For $a = b = -1$, the spatial function is reversed. The frequency axes are also reversed, which, for real images, changes only the phase spectra.

3. Shift theorem. Shifting or translating a spatial function a distance $x = a$ adds a linear phase $\theta = ua$ to the original phase. Conversely, a linear phase filter produces a translation of the image. The magnitude spectrum is invariant to translation.

4. Frequency shift. Multiplying a spatial function by a complex sinusoid translates its spectrum to center at about u_0 rather than zero frequency.

5. Convolution theorem. The convolution of two spatial functions requires the reversal left to right and bottom to top of one of the functions: a translation, multiplication, and summation. Convolution occurs whenever a function is imaged or filtered by a linear, position-invariant imaging system. The frequency effect is simply to multiply the individual spectra. If one of the spatial functions is a unit impulse, its spectrum remains the same.

6. Product. The product of two spatial functions occurs whenever a scene function is illuminated by another function, such as in transmission or reflection. The spectrum of the product is the convolution of the spectra. If one of the spatial functions is the train of impulses, the spectrum is replicated at spacing inversely proportional to the spacing of the spatial impulse train.

7. Correlation theorem. The correlation of two spatial functions corresponds to the product of one spectrum and the conjugate of the other. If the two functions are identical, the spectrum is the magnitude squared or power spectral density.

8. Rotation. Rotating a function through an angle θ rotates the spectrum through an identical angle. Neither magnitude nor phase spectra are invariant to rotation, although invariant functions may be derived.

9. Edge detection.   Differentiating a spatial function in any direction corresponds to a form of high-pass filtering with the filter function shown. High-pass spatial frequency filtering characteristically "sharpens" the image.

10. Blurring.   Integrating a spatial function in any direction corresponds to a form of low-pass filtering with the filter shown. Low-pass spatial frequency filtering characteristically "blurs" the image.

Chapter 6

CHANNEL CHARACTERIZATION

6.1 INTRODUCTION

In the following, the reflective optical videodisc system will be given the most attention because it is the most complex. The greatest portion of this is valid for other systems. The principal characteristics of other systems will also be noted.

The signal in optical videodisc systems takes form in three domains: electrical, optical, and mechanical [1]. In each of these domains, there are many sources of impairment. As in any other communication system, information is transferred from the information source to the user. Because of system impairments, the output signal can differ from the input signal. In a general sense, these impairments introduce unwanted signals, random or deterministic, which interfere with the reproduction of the desired signal. These unwanted signals are due to a variety of sources and can be classified as human-made or naturally occurring.

The human-made type of interference includes inadequate power supply filtering, electromagnetic pickup of radiating signals, terms arising from improper sampling (if this exists in the system), mechanical vibrations resulting in electrical disturbances, and so on. The effects of the human-made sources of noise can be eliminated (or at least minimized) by careful design and practice. We will not dwell on them.

A simplified block diagram of the videodisc system is shown in Fig. 6–1. The videodisc itself is a channel and contains some impairments introduced during the mastering process. The electro-optical transducers (optomodulator and photodiode), light sources (laser, laser diodes), light controllers (lenses, mirrors, beam splitters), and the disc make an optomodulation or optical channel. In this part of the system, the information has a nonelectrical form. Finally, the signal or electromodulation

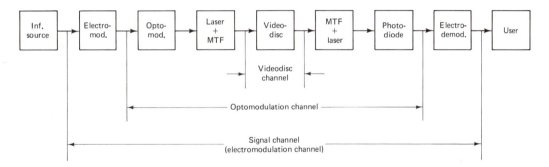

Figure 6–1   Videodisc system.

channel includes the optical channel as well as the modulator/demodulator of the electrical signal. The modulator and the demodulator are taken in a general sense. For example, if the message source generates a binary signal, it is necessary to split the modulator into an encoder and a modulator; the demodulator is split into a demodulator and a decoder. In general, impairments are referred to either as disc or electrical impairments.

All three channels will be considered here. In any event, the channel is analog and its function is to reproduce at its output the message presented to it at its input. A real channel accomplishes this only approximately. Its output typically differs from its input in two ways. First, the channel may modify the message in a deterministic, although not necessarily known fashion: for example, frequency offset and nonlinearities. The channel can also corrupt the transmitted (recorded) signal statistically. To this category belong various types of additive noise, such as thermal and impulse noise, and multiplicative noise.

6.2 VIDEODISC CHANNEL

Generally, two kinds of impairment can be distinguished in a videodisc channel (VDC). The mastering process can corrupt the message as well as the disc itself [2,3]. For some impairments, it is almost impossible to detect their sources.

In the first kind, each step in mastering is a potential source of unwanted signals. The substrate material (usually sheet glass) is required to have microsize defects of less than one dropout per TV frame in the playback signal. Improved polishing is needed to improve the final SNR. The photoresist film can introduce nonlinearity. Thickness variation (i.e., the variation in depth of the pits), unadjusted exposure, and development parameters can reduce the signal quality. Since exposure energy, resist thickness, developer concentration, resist sensitivity, and so on, can vary from disc to disc, the development time should be reinvestigated for every master disc [4]. Electroforming and replication can produce bumps with shapes varying from almost Gaussian to T shapes, so that phase jitter can be introduced in the signal after molding. Metallization introduces, besides noise (a convolution process), blurring edges of the pits. Pinholes in the metallization cause dropouts.

Another generator of impairments is the videodisc. This includes, for example, refractive index variations, eccentricity, and plastic inhomogeneity (dropouts). Impairments in the videodisc channel itself can be broadly classified as noise or dropouts.

6.2.1 Noise

The disc noise is the dominant noise in the system. This noise originates primarily from the following sources:

- Surface roughness of the disc
- Irregularity of the pit dimensions
- Refractive index variations of the plastic
- Substrate and inhomogeneities in the reflective metal (aluminum layer)

These noise terms are generally expected to be kept small [5].

The surface noise is generated by an overall roughness of the disc reflective layer introduced during the replication process. For the sake of simplicity, consider a single spatial frequency of the surface undulation of the geometrical amplitude (depth) D_n and length L_n as a spatial period (of the noise). The time frequency of the generated noise is $f_n = v/L_n$, where v is the disc linear velocity. The phase amplitude is $A_n = (2D_n)n(2\pi/\lambda) = 4n\pi D_n/\lambda$, where n is the refractive index of the substrate and λ is the light wavelength. To simplify, consider the information track as one dimensional. If $F_0(l)$ denotes the noise-free reflection function which contains the recorded information, the complex amplitude reflection function $F(l)$ of the disc in the case of a single noise frequency is given as

$$F(l) = F_0(l) \exp\left(jA_n \cos\frac{2\pi l}{L_n}\right) \qquad (6\text{--}1)$$

So a small overall undulation is superimposed onto the recorded (original) information.

Although the previous analysis was very simplified, two important conclusions can be drawn:

1. The noise caused by surface roughness is not dependent on the recorded signal since this noise is present whether or not the signal is recorded on the disc.
2. This noise is not necessarily radius dependent. That is, for the given spatial period, L_n, of the noise, the time frequency of the generated noise is, for a constant-angular-velocity disc, proportional to the radius. If the distribution of L_n is uniform in a wide range, after translation on the frequency scale, the distribution will remain roughly uniform for the limited frequency bandwidth of practical interest.

This is in contrast, for example, with the radius-dependent noise, for constant-angular-velocity discs, caused by the material (i.e., plastic) nonhomogeneities, if the dimensions of particles are in a relatively narrow region, and if the particles are uniformly distributed over the disc surface.

The intensity of the spurious signal detected is proportional to the square of the (small) phase changes, and the detected noise in this case will have a baseband frequency of $2f_n$. The signal detected contains the baseband noise (of $2f_n$) as well as intermodulation terms between the recorded (encoded) and noise signals. For example, if only the carrier frequency, f_c, is recorded and only this noise exists, the spectrum of the detected signal would be $2f_n$, f_c, $f_c \pm 2f_n$. The strength of various intermodulation terms depends on the information as well as the noise structures. This should not be confused with the fact that the noise caused by surface roughness is not correlated with the signal recorded. The noise contained in the signal detected, when the nonlinearity of the readout system is changed, can also change.

Stochastic deviations from a chosen pit geometry cause noise in the detected signal, and this noise exists only when and where a signal is recorded. This is the dominant noise in the disc where a signal is recorded.

Variations in pit depth D cause modulation of the amplitude of the detected signal. But the depth of pits is determined primarily by the thickness of the photoresist layer of the master disc, which is a slowly varying function of the recording coordinates [4]. In practice, the effect of this noise on signal quality can be neglected, due principally to the testing procedure and two phenomena involved in the detection of the pits: scattering and phase shift.

Variations in the width of pits cause an imbalance between the light intensities reflected from the bottom of the pit and from the surface between the pits. Variations in the length of the pits result in edge-position variations. The influence of the irregularity of the area depends on the spatial frequency of the data recorded.

6.2.2 Dropouts

This kind of videodisc channel impairment is, as is impulse noise, characterized by long, quiet intervals followed by bursts of much higher amplitude than predicted by, for example, a normal or Gaussian distribution law.

Dropouts could be provoked in any stage of videodisc construction, from the beginning of the mastering to the coating of the protective plastic layer. The experimental results obtained through reading upon mastering, when a metal film is used in mastering, show that a single pit is usually omitted. When mastering is made on a photoresistive material, error lengths varying from one to two pits are dominant. Similar measurements were made for videodiscs subject to all processing stages. In the same manner, measurements show that depending on plastic quality, the total number of errors can vary by several orders of magnitude. In general, the defect density increases at each stage in the manufacturing of the disc.

Figures 6–2 and 6–3 show examples of particulates in plastics for different

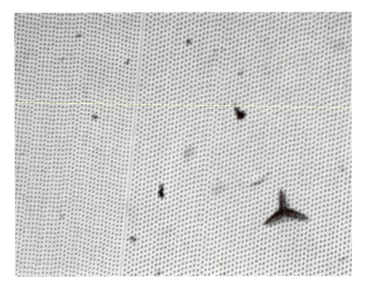

Figure 6–2 Example of the dropout in plastics (×600).

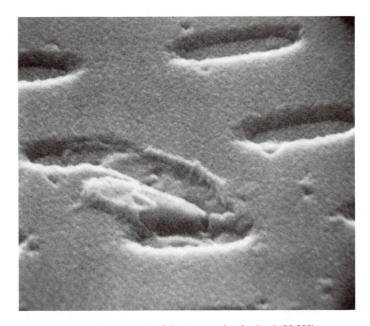

Figure 6–3 Example of the dropout in plastics (×20,000).

values of enlargement. When reading, the particulate in Fig. 6–3 may not cause a dropout, because it does not lie directly in the path. On the other hand, the length of the longest dropout in Fig. 6–2 is approximately six pits (six FM periods when a TV signal is recorded). More examples are given in Appendix 6–2.

6.2.2.1 Dust and scratches on the disc surface.

The high information density on the disc and the resulting microscopically small information details require special measures for the protection of this information. Thus fingerprints and scratches on the disc surface are potential sources of the influence on the amount of reflected light.

Figure 6–4 shows the distribution of the size of dust particles as measured on a disc after laying in a normal living room for two days. Since pit sizes are of the order of a few micrometers only, with no protection a great many dropouts would occur. However, by using the transparent plastic disc itself as a protective layer, dust, fingerprints, and scratches on its surface are out of focus and thus have little influence on the amount of reflected light. The thickness of the protective layer, approximately 1 mm, should be large enough in relation to the focusing depth.

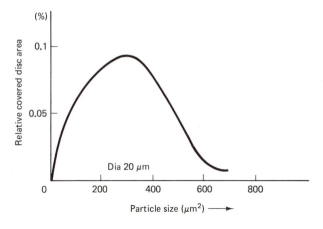

Figure 6–4 Distribution of dust particles on disc after two days in normal living room. (Reported by Philips.)

6.2.3 Noise on a Disc Made of Photographic Film

In general, a videodisc can be made of photographic film. The most common material for photographic recording is silver halide emulsion. A modern black-and-white film consists of the following layers and components:

- A supercoat of gelatine; protects against scratches and abrasion marks
- An emulsion layer: silver halide crystals
- A substrate layer; promotes adhesion of the emulsion to the film base
- The film base or support: cellulose triacetate or a related polymer
- A backing layer; prevents curling

When the film is exposed to the light, an electrochemical conversion process occurs and part of the grain is converted to metallic silver. After chemical developing, a negative is obtained by this nonreversal process and the silver density is inversely proportional to the exposing light.

The exposure and development are not entirely deterministic; silver grains are randomly distributed over the surface of the film. When two films are exposed to light with equivalent exposure, silver grains of the same size and shape will not be obtained in both films. This inherent randomness in silver grain formation, called film-grain noise, leads to a randomness or uncertain amount of light passing through a transparency or reflected from a print [6,7].

Typically, film-grain noise is modeled as a Gaussian distribution:

$$p[D(x, y)] = \frac{1}{\sqrt{2\pi}\sigma_D(x, y)} \exp\left\{ -\frac{[D(x, y) - \mu_D(x, y)]^2}{2\sigma_D^2(x, y)} \right\} \tag{6-2}$$

where $D(x, y) =$ exposed film density
$\mu_D(x, y) =$ a mean density
$\sigma_D(x, y) =$ a variance
The standard deviation is modeled as [8]

$$\sigma_D(x, y) = \alpha[\mu_D(x, y)]^\beta \tag{6-3}$$

where β is between $\frac{1}{3}$ and $\frac{1}{2}$. However, because optical density is a nonnegative quantity, the Gaussian assumption cannot be valid over the full dynamic range of the film. Experimental analysis indicates that film-grain noise may be considered to be a white noise process [8].

Although film-grain noise is signal dependent, it is additive in the density domain. In the intensity domain it is a multiplicative noise process [9]. Thus film-grain noise can be modeled by the technique that lends itself better to the particular use [10]. Film-grain noise can be modeled as additive noise. However, this noise is signal dependent by virtue of its variance. Alternatively, the additive nature of the noise can be sacrificed to eliminate the signal dependence. The difference is that in one model we have additive signal-dependent noise, whereas in the other model we have nonlinear observations with signal-independent noise.

6.2.4 Capacitive Disc Noise

In the capacitive videodisc systems, the disc noise is also the limiting noise. The most significant disc noise is additive in nature. It is related to surface texture and as in other videodisc systems, is a function of all processing and replication steps. Multiplicative noise, although less significant than additive, also exists in the system. It arises from improper mixing of the conductive compound in the disc and it is wideband noise. Other multiplicable noises, for the most part limited in bandwidth, come from stylus elevation changes, which in turn change the detected carrier strength. If these modulations are not severe, they are greatly reduced in the FM limiting process [11].

The carbon loading level is fundamental to the compound's performance and determines the resistivity, melt viscosity, physical characteristics, and surface quality of the disc. The resistivity has a direct bearing on the performance characteristics of the conductive disc [12]. The carrier output may be expressed as

$$C = 20\rho^{-2/3} \qquad (6\text{--}4)$$

where C = carrier output in millivolts peak-to-peak, and ρ = disc resistivity in Ω-cm at 915 MHz, and

$$CNR \sim \rho^{-3/10} \qquad (6\text{--}5)$$

where CNR = 5 MHz carrier-to-noise ratio.

6.2.5 Other Impairments

A number of other distortions in the videodisc channel, apart from noise and dropout, can result in the output of the system being different from the input. Among the miscellaneous impairments are disc eccentricity, time-base error (recorded), nonlinearities, and edge blurriness [1].

Eccentricity is basically a human-made type of interference and its effect can be minimized by careful practice (punching of holes). It causes a time-base error, which can be electronically compensated during reading. A time-base error can also be generated during mastering if the angular speed of the disc is not constant. This can also be minimized and compensated for electronically during reading.

Nonlinearities are always present in an electronic communication system. Significant nonlinearities in the videodisc channel are produced by the photoresist. Photoresists are light-sensitive organic materials with a nonlinear transfer characteristic. The transfer characteristic here is the curve of the amplitude transmittance versus exposure. The deviation from linearity depends to a large degree on the magnitude of the variations of exposure to which the medium is subjected. Thus a baseband signal will be generated because the duty cycle is modulated independent of the signal.

There are many sources of edge blurriness (and jitter, too) in the videodisc channel. If the duty cycle changes, an asymmetry is introduced. Even if, after molding, pits have ideal edges, after metallization the edges will be less ideal (Fig. 6–5). This is equivalent to the convolution of the signal with the pulse function or passing the signal through a low-pass filter. If, during metallization, edges are not blurred equally, jitter will be generated. The influence of metallization may be different for reading from the metallization side or through the plastic, depending on the metal layer thickness, for example, and the process of metallization.

Al → | ← Ideal
edge

Figure 6–5 Edge blurriness by reflective layer.

6.3 OPTICAL MODULATION CHANNEL

The optical modulation channel (OMC) is that portion of the videodisc system which includes the videodisc channel, optical system, and electro-optical transducers. Besides the VDC, the photodetector and lens optical transfer function (OTF) are primary sources of impairments in the OMC.

6.3.1 Photodetector

Any device that reveals the presence of incident radiant energy and transforms it to electrical energy can be called a photodetector, also known as quantum or photon detector. Photon detectors can be divided into photoelectric detectors, sensitive principally in the visible and ultraviolet wavelengths, and semiconductor detectors, most useful in infrared. Obviously, there is considerable overlap in the vicinity of the visible–infrared boundary [13, p. 289].

In videodisc systems, semiconductor detectors are used. An optical system with properly positioned stops can limit the radiant power reaching the detector to power that originates from a specified target [13, p. 286]. However, these very stops, baffles, and other structures surrounding the detector form an enclosure which is, in effect, a background of the target and the detector and background radiate to each other, even though the average net incident power is a random event. The fluctuation in the photon arrival or emission rate is a statistical quantity distributed according to the Bose–Einstein formula. The variance or the mean square (σ_λ^2) of the fluctuations about the mean of the arrival rate (M_λ) at the wavelength is

$$\sigma_\lambda^2 = \overline{M}_\lambda \frac{e^x}{e^x - 1} \tag{6–6}$$

where $x = (hc/k)T$
h = Planck's constant = 6.625×10^{-4} J-s
c = speed of light
k = Boltzmann's constant = 1.38×10^{-23} J/K
T = temperature of the medium, kelvin
The standard deviation is σ_λ.

The photon noise is barely significant in the infrared range. Where the photon noise forming the background is predominant, that is, when all other noise in a system has been reduced, the system is said to be background limited [13, p. 286].

Noise, N, in semiconductor photodetectors is composed of the thermal (Johnson–Nyquist) noise, N_t; generation–recombination noise, N_{gr}; shot noise, N_s; and excess or low-frequency noise, N_f. The noise from these sources is uncorrelated and the powers add:

$$N = N_t + N_{gr} + N_s + N_f \tag{6–7}$$

Thermal noise is generated by the random motion of charged carriers moving about at thermal velocities. It is a white noise; that is, the noise per unit bandwidth is constant with frequency up to the frequencies of the order of the reciprocal of the electron relaxation time (10^{12} to 10^{14} Hz). The average of the squared thermal open-circuit noise voltage is

$$\left(\overline{\frac{V^2}{T}}\right) = 4kTR\,\Delta f \tag{6-8}$$

where Δf is the noise bandwidth in hertz, and R is the resistance of a detector element. Thermal noise is present even in the absence of current flow.

Generations and recombinations of carriers in a semiconductor are statistical in nature; that is, individual generations and recombinations are independent of one another. Equilibrium in a semiconductor, even in the absence of illumination, is dynamic and the free-carrier densities are continuously fluctuating about their mean values. These random generation and recombination processes and the consequent fluctuations in conductivity give rise to noise voltages in the presence of current flow.

At steady state the total number of generations must equal the total number of recombinations over a long interval of time, but an instantaneous fluctuation in the total number of carriers available in a given volume of semiconductor can occur. The fluctuation in the carrier concentration directly affects, for instance, the filament resistance in that volume of the semiconductor. Under an applied bias current, the drop in the potential across that element of the semiconductor will fluctuate about some average value. The generation–recombination current noise per unit bandwidth has been derived by Van Vliet [14]:

$$N_{gr} = \frac{I_{dc}4\tau\,\Delta f(1 - p_i)}{N(1 + (\omega\tau)^2)} \tag{6-9}$$

where I_{dc} is the bias current, τ is the carrier lifetime, N is the total number of free carriers, ω is $2\pi f$, and p_i is the probability that the donor and acceptor levels are ionized. The value p_i is very nearly zero for any photoconductor in a background-limited condition.

The generation–recombination noise is flat, or white, up to frequency of the order $1/\tau$, where the noise current begins to roll off with frequency. Decreasing the background radiation on a background-limited photoconductor will decrease the applied electric field required for optimum operation, and this, in turn, will require less bias current. Generation–recombination noise reduction is best approached by writing Eq. (6–9) in terms of the electronic field E across the background-limited detector:

$$I_{gr}^2 = \frac{4E^2w\mu^2q^2J_rN\sigma^2\,\Delta f}{l[1 + (\omega\tau)^2]} \tag{6-10}$$

where w is the width of the detector element, μ is the carrier mobility, q is the charge on the electron, J_r is the photon flux density for background photons within the bandgap for the detector, N is the quantum efficiency, and l is the length of the detector element. One cannot, unfortunately, reduce the noise simply by reducing all the parameters in the numerator while increasing the length l of the decoder element. The analysis of the signal shows that these same parameters are important in the expression for the signal strength.

Shot noise, like thermal noise, is due to the discrete nature of physical matter. It arises in physical devices when a charged particle moves through a potential gradient without collisions and with a random starting time. Averaging over many such particles yields an average flow, but there will be fluctuations about this average.

Shot noise in semiconductor devices arises as a result of a random diffusion of minority carriers. The shot noise current per unit bandwidth is given by [15]

$$\frac{N_s}{\Delta f} = 2qI_{\text{dc}} \qquad (6\text{--}11)$$

Shot noise in semiconductor diodes is white and it is a dominant noise in background-limited photovoltaic detectors. It follows the Gaussian distribution or Poisson distribution and it occurs only where the light level is extremely low. Shot noise differs from thermal noise in the sense that the mean statistical value of shot noise is not equal to zero; it is a constant direct current.

Excess noise, also referred to as low-frequency noise and as $1/f$ noise, is the least understood of all noise mechanisms. No adequate explanation of this noise has yet been given, but it seems to have its source in photodetectors either in surface effects or possibly in the electrical contacts [16]. It follows the general empirical formula

$$\frac{N_f}{\Delta f} = \frac{kI_{\text{dc}}^2}{fAd} \qquad (6\text{--}12)$$

where k is a constant, I_{dc} is the total current through the detector element, f is the frequency, A is the detector area, and d is the detector thickness. In general, $1/f$ noise is negligible at frequencies above a few hundred hertz, but has occasionally been found still to be dominant at 10 kHz [17, p. 170].

The photoconductor noise has the form

$$\frac{N}{\Delta f} = \frac{4kT}{R} + \frac{4I^2}{N(1 + (\omega\tau)^2)} + \frac{kI^2}{f} \qquad (6\text{--}13)$$

and the second term, N_{gr}, predominates for the background-limited photodetector; the shot noise term is usually not observed.

The photodiode (the photovoltaic detector) noise has the form

$$\frac{N}{\Delta f} = \frac{4kT}{R_z} + 2qI_{\text{dc}} + \frac{K_f I^2}{f} + \frac{K_r I^2}{f} \qquad (6\text{--}14)$$

where R_z is the real part of the parallel impedance for the detector preamplifier combination taken at its operating point, and K_f and K_r are constants of proportionality for the forward and the reverse current in the photodiode. The N_{gr} term from the photoconductor does not appear. The shot noise across the photodiode as a result of the photo-induced current predominates for background-limited photodiodes. In practice, the photodiode noise is 10 to 20 dB (in voltage) below noise of the disc.

6.3.2 Optical Readout

The weakest link in the entire system is the optical readout. First, the numerical aperture (NA) should be chosen to compromise the size of the smallest resolving element $d = \lambda/2NA$, defining the smallest pit, and the focus depth $D = \lambda/(NA)^2$. Second, the optical modulation transfer function (OMTF) changes the amplitude and the phase of the coded signal: for example, if the signal spectrum contains only three components, f_c and $f_c \pm f_m$, for small f_m it is

$$K(f_c - f_m) : K(f_c) = K(f_c) : K(f_c + f_m) \tag{6-15}$$

where $K(f)$ is the corresponding factor with which each spectral component has to multiply the OMTF. For larger f_m, the left-hand side of Eq. (6–15) is small.

As an example, suppose that a one-tone narrow-band FM (modulation index $\beta \leq 0.2$) signal, $s(t)$, is recorded:

$$s(t) = \cos \phi(t) = \cos (\omega_0 t + \beta \sin \omega_m t) \tag{6-16}$$

Because of the lens influence, the readout signal would be

$$s^*(t) = \cos \omega_0 t - a\frac{\beta}{2} \cos (\omega_0 - \omega_m)t + b\frac{\beta}{2} \cos (\omega_0 + \omega_m)t \tag{6-17}$$

(For a flat amplitude characteristic, $a = b$, a signal would not suffer from this kind of distortion.)

The phase of the readout signal, $\phi^*(t)$, is thus

$$\phi^*(t) = \arctan \frac{[(a + b)/2]\beta \sin \omega t}{1 + [(b - a)/2]\beta \sin \omega t} \tag{6-18}$$

After demodulation, a distorted signal is obtained.

Third, for constant angular velocities the OMTF is radius dependent. The worst situation is at the inside radius.

Also, due to the finite size of the reading spot (Fig. 6–6), the output sig-

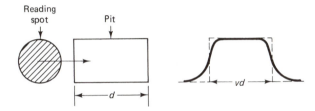

Reading spot   Pit

d

vd

Figure 6–6 Edge blurriness by the reading spot.

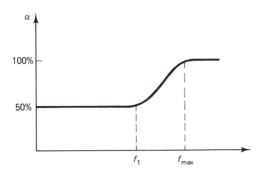

Figure 6–7 Duty-cycle variations with frequencies.

nal would have blurred edges even if recorded pits were ideal. The output signal $f(x, y)$ is

$$f(x, y) = g(x, y) * p(x, y) \qquad (6\text{–}19)$$

where $g(x, y)$ would be the output signal if the reading pulse were ideal, $p(x, y)$ is the reading pulse function (usually approximated by a Gaussian curve), and $*$ is the convolution operator.

If the input signal to the OMC has a constant duty cycle, say 50%, the output duty cycle, α, will change with frequency. This is significant for high frequencies. For some $f = f_{max}$, although the input signal has high and low levels, the output signal would have a constant value (Fig. 6–7).

6.3.3 Other Impairments in the OMC

The laser and optomodulator are also possible sources of noise. During recording, the power of the laser beam should be properly controlled. Too high a power level can destroy a part of the recording area. Too low a power level can cause missing pits. A special servo system is used to control the laser power. For the reading laser, the power should be high enough to eliminate the importance (influence) of the photon (background) noise. Good He-Ne lasers have a white noise spectrum within 10 dB in the frequency range from dc to 20 MHz.

Noise properties of diode lasers are involved, often showing temperature-dependent bursts, but some manufacturers claim to have solved this problem. Advantages of the diode laser are small volume and low drive voltage. Disadvantages of this laser are its oblong transverse mode pattern and the relatively long wavelength (780 to 850 nm) at which it operates.

The laser beam power is not uniformly distributed. It is more or less Gaussian (Fig. 6–8), and the lens can shape it more. In the writing process this can influence the pit geometry. When reading, the beam shape can influence the signal spectrum. Approximately half of the light flux lies inside a circle of diameter d:

$$d = \frac{\lambda}{2a} \qquad a \text{ numerical aperture} \qquad (6\text{–}20)$$

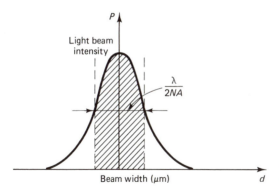

Figure 6-8 Cross section of the laser beam.

This circle is also approximately the contour where the intensity of the central area falls to one-half of its peak value.

The acousto-optical modulator is usually used as a transducer for the electrical-to-light signal transform. If a plane optical wavefront is incident on an acousto-optical modulator, the transmitted wavefront will show periodic waviness corresponding to the periodic changes in the refractive index induced in the modulator by the acoustic wave.

6.4 SIGNAL CHANNEL

The signal channel (SC) is that portion of the videodisc system between the information source and the information destination or information sinc or user (Fig. 6–1). This is a bandlimited channel. A (real) signal $g(t)$ is said to be strictly bandlimited $(-f_m, f_m)$ if the Fourier transform $F(f)$ has the property [16, p. 419]

$$F(f) = 0 \qquad |f| > f_m \qquad (6\text{–}21)$$

The SC includes the OMC and the part of the system where the information is carried by electrical signals. The typical impairments encountered in the electrical signal are noise and linear distortion.

6.4.1 Noise

The dominant noise in the electronic components is thermal noise. This noise is white below frequency kT/h, but it is not white for higher frequencies. However, these frequencies are so high that it can be safely assumed that a thermal noise is white in videodisc systems. For example,

$$kT = 6000 \text{ GHz} \qquad \text{for } T = 290\text{K} \qquad (6\text{–}22)$$

In practice, resistors may actually produce slightly more thermal noise than that indicated by the spectral density, which follows from thermodynamic and quantum mechanical considerations:

$$S_n(f) = 2kT \text{ watts per hertz} \qquad \text{for } f << \frac{kT}{h} \tag{6-23}$$

Thermal noise corrupts the desired signal in an additive fashion and its effects can be minimized by appropriate modulation techniques.

Usually, the input stage of the player (preamplifier) is realized with an FET stage, because the most widely available lowest-noise amplifier device given a high source resistance is the silicon FET. Typical FETs have very large input impedance $(R_i \rightarrow \infty)$ and very low noise $(S_i \rightarrow 0, S_i = \text{spectral density})$. The input capacitance, including the photodiode and FET, is a few picofarads; it can tailor the noise, too.

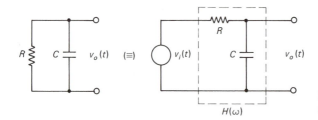

Figure 6-9 RC network and its equivalent-noise model.

An ideal capacitor has no thermal noise source because there are no free electrons present in an ideal dielectric. But as a storage component, the capacity affects noise through the bandwidth of the system. Figure 6-9 shows a capacitor C (e.g., preamp input capacity) connected in parallel with a noisy resistor, R, and the equivalent noise model. Resistor R is replaced by a noise-free resistor in series with a voltage source with power spectral density: $S_n(\omega) = 2kTR (V^2/\text{Hz})$. The mean-square value of the output is given by

$$\overline{v^2(t)} = \frac{1}{2\pi} \int_{-\infty}^{\infty} S_n(\omega) |H(\omega)|^2 \, d\omega \tag{6-24}$$

where $H(\omega)$ is the transfer function. The root-mean-square (rms) output noise voltage is

$$\sqrt{\overline{v^2(t)}} = \sqrt{\frac{kT}{C}} \tag{6-25}$$

The result is independent of R although R is the source of the noise, because of the cancellation of two effects: the mean-square noise voltage is proportional to R and the bandwidth of the low-pass filter is inversely proportional to R.

The second type of noise is impulse noise. This type of noise has many natural and human-made sources. Because of the long, quiet intervals inherent to impulse noise, the speed limitation of the channel is determined by the Gaussian background noise rather than by the more spectacular bursts of impulse noise [18, p. 10].

The main contributions to noise in radiation-detecting systems are [5] shot noise due to the dark current (I_o) of the photodiode, thermal noise (i_R) of the load resistance

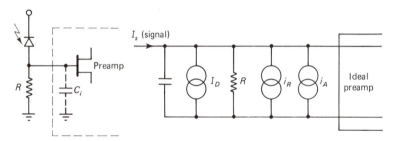

Figure 6–10   Detector circuit and its noise model.

R, and noise of the model based on the equivalent current sources, shown in Fig. 6–10. The influence of the thermal noise of the load resistance can be reduced by increasing the value of R and simultaneously compensating for the poor frequency response with an equalizing network. The other possible way is to use a detector with internal gain (e.g., an avalanche photodiode or a photomultiplier).

6.4.1.1 Capacitive videodisc.   As in the optical videodisc system, the major contribution to the noise level comes from the input circuit. The input circuit contains a detector, first amplification stage, and an oscillator. By careful design, the oscillator noise can be significantly reduced. This includes a very high Q resonant cavity. The signal into the detector, and thus the signal-to-noise ratio, is limited in amplitude due to voltage breakdown of the dielectric layer. The peak voltage across the dielectric is limited to approximately 10 V. The detector and amplification under these conditions are such that the noise from the pickup circuitry is typically 10 dB below the noise detected from the disc surface itself [11].

As mentioned earlier, stylus elevation changes can introduce a multiplicative noise through the detector carrier amplitude change. This can be greatly reduced in the FM limiting process.

6.4.2 Linear Distortion

All real channels exhibit some form of time dispersion. This dispersion can be attributed to the imperfect transfer characteristics of the system or to multipath transmission. The derivative of the phase characteristic, the envelope delay, represents the relative time of arrival of various frequency components of the input signal. The input signal suffers a distortion in passage through the channel because neither the envelope delay nor the attenuation is generally constant with frequency. This distortion takes the form of an overlap in time (e.g., in data transfers) between successive symbols, known as intersymbol interference.

Intersymbol interference (ISI) represents a kind of deterministic channel impairment. In theory, it is usually possible to remove ISI if the channel characteristics are known. In reality, it is impossible to eliminate completely the effects of ISI.

6.4.3 Other Impairments

A number of causes, other than noise and linear distortion, can force the output of the signal channel to be different from the input. Among these are nonlinearities, frequency offset, and phase jitter.

Truly, linear amplifications or filtering is impossible to achieve and nonlinearities are always present in the system. An example is an overloaded amplifier operating in a highly nonlinear region.

Frequency offset and phase jitter result whenever a carrier is used in the system (which it almost always is). When the frequency band is shifted in frequency one or more times, and then shifted back, a baseband can be frequency-shifted typically by a few cycles. This is because the reference carrier may differ in frequency and phase from the modulating carrier. If the demodulation process modifies noise, the preemphasis/deemphasis technique can be used.

6.5 EXPERIMENTAL RESULTS: THE NOISE MEASUREMENTS

Since noise ultimately determines the performance of a system, the study of noise is important and it is necessary to evaluate its limiting effect. As an example, some experimental results are presented for a system with an optical pickup. Both background noise and impulse noise (dropout) are considered. The experimental work was done at DiscoVision Associates, Torrance and Costa Mesa, California.

6.5.1 Background Noise

To determine the amplitude distribution of disc noise, the following procedure is used. First, a noise signal, as read from the disc, is digitized into 256 equal quanta; the sampling frequency used was never less than 20 MHz. Then the digitized noise was processed by a PDP 11, and a plot of the amplitude probability density (APD) function was obtained. Experiments were done with a direct-read-after-write system. The frequency spectrum was obtained by a spectrum analyzer with a scanning filter width of 30 kHz.

Three cases were investigated:

- *Case 1*: The wideband output of the writer while reading a blank disc
- *Case 2*: The bandlimited (7 MHz) output of the writer while reading a blank disc
- *Case 3*: The bandwidth-limited (7 MHz) output of the writer while reading the disc that recorded the carrier (7.5 MHz)

For all cases, an experimental setup was built such that the noise from the disc was dominant. This was checked by comparing with the receiver noise when no light is detected. In this way, the laser noise is added to the disc background

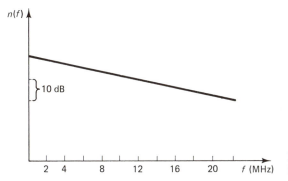

Figure 6–11 Frequency spectrum of the blank disc noise-wide band output.

noise, but it can be considered to be relatively low. The light level is high enough so that the noise figure is determined by the disc noise and by intrinsic laser noise. At low light levels the photon noise can be a fundamental limit.

In Fig. 6–11, the frequency spectrum of the blank disc noise is shown; the preamp bandwidth is around 23 MHz. The slope of the curve can be considered to be the result of the MTF of the read lens. The plot of the corresponding amplitude probability density function is shown in Fig. 6–12. The normalized Gaussian curves are shown for reference only. After the noise was bandlimited to 7 MHz (Fig. 6–13), the APD function obtained was similar to that in Fig. 6–12.

For case 3, a 7.5-MHz carrier was recorded on the disc and then read. The

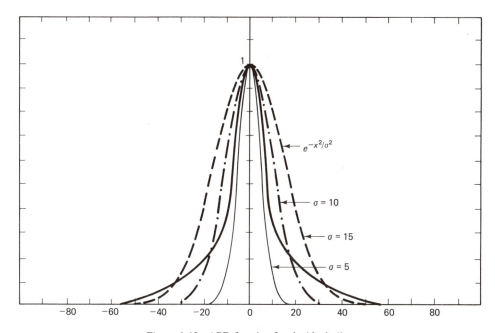

Figure 6–12 APD function for the blank disc.

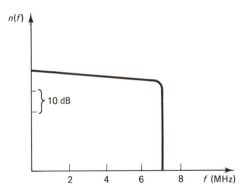

Figure 6–13 Frequency spectrum of the blank disc bandlimited noise.

frequency spectrum of this is shown in Fig. 6–14. This was done to determine if the system nonlinearities would produce intermodulation products between the noise and carrier that would alter the shape of the amplitude distribution. The low-pass filter was used to remove the carrier so that only the distribution of noise up to 7 MHz would be calculated.

The SNR of the carrier (sinusoidal) signal (Fig. 6–14) is about 60 dB; the noise is measured over a bandwidth of $B_1 = 30$ kHz. It should be mentioned that if the noise bandwidth is taken equal to some bandwidth B, then SNR_B would be

$$\left(\frac{S}{N}\right)_B (dB) = \left(\frac{S}{N}\right)_{B_1} - 10 \log \frac{B}{B_1} \qquad (6\text{--}26)$$

For the example with $B = 10$ MHz, the corresponding SNR is around 35 dB.

After the carrier was passed through the LPF with a bandwidth of 7 MHz (Fig. 6–15), the APD function (Fig. 6–16) is similar to that for the blank disc in Fig. 6–12. The main difference is that the curve is now somewhat wider than for the blank disc, which means a larger standard deviation. For reference, the normalized Gaussian curves are shown.

Many experiments with the optical readout were performed; Figs. 6–11 and 6–16 show the results of two of them. At first glance, the amplitude probability

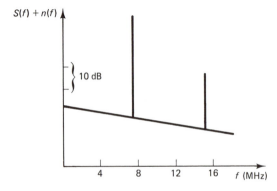

Figure 6–14 Frequency spectrum of the carrier plus noise.

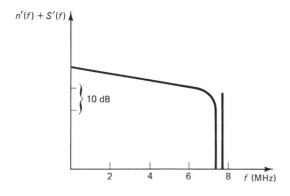

Figure 6–15 Frequency spectrum of the bandwidth-limited output carrier plus noise.

density function (Figs. 6–12 and 6–16) seems to be quite close to a Gaussian distribution; however, detailed analysis shows that the Gaussian distribution may be taken only as the first crude approximation to the experimentally determined probability density function. The power spectral density is almost uniform (i.e., the noise in the system is white). The slope of this function, as shown in Figs. 6–11 and 6–13 to 6–15, can be considered to be the result of the MTF of the read lens, and the slope is somewhat lower if the MTF compensation circuit is included in the player. It may also be noted that the experimental work does not show a clear difference in disc noise with or without information recorded on the disc.

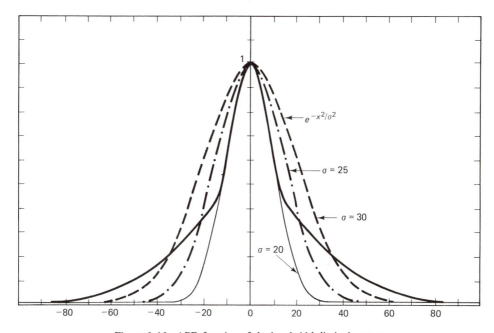

Figure 6–16 APD function of the bandwidth-limited output.

6.5.2 Dropouts

A dropout is a lapse in the (FM) signal played back from a photoresist master, a stamper, or a plastic replica, either clear or metallized, single side (1×), or 2× (finished disc consisting of two single sides bonded together). Impulse noise can be, for example, generated by particles in the substrate or scratches on the surfaces. The first kind of impairment will be considered here.

For experimental measurements, two kinds of special discs were recorded. In the first group, an 8-MHz carrier signal lasting 100 s was recorded, preceded and followed by 20 s of a 10-MHz carrier used as a marker. Measurement starts when the frequency switches from 10 MHz to 8 MHz. This was done three times, for the inside radius, the middle, and the outside radius. In this way it is ensured that when the measurement is repeated, it is done always on the same area on the disc [19].

In the second group, a TV-type signal was recorded: regular sync pulses like an ordinary TV signal, plus the frame number, plus an 8-MHz carrier instead of the FM video signal. In this way, the experiment can be repeated many times. The measurements were performed during the time intervals when the 8-MHz signal was expected.

In Fig. 6–17, the dropout influence is shown, both in the recovered electrical (Fig. 6–17a), and in the signal proportional to the instant frequency (Fig. 6–17b). Whenever the instant frequency is lower than some fixed frequency (e.g., 7.3 MHz),

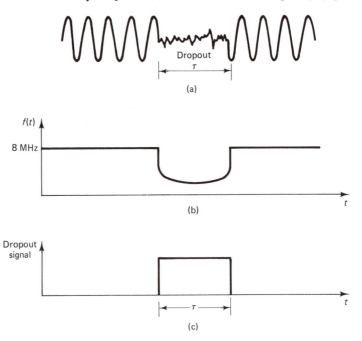

Figure 6–17 Dropout statistic measurements: (a) recovered electrical signal; (b) instant frequency; (c) output of dropout detector.

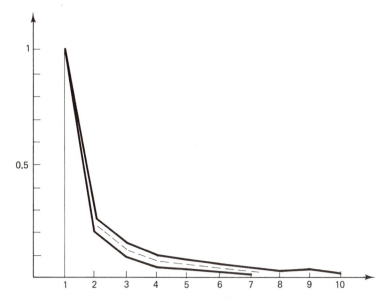

Figure 6–18   Normalized number of dropouts versus their duration.

the dropout detector generates a dropout pulse (Fig. 6–17c). The statistics of primary interest are the number of dropouts (versus dropout duration), and the dropout distribution (in time and on the disc surface).

Figure 6–18 shows the distribution of the normalized number of dropout durations. This diagram is somewhat typical [20], although the total number of errors, depending on the plastic quality and manufacturing cleanliness, can vary by several orders. When we tried to extrapolate the best-fitting curve, using the mean-square error as a criterion, the corresponding curve expression varied. Some of the functions used for particular cases were $y = ae^{-bx}$, $y = a/x^2$, and $y = ae^{-b\sqrt{x}}$.

The variation with radius is caused primarily by a constant angular velocity in the system: the same-size particle will cause a larger dropout for the inside radius than for the outside radius. In general, no clear difference in disc surface dropout distribution is observed.

6.5.3 Polar Figure of the Dropouts and Disc Noise

Sometimes it is convenient to have a two-dimensional presentation of the dropouts, noise, or defects in general, during a stage of production between the glass master and the final replica. This can be obtained, for example, on the CRT of a storage oscilloscope if the following three control signals are used:

1. Horizontal deflection plates: $v_x = A \cos \theta$
2. Vertical deflection plates: $v_y = B \sin \theta$
3. Gate signal: $v_z = s(t)$

where $A = B = kR$ (R is the instantaneous radius)
   θ = angle between instant radius and arbitrary reference
   $s(t)$ = test signal indicates distortions (dropouts, noise, etc.)

Then a hard copy can be obtained for defects of the surface produced at various stages in the manufacturing of the disc. The test signal, $s(t)$, can be obtained optically [21,22] by using specially designed defect detectors, or, for example, an electronic dropout detector can be used.

6.5.4 Discussion

Understanding the origin, characteristics, and interplay of various noise sources is essential to the design and evaluation of a videodisc system. Whenever a particular accuracy is desired, measurements must be made; the various expressions obtained serve mainly to estimate the influences of some parameters.

On the basis of the results of many experiments with optical readout systems, it seems that a fairly good model for the output noise in the system is Gaussian amplitude distribution and uniform spectral density (i.e., Gaussian white noise). Gaussian amplitude distribution could be expected based on the central limit theorem, because in a videodisc system there are many sources of noise. The power spectral density decreases with frequency, which can be considered to be the result of the MTF of the read lens.

This noise model—Gaussian amplitude distribution and uniform spectral density—can be accepted as an initial approximation in order to make preliminary estimations of the system parameters. In a particular case, it is necessary to carry out corrections on the basis of measured results.

Although during the experimental work a variety of conditions were changed (e.g., normal videodisc system, reading after writing, photoresist or metal substrate for the master disc, different photoresist) the overall system noise did not change too much. Also, no clear difference in disc noise was observed in experimental work, whether or not information was present on the disc.

6.6 CATASTROPHIC FAILURES AND THEIR HISTORY THROUGH VIDEODISC PROCESS PRODUCTION

One catastrophic defect per disc, if it is not at the end, can be enough to make a disc unplayable. For the grooved capacitive disc, the groove keeps the stylus in place. In the optical videodisc system a player can fail to focus when starting to play. In these cases "push forward" methods can help. In a certain class of discs a single critically placed dropout on the disc can cause a catastrophic failure (e.g., a return to start). Minimization of the number of dropouts is important not only for improvement of visual quality, but also to eliminate catastrophic defects and their effects on the program material [23].

The glass substrate is visually inspected and serious scratches are rejected. A tangential scratch lasts longer and can be serious, but is visible on only a few frames. Further, minor scratches which are easily visible on the glass surface produce little effect at the PRM and stamper stages, and still less effect in the finished disc.

The dropout rates of the stamper can be greater than those on the PRM master by a factor of 2 to 9. If a thin metal (titanium) coating is applied to the glass followed by a layer of photoresist, the noise and dropout characteristics of the master recording are also dependent on the metal photoresist property. It was noticed in many experiments that the most serious problems, at least as far as dropouts are concerned, come after molding. Thus even the ability to produce absolutely perfect stampers might have only a marginal impact on the proportion of discs passing the final test. But a bad stamper definitely means bad final discs. Thus improvements in the glass/ PRM area may slightly increase the nominal yield.

The dropout density of the replicas can be a few times (four, for example) that of the stamper. In some cases the plastic replicas are unusable although the picture from the stamper appears nearly perfect. This can happen if on the stamper the clusters of dropouts are composed of isolated, widely separated defects. On the replicas, the defects in these areas are frequently long, multiple-line dropouts which are subjectively objectionable because they cannot be compensated for by the player's dropout compensator and because they frequently cause the picture to break up.

Ideally, the medium on which information is recorded should be smooth and featureless prior to recording. Small deviations from this uniformity scatter light in a matter similar to the scattering of the laser light used to read the information bumps on the videodisc.

The dropout density on a metallized disc is greater than that on a clear disc. But the many common features of the clear disc and metallized 1× dropout statistics suggest that, to some extent, the regions of poor metallization are determined before metallization. Also, the initial dropout statistics of the replica (clear disc) are enhanced by the added contamination before metallization.

Generally, metallized 1× replicas have many more dropouts than stampers, and with patterns that differ randomly from replica to replica. The regions of higher dropouts correspond to poorer metallization. Poor metallization can be indicated by light transmitted through the metallized 1× replica. These pinholes in the metal layer look like dark flakes when viewed with reflected light. Depending on the playing mode (CAV, CLV), a uniform spatial density of dropouts of the same physical size over the disc results in a rate of dropouts proportional to the radius.

The average dropout rate increases from 1× to 2× replicas. The added dropouts are largely in the form of scratches and other concentrated and repetitive features.

REFERENCES

1. J. Isailović, "Channel characterization of the optical video disc," *Int. J. Electron.*, Vol. 54, No. 1, 1983, pp. 1–20.

2. R. Adler, "An optical videodisc player for NTSC receivers," *IEEE Trans. Broadcast Telev. Receivers*, Vol. BTR, 20, No. 3, Aug. 1974, pp. 230–234.

3. J. S. Winslow, "Mastering and replication of reflective videodiscs," *IEEE Trans. Consum. Electron.*, Nov. 1976, pp. 318–326.

4. B. A. J. Jacobs, "Laser beam recording of video master discs," *Appl. Opt.*, Vol. 17, No. 13, July 1, 1978, pp. 2001–2006.

5. J. P. J. Heemskerk, "Noise in a videodisc system: experiments with an (A1Ga) As laser," *Appl. Opt.*, Vol. 17, No. 13, July 1, 1978, pp. 2007–2012.

6. C. E. Mills, *The Theory of Photographic Process*, Macmillan, New York, 1966.

7. E. W. Thomas, *SPSE Handbook of Photographic Science and Engineering*, Wiley-Interscience, New York, 1973.

8. W. K. Pratt, *Digital Image Processing*, Wiley-Interscience, New York, 1978.

9. B. R. Hunt, Digital Image Processing, *Proc. IEEE*, Vol. 63, No. 4, Apr. 1975, pp. 693–708.

10. F. Naderi and A. A. Sawchuk, "Estimation of images degraded by film-grain noise," *Appl. Opt.*, Vol. 17, Apr. 15, 1978, pp. 1228–1237.

11. J. K. Clemens, "Capacitive pick-up and the buried subcarrier encoding system for the RCA videodisc," *RCA Rev.*, Vol. 39, No, 1, Mar. 1978, pp. 33–59.

12. L. P. Fox, "The conductive videodisc," *RCA Rev.*, Vol. 39, No. 1, Mar. 1978, pp. 116–135.

13. C. S. Williams, and O. A. Becklund, *Optics: A Short Course for Engineers and Scientists*, Wiley-Interscience, New York, 1972.

14. K. M. Van Vliet, *Proc. IRE*, Vol. 46, 1958, p. 1004.

15. A. Van der Ziel, *Proc. IRE*, Vol. 46, 1958, p. 1019.

16. *Electronics Engineers Handbook*, ed. D. G. Fink and A. A. McKenzie, McGraw-Hill, New York, 1976.

17. T. S. Moss, G. J. Burrell, and B. Ellis, "*Semiconductor Optoelectronics*, Butterworths, London, 1973.

18. R. Lucky, W. J. Salz, and E. J. Weldon, *Principles of Data Communication*, McGraw-Hill, New York, 1968.

19. J. Isailović, "The optimization of error correcting technique for optical memories," Colloquium on Microwave Communication, Budapest, 1978.

20. S. Itoya, M. Nakada, and T. Kubo, "Development of the PCM laser sound player," *IEEE Trans. Consum. Electron.*, Vol. CE-24, No. 3, Aug. 1978, pp. 443–452.

21. I. Gorog, "Optical techniques developed for the RCA videodisc," *RCA Rev.*, Vol. 35, No. 1, Mar. 1978, pp. 162–185.

22. W. R. Roach, C. B. Carroll, A. H. Firester, I. Gorog, and R. W. Wagner, "Diffraction spectrometry for RCA videodisc quality control," *RCA Rev.*, Vol. 39, No. 3, Sept. 1978.

23. Private discussion with W. V. Smith, J. Winslow, R. Dakin, R. Wilkinson, and J. Mosher, all DVA.

APPENDIX 6.1 MATHEMATICAL PRESENTATION OF NOISE

In the analysis of a videodisc system, we often encounter random signals, signals whose behavior cannot be predicted exactly. For deterministic signals it is implicitly assumed that it is possible to write an explicit time function. For random signals it is not possible to write such expression. The main objective of this appendix is to develop probabilistic models for random signals, noise in particular.

A6.1-1 RANDOM PROCESSES

The outputs of information sources and noise are random in nature. The concept of random variables provides us with a probabilistic description of the numerical values of a variable. A similar model, called a random process or a stochastic process, can be used as a probabilistic description of functions of time. Random processes are used to model message waveforms as well as noise waveforms that are encountered in a videodisc system.

Data representing random phenomena cannot be described by an explicit mathematical relationship because each observation of the phenomenon will be unique. In other words, any given observation will represent only one of many possible results that might have occurred. For example, assume that the voltage from output noise in the videodisc system is recorded as a function of time. A specific voltage-time-history record will be obtained, as shown in Fig. A6.1–1. But if we repeat the experiment with the same or a different disc, a different voltage-time-history record would result. Hence the voltage time history for any one noise generator is merely one example of an infinitely large number of time histories that might have occurred (Fig. A6.1–1)

Each time history representing a random phenomenon is called a sample function or sample record or realization of the process. The collection of all possible sample functions that the random phenomenon might have produced is called a random process or a stochastic process. Hence a sample record of data for a random physical phenomenon may be thought of as one physical realization of a random process.

A random process may be categorized as being either stationary or nonstationary. A random process is said to be stationary if its statistical properties are invariant to time translation. This invariance implies that the underlying physical mechanism producing the process is not changing with time. Many of the important properties of stationary processes commonly encountered are described by first and second moments. Consequently, it is relatively easy to develop a simple but useful theory, spectral theory, to describe these processes. The noise in a videodisc system can be considered to be stationary.

Stationary random processes may be further categorized as being either ergodic or nonergodic. The random process is said to be ergodic if time averages of sample functions of the process can be used as approximations to corresponding ensemble averages.

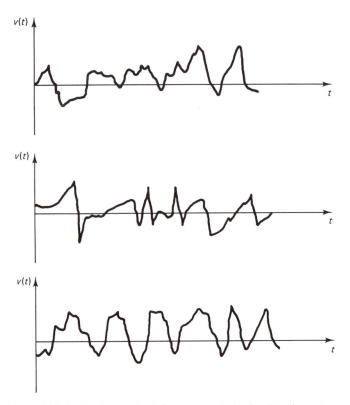

Figure A6.1–1   Simple records of the output noise in the videodisc system.

A6.1–2 PROBABILISTIC MODELS (PROBABILITY DENSITY FUNCTIONS)

This section introduces specific models of probability density functions employed in the analysis of random process in videodisc systems.

A6.1–2a Gaussian Probability Density Function

Our primary interest in studying the Gaussian probability density function (pdf) is from the viewpoint of using it to model (output) noise in the videodisc system. The most naturally occurring distribution is the Gaussian or random distribution, which applies to many physical phenomena (e.g., thermal noise, errors in practical measurements, quality control of components, etc.) and to many applications because of a remarkable phenomenon called the central limit theorem. This theorem implies that a random variable that is determined by a large number of independent causes tends to have a Gaussian probability distribution. In other words, the pdf of the sum of

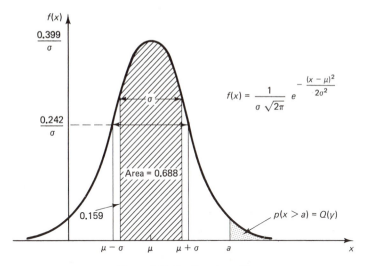

$$f(x) = \frac{1}{\sigma\sqrt{2\pi}} \, e^{-\frac{(x-\mu)^2}{2\sigma^2}}$$

Figure A6.1–2   Gaussian pdf.

a large number of random variables, where each of them can have any distribution, always tends to be a Gaussian distribution.

For a continuous random variable x, the Gaussian pdf $f(x)$ is given by

$$f(x) = \frac{1}{\sigma\sqrt{2\pi}} \exp\left[-\frac{(x-\mu)^2}{2\sigma^2}\right] \qquad -\infty < x < \infty$$

This is illustrated in Fig. A6.1–2. The bell-shaped graph of $f(x)$ is called the random curve; its evaluation was derived by De Moivre in (1733). It is also called Gaussian in honor of Gauss (1777–1855), who also derived the equation in his work concerning errors in repeated measurements.

There are only two parameters in the expression; these will be discussed later. The area between two finite points cannot be evaluated in closed form, due to the nature of $f(x)$, and requires numerical evaluation.

Often there is interest in probabilities such as

$$P(X > a) = \int_a^\infty f(x)\,dx$$

After a change of variables $z = (x - \mu)/\sigma$, the preceding integral can be reduced to

$$P(X > a) = \int_{(a-\mu)/\sigma}^\infty \frac{1}{\sqrt{2\pi}}\, e^{-z^2/2}\,dz$$

Sometimes the Marcum Q function, defined as

$$Q(y) = \frac{1}{\sqrt{2\pi}} \int_y^\infty e^{-z^2/2} \, dz \qquad y > 0$$

is tabulated.
Since

$$\int_{-\infty}^\infty e^{-\alpha x^2} \, dx = \sqrt{\frac{\pi}{\alpha}}$$

or more precisely,

$$\int_{-\infty}^\infty e^{-x^2/2} \, dx = \sqrt{\frac{\pi}{2}}$$

the Q function can be expressed as

$$Q(y) = \frac{1}{2}\left[1 - \frac{1}{2}\operatorname{erf}\left(\frac{y}{\sqrt{2}}\right)\right]$$

where erf (x) is the error function:

$$\operatorname{erf}(x) = \frac{2}{\sqrt{\pi}} \int_0^x e^{-x^2} \, dx$$

Sometimes, erf (x) is defined as

$$\operatorname{erf}(x) = \frac{1}{\sqrt{2\pi}} \int_0^x e^{-u^2/2} \, du$$

Then

$$Q(y) = \frac{1}{2} - \operatorname{erf}(y)$$

A6.1–26 Rayleigh's Distribution

This probability distribution occurs in the study of bandlimited Gaussian noise. It is also used in modeling short-term fading due to tropospheric scattering. In the ballistics field, it gives the probability of hitting a target area, because it is associated with the probability distribution in two dimensions.

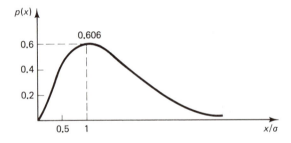

Figure A6.1–3 Rayliegh pdf.

For a continuous random variable x, the Rayleigh pdf $f(x)$ is given by

$$f(x) = \frac{x}{\sigma^2} e^{-x^2/2\sigma^2} \qquad x \geq 0$$

This is illustrated in Fig. A6.1–3.

A6.1–2c Uniform, Laplace, Cauchy, and Maxwell pdf

We now look at some other useful models for continuous random variables.
Uniform (rectangular) pdf (Fig. A6.1–4a):

$$f(x) = \begin{cases} \dfrac{1}{b-a} & a \leq x \leq b \\[2mm] 0 & \text{otherwise} \end{cases}$$

Laplace pdf (Fig. A6.1–4b):

$$f(x) = \frac{\alpha}{2} e^{-\alpha x}$$

Cauchy pdf (Fig. A6.1–4c):

$$f(x) = \frac{\alpha/\pi}{\alpha^2 + x^2}$$

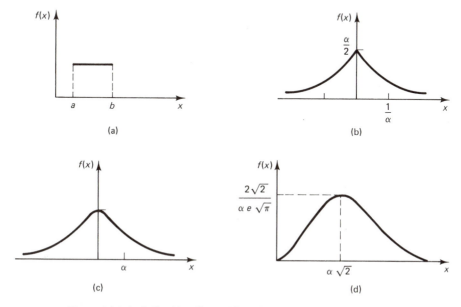

Figure A6.1–4 Pdf's: (a) uniform; (b) Laplace; (c) Cauchy; (d) Maxwell.

Maxwell pdf (Fig. A6.1–4d):

$$f(x) = \frac{\sqrt{2}}{\alpha^3\sqrt{\pi}} x^2 e^{-x^2/2\alpha^2} \qquad x \geq 0$$

Assuming an original $f_1(x)$ defined over $(-\infty, \infty)$, the so-called truncated pdf can be defined as

$$f(x) = \begin{cases} Cf_1(x) & a \leq x \leq b \\ 0 & \text{otherwise} \end{cases}$$

where

$$\int_{-\infty}^{\infty} f(x)\, dx = C \int_{a}^{b} f_1(x)\, dx = 1$$

A6.1–3 DISCRETE MODELS

In general, a discrete pdf is given by

$$f(x) = A\delta(x - a) + B\delta(x - b) + \cdots + N\delta(x - n)$$

where $A + B + \cdots + N = 1$.
The binomial distribution (Fig. A6.1–5a) is

$$f(x) = \sum_{k=0}^{n} \binom{n}{k} p^k q^{n-k} \delta(x - k) \qquad q = 1 - p$$

The Poisson distribution (Fig. A6.1–5b) is

$$f(x) = e^{-a} \sum_{k=0}^{\infty} \frac{a^k}{k!} \delta(x - k)$$

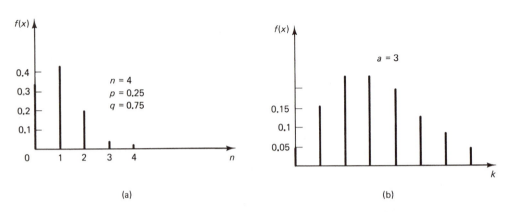

Figure A6.1–5   Discrete pdfs: (a) binominal; (b) Poisson.

The binominal pdf can be used in estimating digital errors such as those transmitted by a PCM source. It can also be used to determine the reliability of components and is associated with the binominal theorem. The Poisson pdf can be applied to various random phenomena, such as shot noise in a diode, the transmission of telegraph signals, and radioactive decay.

A6.1–4 NOISE AVERAGING: EXPECTED VALUE, DISPERSION, AND MOMENTS

When examined over a long period, a random signal may exhibit certain regularities that can be described in terms of probabilities and statistical averages. Descriptions and analyses of messages and noise using concepts that have been borrowed from probability theory are very useful in the videodisc systems.

In forming averages of any signal, random or nonrandom, we find parameters that tell us something about the signal. Much of the detailed information about the signal, of course, is lost in the process.

The averaging can be performed in the time domain or in the amplitude domain. If the results obtained in both cases are the same, the random stationary process is said to be ergodic.

The time average of the random process $x(t)$ is the mean value μ:

$$\mu = \overline{x(t)} \lim_{T \to \infty} \frac{1}{T} \int_{-T/2}^{T/2} x(t) \, dt$$

For an electrical signal the parameter μ is often called the dc component.

The mean-square value, $\overline{x^2(t)}$, is

$$\overline{x^2(t)} = \lim_{T \to \infty} \frac{1}{T} \int_{-T/2}^{T/2} |x(t)|^2 \, dt$$

The square root of $\overline{x^2(t)}$ is called the root-mean-square (rms) value of $x(t)$.

The ac, or fluctuation, component of $x(t)$ is that component which remains after the mean value $\overline{x(t)}$ has been taken out:

$$y(t) = x(t) - \overline{x(t)}$$

The mean or expected value of a random process $x(t)$ is defined as

$$\overline{x(t)} = E\{x\} = \int_{-\infty}^{\infty} x f(x) \, dx$$

where $E\{\cdot\}$ is the mathematical expectation. In the same way, that is, averaging per ensemble, the mean-square value $\overline{x^2(t)}$ is

$$\overline{x^2(t)} = E\{x^2\} = \int_{-\infty}^{\infty} x^2 f(x) \, dx$$

For the discrete variables in previous expressions, the integral should be substituted by the corresponding sums.

The variance or dispersion, σ^2, is defined by

$$\sigma^2 = E\{(x - \mu)^2\} = \int_{-\infty}^{\infty} (x - \mu)^2 f(x)\, dx$$

We note that

$$\sigma^2 = E\{x^2\} - E^2\{x\}$$

The positive square root is called the standard deviation. The standard deviation is important in statistics because it gives the spread of values about the mean, since the smaller σ is, the closer the values are to the mean. In ac theory, σ corresponds to the rms voltage or current in the circuit.

If two random variables x and y have standard deviations σ_x and σ_y, respectively, the variance, σ_z, of their sum, $z = x + y$, is

$$\sigma_z^2 = \sigma_x^2 + \sigma_y^2$$

For the pdf's in this case the convolution relation is

$$f_z(z) = f_x(x) * f_y(y) = \int_{-\infty}^{\infty} f_x(z - u) f_y(u)\, du$$

A more complete specification of the statistics of x is possible if one knows its moments m_k defined by

$$m_k = E\{(x^k\} = \int_{-\infty}^{\infty} x^k f(x)\, dx$$

Clearly, $m_0 = 1$ and $m_1 = \mu = E\{x\}$.

The central moments are given as

$$\mu_k = E\{(x - \mu)^k\} = \int_{-\infty}^{\infty} (x - \mu)^k f(x)\, dx$$

Obviously, $\mu_0 = 1$, $\mu = 0$, and $\mu_2 = \sigma^2$.

The general moments are

$$m_k^a = E\{(x - a)^k\}$$

A summary of a statistical average appears in Table A6.1–1 with more details.

In videodisc systems, we are always trying to model physical processes, with moments up to the second order (second-order statistics).

For the Gaussian pdf

$$\overline{x(t)} = \mu$$

TABLE A6.1–1 SUMMARY OF STATISTICAL AVERAGES

Name	Definition	Indicated operation	
		Discrete case	Continuous case
Mean = m. Also called average, expectation, ensemble average	$E[X]$	$\sum_{i=1}^{N} x_i P(x_i)$	$\int_{-\infty}^{\infty} xp(x)\, dx$
Variance = σ^2 (standard deviation = σ)	$E[(X - m_x)^2]$	$\sum_{i=1}^{N}(x_i - m_x)^2 P(x_i)$	$\int_{-\infty}^{\infty}(x - m_x)^2 p(x)\, dx$
nth moment	$E[X^n]$	$\sum_{i=1}^{N} x_i^n P(x_i)$	$\int_{-\infty}^{\infty} x^n p(x)\, dx$
nth central moment or nth moment about the mean	$E[(X - m_x)^n]$ (= 0 for n = 1)	$\sum_{i=1}^{N}(x_i - m_x)^n P(x_i)$	$\int_{-\infty}^{\infty}(x - m_x)^n p(x)\, dx$
Mean of a function g(x)	$E[g(X)]$	$\sum_{i=1}^{N} g(x_i) P(x_i)$	$\int_{-\infty}^{\infty} g(x) p(x)\, dx$
(n + k)th joint moment	$E[X^n Y^k]$	$\sum_{i=1}^{N}\sum_{j=1}^{M} x_i^n y_j^k P(x_i, y_j)$	$\int_{-\infty}^{\infty}\int_{-\infty}^{\infty} x^n y^k p(x, y)\, dx\, dy$
(n + k)th joint central moment	$E[(X - m_x)^n(Y - m_y)^k]$	$\sum_{i=1}^{N}\sum_{j=1}^{M}(x_i - m_x)^n(y_j - m_y)^k P(x_i, y_j)$	$\int_{-\infty}^{\infty}\int_{-\infty}^{\infty}(x - m_x)^n(y - m_y)^k p(x, y)\, dx\, dy$
Covariance	$E[(X - m_x)(Y - m_y)]$	$\sum_{i=1}^{N}\sum_{j=1}^{M}(x_i - m_x)(y_j - m_y) P(x_i, y_j)$	$\int_{-\infty}^{\infty}\int_{-\infty}^{\infty}(x - m_x)(y - m_y) p(x, y)\, dx\, dy$
Mean of a function g(x, y)	$E[g(X, Y)]$	$\sum_{i=1}^{N}\sum_{j=1}^{M} g(x_i, y_j) P(x_i, y_j)$	$\int_{-\infty}^{\infty}\int_{-\infty}^{\infty} g(x, y) p(x, y)\, dx\, dy$

For the Rayleigh pdf

$$\overline{x(t)} = E\{x\} = \sigma \sqrt{\frac{\pi}{2}}$$

and

$$E\{x^2\} = 2\sigma^2$$

Hence

$$\sigma_x^2 = \left(2 - \frac{\pi}{2}\right)\sigma^2$$

For the Maxwell pdf

$$\overline{x(t)} = E\{x\} = 2\alpha \sqrt{\frac{2}{\pi}}$$

and

$$E\{x^2\} = 2\alpha^2$$

For the uniform pdf

$$\mu = \frac{b + a}{2} \quad \text{and} \quad \sigma^2 = \frac{(b - a)^2}{12}$$

Now, we can summarize the meaning of various time averages for an erodic random process:

- The mean $\overline{x(t)}$ is the dc component.
- The mean-square value $\overline{x^2(t)}$ is the total average power.
- $[\overline{x(t)}]^2$ is the dc power.
- The variance $\overline{x^2(t)} - [\overline{x(t)}]^2$ is the ac power.
- The standard deviation is the rms value.

The power spectral density function, $S(\omega)$, of a random process is also given through the averaging process:

$$S(\omega) = \lim_{T \to \infty} \frac{1}{T} \overline{|F(\omega)|^2}$$

where $F(\omega)$ is the Fourier transform of $x(t)$ and $\overline{|F(\omega)|^2}$ is the ensemble average.

A6.1–5 THE WIENER–KHINTCHIN THEOREM

The autocorrelation function $R(\tau)$ of a function $f(t)$ is given by

$$R(\tau) = \lim_{T \to \infty} \frac{1}{T} \int_{-T/2}^{T/2} f(t)f(t - \tau) \, dt$$

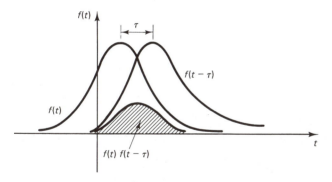

Figure A6.1–6 Autocorrelation function for a single value of τ.

A single value of $R(\tau)$ is represented by the shaded area in Fig. A6.1–6.

Autocorrelation is not a reversible process; it is not possible to get back from an autocorrelation function to the original function from which it was derived. Autocorrelation thus involves a loss of information. For example, the autocorrelation function does not reveal the phase of a harmonic function.

An important relationship between the autocorrelation function $R(\tau)$ and the spectral density function $S(f)$ is given by the Wiener–Khintchin theorem, which states that $R(\tau)$ and $S(f)$ form a set of Fourier transform pairs given by

$$S(f) = \int_{-\infty}^{\infty} R(\tau)e^{-j\omega\tau}\, d\tau$$

$$R(\tau) = \int_{-\infty}^{\infty} S(f)e^{j\omega\tau}\, df \qquad \omega = 2\pi f$$

Obviously,

$$R(0) = \int_{-\infty}^{\infty} S(f)\, df$$

is the average power of the network. This theorem permits the determination of the spectral density $S(\omega)$ from a given correlation function $R(\tau)$, and vice versa.

Consider a random binary waveform that consists of a sequence of pulses with amplitude ± 1 and pulse duration T. If the starting line t_0 of the first pulse is equally likely to be anywhere between 0 and T, the random binary waveform can be written as

$$x(t) = \sum_{k=-\infty}^{\infty} a_k p(t - kT - t_0)$$

where a_k is a sequence of independent random variables with $p(a_k = 1) = p$ $(a_k = -1) = \frac{1}{2}$, $p(t)$ is a unit amplitude pulse of width T, and t_0 is a random variable having a uniform pdf in the interval $[0, T]$ (see Fig. A6.1–7).

The mean value is

$$\mu = \overline{x(t)} = E\{x(t)\} = 0$$

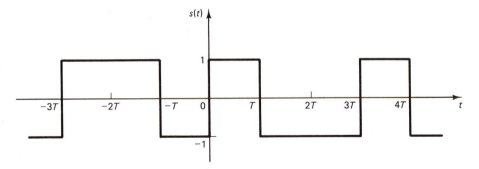

Figure A6.1–7   Random binary waveform, $t_0 = 0$.

Let $A(\tau)$ be a random event with a value of t_0 such that the first pulse starts during the interval τ. The probability for that event is

$$p(A) = \frac{\tau}{T}$$

and

$$P(\bar{A}) = 1 - \frac{\tau}{T}$$

The autocorrelation is

$$R(\tau) = E\{x(t)x(t - \tau)\} = \begin{cases} 1 - \dfrac{\tau}{T} & |\tau| < T \\ 0 & |\tau| > T \end{cases}$$

By taking the Fourier transform of $R(\tau)$, we obtain the psd of $x(t)$ as

$$S(f) = T\left(\frac{\sin u}{u}\right)^2 \qquad u = \pi f T$$

It is of practical interest to know the bandwidth of the signal, B, where, for example, 95% of the signal power is contained in the interval $-B$ to B. It can be obtained by numerical integration:

$$B = \frac{1.55}{T}$$

This is the minimal bandwidth needed to pass through a random binary signal.

A6.1–6 LINEAR SYSTEM RESPONSE

A system is defined to be a mapping of a set of input functions, $x(t)$, into a set of output functions, $y(t)$:

$$y(t) = O\{x(t)\}$$

A system is said to be linear if the following superposition property is obeyed:

$$O\{a_1x_1(t) + a_2x_2(t)\} = a_1O\{x_1(t)\} + a_2O\{x_2(t)\}$$

for any a_1, a_2, $x_1(t)$, and $x_2(t)$.

A system $O\{\cdot\}$ is time invariant if

$$O\{x(t - \tau)\} = y(t - \tau)$$

The impulse response of the system, $h(t)$, is the response of $O\{\cdot\}$ to the delta function $\delta(t)$:

$$h(t) = O\{\delta(t)\}$$

The system is said to be casual if $h(t) = 0$ for $t < 0$.

The response $g(t)$ of a linear system to an arbitrary input $f(t)$ is given by the convolution integral:

$$g(t) = \int_{-\infty}^{\infty} f(\tau)h(t - \tau)\, d\tau = f(t)*h(t)$$

where $*$ denotes convolution of $f(t)$ and $h(t)$.

The system transfer functions, $H(\omega)$, is the Fourier transforms of the impulse response:

$$H(\omega) = \int_{-\infty}^{\infty} h(t)e^{-j\omega t}\, dt$$

From the last two equations it follows that if $F(\omega)$ and $G(\omega)$ are the Fourier transform of $f(t)$ and $g(t)$, respectively, then

$$G(\omega) = F(\omega)H(\omega)$$

If the input to a linear system is a stochastic process, for example, noise $n_i(t)$, with the mean μ_i, autocorrelation $R_i(\tau)$, and power spectral density $N_i(\omega)$, then the resulting output,

$$n_o(t) = \int_{-\infty}^{\infty} n(t - \alpha)h(\alpha)\, d\alpha$$

is a stochastic process with the mean

$$\mu_o = E\{n_o(t)\} = \mu_i H(0)$$

the autocorrelation function

$$R_o(\tau) = R_i(\tau)*h(\tau)*h^*(-\tau)$$

and the power spectral density

$$N_o(\omega) = N_i(\omega)|H(\omega)|^2$$

If the pdf of the input signal is $p_i(x)$ and the output pdf is $p_o(y)$, where the transformation $y = T(x)$ can be expressed algebraically and inverse $x = T^{-1}(y)$

readily found, then a simple relation between $p_y(y)$ and $p_x(x)$ can be developed. When x varies between x_1 and x_2, y varies between $y_1 = T(x_1)$ and $y_2 = T(x_2)$ with the same probability; thus

$$\int_{x_1}^{x_2} p_i(x)\,dx = \int_{y_1}^{y_2} p_o(y)\,dy$$

Assuming that $x = T^{-1}(y)$ is a single-valued function of y, the left side can be changed:

$$\int_{y_1}^{y_2} p_i[T^{-1}(y)]\,[T^{-1}(y)]'\,dy = \int_{y_1}^{y_2} p_o(y)\,dy$$

Comparing the two integrals, we get the relation

$$p_o(y) = p_i[T^{-1}(y)]\,[T^{-1}(y)]'$$

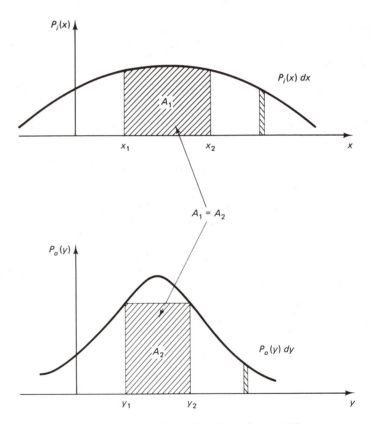

Figure A6.1–8 Transformation of a random variable.

or

$$p_o(y) = \frac{P_i(x)}{dT(x)/dx} = \left[p_i(x)\frac{dx}{dy} \right]_{x = T^{-1}(y)}$$

Basically, this result is obtained by equating the two shaded areas in Fig. A6.1–8. In particular, when $x_1 \to x_2$, it is

$$p_o(y)\, dy = p_i(x)\, dx$$

Table A6.1–2 lists several examples. The last case illustrates the Gaussian distributed noise passed through a square-law detector. Because $p_i(x)$ is double-valued function,

$$p_o(y) = 2p_i(x)x'$$

This follows from the equality of the two corresponding areas.

TABLE A6.1–2 EXAMPLES OF THE TRANSFORMATIONS OF A RANDOM VARIABLE

	$P_i(x)$	$y = T(x)$	$P_o(y)$
1	$2 - 2x$ $0 \leq x \leq 1$	$-x^2 + 2x$	1 $0 \leq y \leq 1$
2	1 $0 \leq x \leq 1$	x^2	$\dfrac{1}{2\sqrt{y}}$ $0 \leq y \leq 1$
3	e^{-x} $0 \leq x \leq \infty$	x^2	$\dfrac{e^{-\sqrt{y}}}{2\sqrt{y}};\ \ 0 \leq y \leq \infty$
4	$\dfrac{1}{\pi}$ $-\pi/2 \leq x \leq \pi/2$	$a \sin x$	$\dfrac{1}{\pi\sqrt{a^2 - y^2}};\ \ -a \leq y \leq a$
5	$\dfrac{1}{\sigma\sqrt{2\pi}}e^{-x^2/2\sigma^2}$ $-\infty < x < \infty$	x^2	$\dfrac{1}{\sigma\sqrt{2\pi y}}e^{-y^2/2\sigma^2};\ \ 0 \leq y \leq \infty$

A6.1–7 BANDLIMITED WHITE NOISE

In analogy to white light, a random signal whose power spectral density is constant, independent of frequency, is referred to as white noise. Strictly speaking, the white-noise model cannot be used to describe any physical process because it implies an infinite amount of power.

For a constant power spectral density of N_0 watts per hertz, measured over the positive frequency and if $n(t)$ has zero mean value, the spectrum of white noise is

$$S(\omega) = \frac{N_0}{2} \quad \text{for all } \omega$$

The factor of ½ is necessary to have a two-sided power spectrum for mathematical purposes. The autocorrelation function is

$$R(\tau) = \int_{-\infty}^{\infty} \frac{N_0}{2} e^{j\omega\tau} df = \frac{N_0}{2}\delta(\tau)$$

where $\delta(\tau)$ is the Dirac delta function shown in Fig. A6.1–9b. Since $R(\tau)$ has a value at $\tau = 0$ only, there is no correlation between two samples of white noise separated by an interval $\tau > 0$ and they are therefore statistically independent.

If the bandwidth of the system, B, is narrower than the bandwidth limitations of the physical process being observed, it is more practical to consider the results of passing white noise through a filter with the same bandwidth B. The output noise

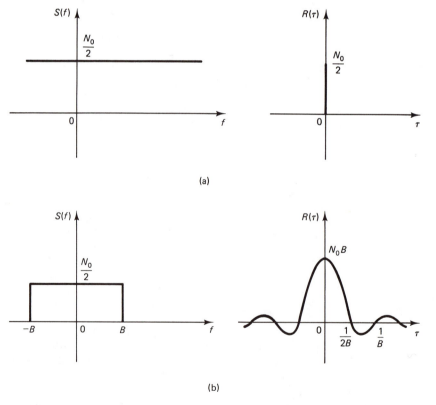

Figure A6.1–9 Spectra and autocorrelations: (a) white noise; (b) bandlimited white noise.

is then called bandlimited white noise or colored noise. The average power is given by

$$P_{av} = 2 \int_0^\infty \frac{N_0}{2} \, df = N_0 B$$

If the transfer function $H(\omega)$ of the linear network is not constant, the equivalent noise bandwidth, B_n, of a network can be defined as a bandwidth of an ideal rectangular network having the same maximum gain and passing the same average power from the white source as the actual network under consideration:

$$B_n = \frac{1}{|H(\omega_o)|^2} \int_0^\infty |H(\omega)|^2 \, df$$

The autocorrelation function $R(\tau)$ of the filtered (bandlimited) white noise is

$$R(\tau) = \int_{-\infty}^\infty \frac{N_0}{2} |H(\omega)|^2 e^{j2\pi ft} \, df = \frac{N_0}{2} \int_{-\infty}^\infty e^{j2\pi ft} \, dt$$

$$= N_0 B (\sin x / x)$$

where $x = 2\pi B\tau$ and is shown in Fig. A6.1–9b. Since the shape of $R(t)$ is a sin x/x function, there is a correlation on either side of $\tau = 0$, and correlation is periodical at intervals of $\tau = 1/(2B)$. Hence, filtering uncorrelated white noise produces correlated, bandlimited white noise or color noise. By increasing the (filter) bandwidth B, the output noise power is increased, but the correlation between $n(t)$ and $n(t + \tau)$, for a fixed τ, is decreased, or in other words, the rate of variance of noise at the output is increased.

A6.1–8 NARROW-BAND NOISE REPRESENTATION

A random process is said to be a narrow-band random process if the width $\Delta \omega$ of the significant region of its spectral density is small compared with the central angular frequency ω_c of that region [10, p. 159].

If the input in an ideal bandpass filter, with a central frequency f_c and bandwidth Δf, is a zero mean stationary Gaussian white noise with a spectral density $N_0/2$, the output of the filter will be a zero mean stationary Gaussian process with the spectral density, shown in Fig. A6.1–10a given by

$$S_o(f) = \begin{cases} \dfrac{N_0}{2} |H(f)|^2 & |f - f_c| < \dfrac{\Delta f}{2} \\ 0 & \text{elsewhere} \end{cases}$$

The output can be considered as a sample function of such a random process. That noise $n(t)$, although random, will be oscillating on the average, with frequency f_c.

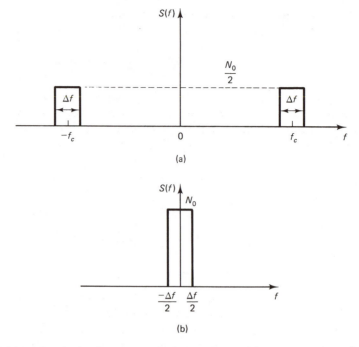

(a)

(b)

Figure A6.1–10 (a) Spectral density of narrow-band white Gaussian noise; (b) the low-frequency translation.

For the narrow band (Fig. A6.1–11) the noise can be expressed as

$$n(t) = V(t) \cos [\omega_c t + \phi(t)]$$

the envelope $V(t)$ and the phase $\phi(t)$ varying in a random fashion, roughly at the rate Δf hertz.

It is often convenient to express narrow-band noise in a quadrature form:

$$n(t) = n_c(t) \cos \omega_c t - n_s(t) \sin \omega_c t$$

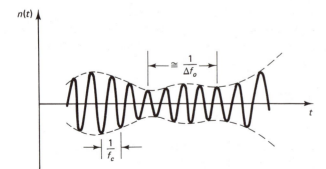

Figure A6.1–11 Narrow-band noise.

where

$$V(t) = \sqrt{n_c^2(t) + n_s^2(t)}$$

and

$$\tan\phi(t) = \frac{n_s(t)}{n_c(t)}$$

If the narrow-band noise process $n(t)$ is approximated by the Fourier series:

$$n(t) = \sum_{n=1}^{\infty} \sqrt{2S'(f_n)\,\Delta f} \cos{(\omega_n t + \theta_n)}$$

where $S'(f_n)$ denotes the noise power spectral density, one-sided, and θ_n is a uniformly distributed random variable, then

$$n_c(t) = \sum_{n=1}^{\infty} \sqrt{2S'(f_n)\,\Delta f} \cos{[(\omega_n - \omega_c)t + \theta_n]}$$

$$n_s(t) = \sum_{n=1}^{\infty} \sqrt{2S'(t_n)\,\Delta t} \sin{[(\omega_n - \omega_c)t + \theta_n]}$$

The beat frequency, $\omega_n - \omega_c$, is now low frequency.

The spectral density (low-pass) is (Fig. A6.1–10b)

$$S_{ns}(\omega) = S_{nc}(\omega) = [S_n(\omega - \omega_o) + S_n(\omega + \omega_o)]_{LP}$$

The $n_c(t)$ and $n_s(t)$ are zero mean Gaussian random processes and they are uncorrelated. It is the random nature of the noise that tends to distribute the noise components over both the cosine and sine terms. This is in direct contrast to a deterministic signal in which we can control the phase to give cosine and sine terms. The mean-square values are

$$\overline{n^2(t)} = \overline{n_c^2(t)} = \overline{n_s^2(t)} = N_0$$

Also, the mean noise power is divided equally between the cosine and sine terms:

$$\overline{n^2(t)} = \tfrac{1}{2}\,\overline{n_c^2(t)} + \tfrac{1}{2}\,\overline{n_s^2(t)}$$

The pdf's of $V(t)$ and θ are:

$$p_v(V) = \frac{V}{N_0} \exp\left(\frac{-V^2}{2N_0}\right) \qquad V > 0$$

and

$$p_v(\phi) = \frac{1}{2\pi} \qquad -\pi < \phi < \pi$$

In modulation theory there is a special interest in the envelope of the sum of sine wave and narrow-band Gaussian noise:

$$z(t) = A_c \cos{(\omega_c t + \phi)} + n(t)$$

This can be expressed as

$$z(t) = A_c \cos \omega_c t + n_c(t) \cos \omega_c t - n_s(t) \sin \omega_c t$$

where $\phi = 0$ is taken for the sake of simplicity, or

$$z(t) = [A_c + n_c(t)] \cos \omega_c t - n_s(t) \sin \omega_c t$$

$$z(t) = R(t) \cos [\omega_c t + \theta(t)]$$

The envelope of $z(t)$ is given by

$$R(t) = \sqrt{Z_1^2 + Z_2^2} = \sqrt{[A_c + n_c(t)]^2 + [n_s(t)]^2} \qquad R(t) > 0$$

where

$$Z_1 = A + n_c(t)$$
$$Z_2 = n_s(t)$$

Z_1 and Z_2 are two independent Gaussian random variables with

$$E\{Z_1\} = A \qquad E\{Z_2\} = 0$$

and

$$\sigma_{z1}^2 = \sigma_{z2}^2 = E\{n^2(t)\} = N_0 = \sigma^2$$

The joint pdf of Z_1 and Z_2 is

$$p_{1,2}(Z_1, Z_2) = \frac{1}{2\pi N_0} \exp\left[-\frac{(Z_1 - A)^2 + Z^2}{2N_0} \right]$$

Then the joint density for $R(t)$ and $\theta(t)$ is

$$p(R, \theta) = \frac{R}{2\pi\sigma^2} \exp[-(R^2 + A_c^2 - 2AR \cos \theta)/2\sigma^2]$$

The pdf for $R(t)$ can be obtained by integrating over θ:

$$p(R) = \int_0^{2\pi} p(R, \theta) \, d\theta$$

and it can be shown that the pdf of the envelope of a sine wave plus narrow-band noise is

$$p(R) = \frac{R}{\sigma^2} e^{-(R^2/2\sigma^2 + x)} I_0\left(\frac{R}{\sigma} \sqrt{2x} \right) \qquad R > 0$$

where x is the input signal-to-noise ratio, $x = A_c^2/2\sigma^2$, and $I_0(\cdot)$ is the modified Bessel function of the first kind of zero order. This pdf is known as the Rice–Nagakami pdf or simply the Rician pdf, and can be simplified for some special cases.

For $x < 1$, the small input carrier-to-noise ratio

$$p(R) \simeq \frac{R}{\sigma} e^{-R^2/2\sigma^2}$$

which is the Rayleigh density function.

For the range of input carrier-to-noise ratio $x >> 1$,

$$p(R) \simeq \frac{1}{\sqrt{2\pi} \ \sigma} e^{-(R - A_c)^2/2\sigma^2}$$

This is illustrated in Fig. A6.1–12.

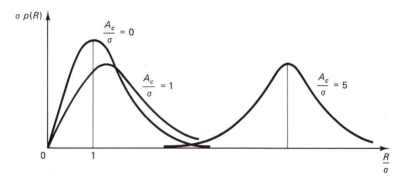

Figure A6.1–12   Pdf of the envelope of narrow-band noise plus sinusoid.

The pdf of the phase θ is

$$p(\theta) = \int_0^\infty p(R, \theta) \ dR$$

After some manipulation it can be shown that

$$p(\theta) = \frac{1}{2\pi} e^{-A_c^2/2\sigma^2} + \frac{A_c \cos \theta}{2\sqrt{2\pi} \ \sigma} e^{-(A_c/2\sigma^2) \sin \theta} \left[1 + \mathrm{erf} \left(\frac{A_c \cos \theta}{\sigma\sqrt{2}} \right) \right]$$

where the error function is defined as

$$\mathrm{erf}(x) = \frac{2}{\sqrt{\pi}} \int_0^x e^{-z^2} \ dz$$

The phase function is shown in Fig. A6.1–13 for several values of the parameter $x = A_c^2/2\sigma^2$ (i.e., the ratio of signal power to noise power).

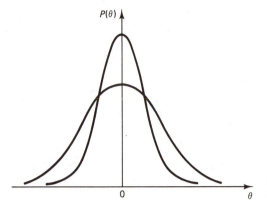

Figure A6.1–13   Pdf for phase of sine wave plus noise.

APPENDIX 6.2 FAILER EXAMPLES

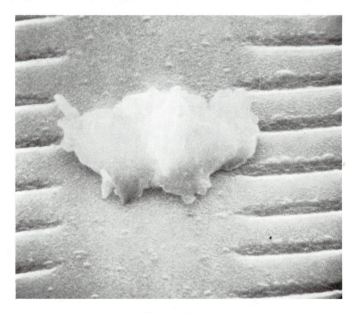

Figure A6.2–1

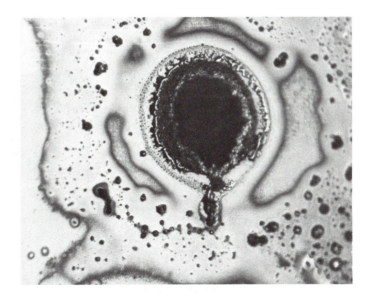

Figure A6.2–2

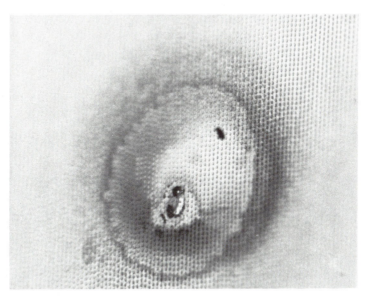

Figure A6.2–3

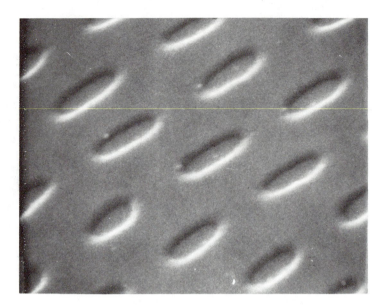

Figure A6.2–4

Chapter 7

OPTICAL MEMORIES

7.1 INTRODUCTION

Videodisc technology as discussed in the preceding chapters is spawned by the home entertainment industry. Optical videodisc technology includes the storage and retrieval of random-access graphics on videodisc as well as storage of digital information. Videodisc can be used as an analog read-only memory (ROM) for video or audio signal storage or digital data storage. Also, the videodisc can be used as a digital ROM, where the disc, itself considered as a channel, is digital; the pit length and the space in between pits form a finite set. One example is a compact disc (CD) for a digital audio. At DiscoVision Associates a double-density Jordan code was used to record a digital data [1] on the videodisc (summer 1977); maximum capacity was 30×10^9 row bits. For the stop audio, based on the digital data, a video-like signal is generated and then recorded on the videodisc [2,3] (summer 1980). Figure 7–1 displays a microscopic photograph of the memorized information in the Jordan code on the disc: 2, 1.5, and 1 bit lengths in pit and no-pits form can be seen on the picture. The microscopic enlargement is 10^4.

Optical data storage systems that utilize a highly focused laser beam to record and instantaneously play back information [4–8] are very attractive in computer mass storage technology. They offer very high storage density with very high data rates, rapid random access to the data, potential archival properties, and a projected low media cost. These systems are also attractive for television broadcasters and other communications professionals who need to store and then retrieve extremely large quantities of video information text, or moving and/or stationary pictures.

There is sometimes a confusion in the names used to refer to those two classes

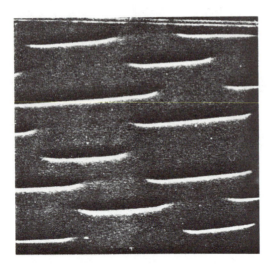

Figure 7–1 One example of the disc layout for the Jordan code.

of optical memory. Both can be used as analog or digital memories; that is, both can be used for video signal and/or digital data recording. Thus names based on the form of the information stored in optical memory are typically not self-explanatory: digital disc(k), videodisc, digital storage, optical digital disc, digital optical memory, and so on. Also, a direct-read-after-write (DRAW) property typically exists in all optical recorders. (*Note*: "Disc" is customarily spelled with a "c" in the entertainment industry and with a "k" ("disk") in the computer industry. Sometimes, one spelling is used for both types of applications.)

Figure 7–2 shows a different type of optical memory system. In a videodisc system (Fig. 7–2a), after the information takes a mechanical form, the mastering process is continued (under clean-room conditions) through the stages of development, nickel sputtering, nickel plating (master–mother–stamper), and so on. Direct-read-after-write during recording can be used for quality control, for example. At the end, read-only ("playback only"—prerecorded signal playback) discs (RODs) are obtained.

In the second type of optical memory ("write once/read many times"), a "programmable" disc is formed before recording and DRAW can be used as noted above. In the recording process, the disc carries its own clean room with it. There are two groups in this class. In the first group (Fig. 7–2b), after recording a playback-only disc is obtained. Actually, this is a programmable ROD (PROD). In the second group of the second type of optical memory (Fig. 7–2c), recorded information can be erased and new information recorded. This is the erasable PROD (EPROD) system.

In this chapter the second type of optical memory, the PROD, and its special case EPROD, are discussed. The term "programmable" or "optical disk" will be used throughout.

As in videodisc systems, in general, the optical disk also suffers from a lack of disk standards [9]. Even the disk radius varies a lot: disks are now being made in 20, 30, and 35 cm sizes, with 13-, 7.5-, and possibly 5-cm disks likely in the

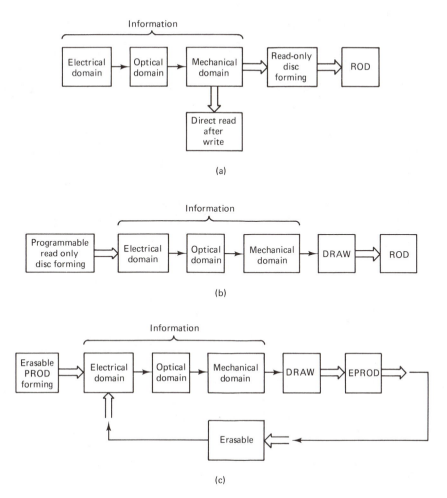

Figure 7.2 Optical memory systems: (a) videodisc ("playback only") systems; (b) programmable read-only disc (k) (PROD) systems; (c) erasable PROD systems.

near future for use in small computers. Problems with stability, archivability (shelf life), data integrity, producibility of the media themselves, and the strong interdependence between the drive and the media, making it difficult to develop one or the other without both [10], are probably the most important reasons for the delay in introduction of optical recording technology and its standard(s).

If the user's information to be recorded requires visual display to the user, the compatibility of the information with a particular display mode will very probably determine the storage formats selected for the application. The display of pictures, maps, images, and graphics lends itself to a simple video approach, as long as sufficient resolution can be provided. In such a case the information could be stored as a video signal and the system could employ the standard videodisc format, commercial video players preferably under computer control, and standard video monitors. This approach suffers from the resolution limitations of the standard video formats (NTSC, PAL, SECAM).

For digital optical disks various formats are selected. Figure 7–3 shows three planar addressing modes. The disk surface can be divided into concentric areas, rings, zones, or tracks (Fig. 7–3a), or into angular sectors (Fig. 7–3b). Those two can be combined (Fig. 7–3c).

The serial format time/space domain is shown in Fig. 7–4. Markers and segment address are added to the data bits. Data bits include data and error-correcting/detecting bits. The DRAW property can efficiently be combined with the error-correcting codes. The playback quality can be monitored during recording. Any difference between the incoming data and the recovered playback signal that approaches the correcting power limit of the code causes the system to rewrite the entire sector in the

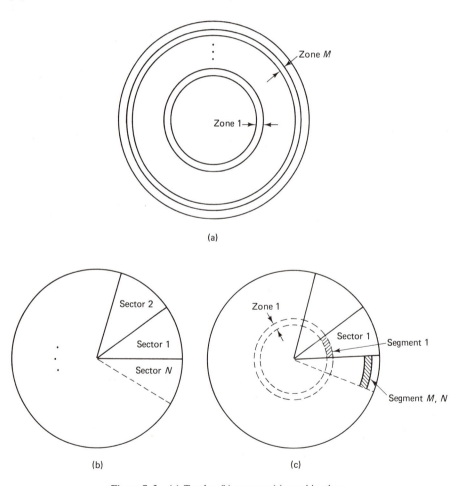

Figure 7–3 (a) Tracks; (b) sectors; (c) combinations.

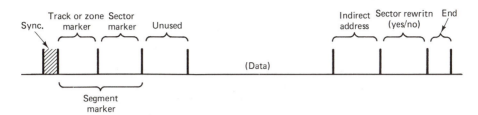

Figure 7-4 Serial format for the digital channel.

next block. When no errors (or correctable errors) are detected, the proper code word is written. Indirect addressing can be implemented by leaving in the sector a gap that may later be recorded.

Use of erasable versus nonerasable optical disks depends on the particular application. Erasability is desirable in some applications; in particular, erasable optical disks could be used in place of video tape or similar recording media, which also suffer from limited reading capability and limited read and write performance [11]. Normally, the main disadvantages of today's erasable optical disks are the low signal-to-noise ratio and the fact that the bit error rate (BER) increases after consecutive reads on the same location. The chief attribute of optical disks is the relatively fast access time to any point from beginning to end. New applications of erasable optical recording should be significant.

Many applications of optical storage devices do not fundamentally require the erasability feature. In some applications nonerasability is an important attribute: for example, in the storage and preservation of legal, financial, medical, and insurance records. Nonerasable optical media have the benefit of not being degraded by large numbers of repeated reads of the same information. This characterisic is generally a consequence of the high pit-formation temperatures that are required [11]. The reading beam, being of very low power, does not raise the media surface to a high enough temperature to cause a permanent change in the integrity of the recording structure. The major applications of nonerasable optical recording involve image and document storage. In image and graphic applications the ability to reread a document many times for long periods without degradation is a principal attribute of the medium.

The major disadvantage of nonerasable optical recording (PROD) is that it is fundamentally incompatible with existing computer communications systems, thus requiring significant changes in system architecture and software.

7.2 OPTICAL DISK DRIVE: RECORDING AND PLAYBACK SYSTEM

The principle of optical recording and playback in programmable optical disk systems is the same one as in videodisc systems. Laser light is used for both recording and reading. Three servo systems—tangential, radial, and focus—control the focused beam

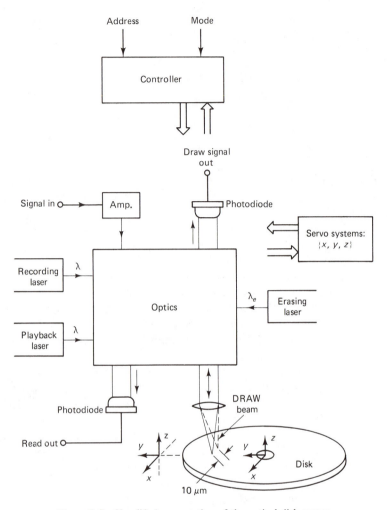

Figure 7–5 Simplified presentation of the optical disk system.

in all three spatial coordinates (x, y, z). In addition, a motor servo is required to maintain constant the disk rotational speed. The controller design depends on whether the disk is used as a digital channel (address, operational mode, etc., are required) or an analog-video channel (frame numbers, synch pulses, etc., are required).

A three-layer system is shown in Fig. 7–5, where separate layers are used for recording, playback, and erasing. Reading and writing are typically accomplished with lasers of the same wavelength (e.g., $\lambda = 830$ nm for a semiconductor laser, $\lambda = 633$ nm for a He-Ne laser, $\lambda = 488$ nm for an argon laser, and $\lambda = 442$ nm for a He-Cd laser), and erasing is done, for example, by a laser of $\lambda_e = 780$ nm wavelength. The incident power of the recording and erasing lasers is between 6 and 20 mW and 3 and 8 mW, respectively. Reading requires less power. Recording and playback can be accomplished with the same laser, with different light power focused on the disk surface. During recording a DRAW beam is focused typically 10 μm away behind the recording beam. In systems with separate recording and playback lasers, the DRAW beam can be obtained from either laser.

A recording and playback system using one laser is shown in Fig. 7–6. The power of the recording laser may directly determine the recording bit rate, disk speed, and speed of the arm (sled) [12]. The light output of the laser is first divided, with most of it (say 90%) going to the channel that records. The recording beam is directed through an optomodulator (electro-optic or acousto-optic), which modulates the light in response to an input electrical signal. The attenuator following the modulator is used to reduce light intensity during playback. The modulator could also perform this function, but because of the potential for accidental erasure, a stable attenuator seems the prudent choice [13]. The beam is then expanded to fill a focusing lens (typically, a 0.40 to 0.85 NA microscope objective). A polarizing beam splitter and a quarter-wave plate transmit nearly all of the incident light toward the disk, while directing nearly all the reflected light toward the DRAW signal detector. The playback beam passes similar devices, except an optomodulator; the separate focusing lens is shown. But optics can be rearranged so that the record and playback beams are focused with the same lens. Since the read beam is slightly skewed to the optical recording axis, the playback spot trails the recording spot by a few micrometers. In this way the recorded signal is revealed shortly after writing. This is a DRAW signal, typically used to detect any difference between the input and recorded signals, for error-free recording. The information can be recorded immediately, if necessary. Thus the read signal can be used as the DRAW signal, during recording, and as a playback signal, after recording.

The depth of focus is typically ± 0.5 μm. When the NTSC video signal is recorded (1800 rpm), the surface undulations generate error frequencies that are multiples of 30 Hz. Since these frequencies are in the audio range, the lens can be moved with a loudspeaker type of voice coil.

The limitations for the total pit capacity of the storage medium and the rate at which the signal can be stored and retrieved are the same as for optical videodisc systems. Data capacity and data rate are also functions of the channel coding method. Data access is a similar operation to "frame search" on the videodisc. Data access time is an important parameter of the optical disk system. A possible procedure for accessing recording data on the disk is: slew the translation stage to the appropriate track location, read the track address and search for the track desired, and find the data desired on the track once the proper track has been located. The slewing can be performed using an air sled with a linear motor, or a ball screw with a rotary motor. A film of air in the air sled provides smooth motion. The ball screw uses a precision-threaded rotating screw to move the translation stage, supported by a set of ball bearings. Both systems move at about 25 cm/s. Multiple recording of the track address around a track reduces access time. Typical data access times are on the order of hundreds of milliseconds [14].

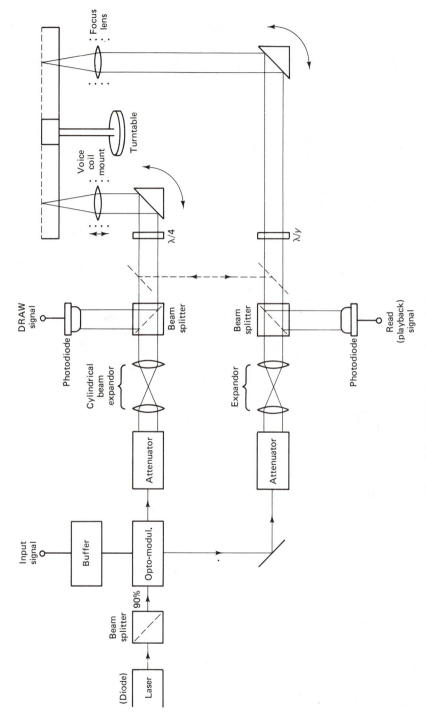

Figure 7–6 Optical disk recording and playback system.

7.3 RECORDING MEDIA

The recording medium is the basis for successful optical disk data storage and retrieval. Desired optical videodisc recording material characteristics, such as high resolution sensitivity, high SNR, and real-time recording and instant playback, discussed in Section 2.3.1.2, are also desired characteristics for optical disk media [15,16]. Two of those characteristics, high sensitivity and high immunity to defects, have, at least in general, a more important role in programmable optical disk recording than in videodiscs.

Until recently, gas lasers (He-Cd, Ar+, and He-Ne) were the only lasers used for optical recording applications. High output power and short wavelengths are their advantages. But for programmable recording systems, extremely compact design, with significant reductions in the size and cost of the records and playback systems, are required. The size, high efficiency, and potential high reliability of diode (semiconductor) lasers made them more favorable for programmable optical systems. Thus high sensitivity of the recording material is a necessity.

The original intention, and still the principal application of the videodisc, is for the video signal recording. The influence of defects (material, pinholes, dust, and other surface contaminants which can prevent the recording of many signal elements, or obscure them after recording) is greatly reduced by the masking effects of the human eye. But the programmable optical disk is expected to be used widely for digital data recording. To be applicable for data storage in electronic data processing systems, a storage peripheral must, in general, be capable of retrieving stored data with a final error rate of 10^{-12} or less [17]. Thus the material must have higher immunity to defects.

In the chain of optical videodisc production (Chapter 2), the recording medium is the only basis for stamper production. For the programmable optical disk the recording medium also serves as a final memory. The recording material should allow permanent recording of the data and should not degrade under ambient conditions or prolonged readout. That is, archive storage is required.

Optical recording materials can be classified different ways according to the criteria used for classification. There is, for example, an 11-group classification [18]: photographic films, photoresists, photopolymers, thermoplastics, photochromics, chalcogenide films, ablative thin films, magneto-optic, photoferroelectric, photoconductive/electro-optic, and electro-optic. For convenience, we classify optical recording materials into only two groups: nonerasable (nonrecyclable) and erasable (recyclable) media.

7.3.1 Nonerasable Optical Recording Media

Nonerasable media include ablative thin films, photographic films, photoresists, and photopolymers. The recording mechanism is generally permanent and there is no erase-recycling process for those materials. In general, replication can be accomplished except for photopolymers.

Further classification is possible as to whether after the material is optically exposed, an additional processing step is required. For ablative thin films, no further processing is required after recording. For other nonerasable optical recording materials, further processing may be required.

7.3.1.1 Ablative thin films.

The reworking mechanism is essentially thermal in nature; pits are recorded through a process of ablation or melting. The threshold laser recording power is determined by the combined optical and thermal efficiencies of the recording structure. The optical efficiency, η_{op}, of the structure in which the recording medium is placed is equal to that fraction of the incident light beam which is absorbed into the recording medium. The thermal efficiency, η_{th}, is the fraction of heat generated within the recording layer that does not diffuse into the materials that are in contact with the layer of recording medium.

Readout is based on the light amplitude modulation by the recorded pattern. Either reflection or transmission of light can be used.

Nominal resolution is 1000 to 1500 cycles/mm. Organic dyes require approximately 10^{-2} J/cm^2; metals (bismuth and rhodium, for example) are 3 to 10 times less sensitive, but are more resistant to damage while handling. The replication can be accomplished by embossing techniques.

To date, a great percentage of the research into materials for optimal memory media has concentrated on a rare nonmetallic element, tellurium (Te), which resembles sulfur and selenium in chemical properties [9]. Tellurium is usually found alloyed with other elements because pure tellurium oxidizes rapidly when in contact with moisture. This is a soft gray material which sublimes at temperatures well below its 452°C melting point and may be readily evaporated in a high-vacuum chamber using a refractory metal coat as a source [19].

Much optical-media research in recent years has concentrated on finding feasible alternatives to tellurium. Among them are silver halide and gold/platinum alloys, amorphous Te-C films [20], suboxide thin films (Sb_2O_3, TeO_2, MoO_3, GeO_2 [21]), tellurium/carbon alloy, tellurium/copper alloy, and so on.

7.3.1.2 Media with postprocessing.

Photographic films, photoresist, and (some) photopolymers require processing after optical exposure, testing for example. This prevents use of the common DRAW error-correction schemes, which rely on the ability to read a spot milliseconds after it has been formed. Some photoresists under development offer the DRAW technique. Photographic films and photoresist were discussed in Chapter 2, and these materials are available commercially. Photopolymers are still in the experimental stage.

Photographic media are very sensitive, with recording thresholds on the order of 10 to 50 μJ/cm^2, about 100 times lower than for tellurium alloys. The resolution is up to 3000 cycles/mm. The recording process is photochemical in nature. The readout is based primarily on the amplitude modulations of the readout beam through the optical density change, although the readout may be based on the index of reflectivity change. Wet chemicals or heat are used after optical exposure. Replication can be accomplished by contact-printing techniques.

Wet chemicals or heat are also required for both positive or negative photoresists. During the development process, soluble areas (unexposed for negative photoresist, exposed for positive) dissolve away. The resulting surface relief can be read by photodetection. The resolution and recording sensitivity are similar to those of oblative metal films. Replication is also by embossing techniques.

Replication is not possible with all photopolymers when recording is performed by utilizing reflective index changes. Hitting or post exposure may be required after optical data recovery. Resolution is up to 5000 cycles/mm.

7.3.2 Erasable Optical Media

Three processes should be clearly distinguished: writing, reading, and erasing. Besides parameters important for the nonerasable media (raw BER, S/N, etc.) a new one is added: the constancy of those parameters after many erasures and even after large numbers of readings.

One way to classify optical reversible or erasable materials is on the basis of storage time: permanent storage or limited storage.

7.3.2.1 Permanent storage.
Typical materials include magneto-optic, thermoplastic, and chalcogenide films. Nominal resolution is 100 to 2000 cycles/mm. Recording sensitivity is 10^{-4} to 10^{-2} J/cm².

Magneto-optical (MO) Recording.   A low playback SNR due to the media noise associated with the microcrystalline nature of the media, poor chemical stability, and a relatively limited exposure latitude for recording are among the major problems (in MnBi, for example) that have reduced media applicability. Lately, significant improvements have been reported using certain rare-earth materials.

Vertically oriented media must be used in MO techniques to achieve the maximum effect [22]. A laser heats the surface of an amorphous magnetic film (MnGaGe, PtCo, etc.) above its Curie temperature (Fig. 7–7). Subsequent cooling in the presence of an external magnetic field alters the magnetization direction of the heated spot in a predictable way. Data are erased by reversing the direction of the magnetic field applied; that is, the erase process is the same as the recording process.

Because the writing is done by the time–space coincidence of a magnetic field and localized heating by an optical beam, the smallest bit size is intrinsically limited by the optical resolution of the system. But the resolution or the size of the written spot also depends on the magnetization reversal process.

There are also materials that have a temperature dependence of coercivity (e.g., CoP) that can be used to store information by applying a suitable field coincident with the laser heating [23].

More recently, attention has focused on amorphous rare-earth transistion-metal alloys, such as Tb-Fe, Gd-Go, Gd-Fe, Gd-Tb-Fe, Tb-Dy-Fe, and Gd-Fe-Bi. All these

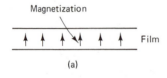

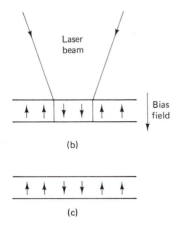

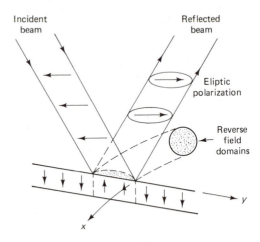

Figure 7-7   Magneto-optical (MO) recording.

Figure 7-8   MO reading: Kerr effect.

materials have acceptable Kerr rotation angles and are easy to move into vertical media, generally by sputtering. In this substance, magnetic moments of the rare earths are antiparallel to those of transition metals. The temperature dependencies of these moments are quite different. The terbium moment is larger than the iron moment at low temperatures. As a result of antiparallel alignment, there is a temperature, below the Curie temperature, where the two will just cancel. This is called the compensation temperature. This temperature can be varied by changing the composition. The writing process does not require heating to the Curie point. Also, cobalt may be added to the material to raise its Curie temperature [22]. This is important because the Kerr effect decreases as the Curie temperature is approached.

Reading is done by using the principles of either the Kerr or the Faraday effect. In the Kerr effect (Fig. 7–8; incident and reflected beams are space separated for clarity), a linear polarized beam reflected off a vertically magnetized surface will have its polarization partially rotated. An elliptic polarization is created, with the axis of the ellipse depending on the direction of magnetization. In the Faraday effect, the polarized beam passes through the material rather than being reflected from its surface.

The magneto-optic readout signal in an isotropic medium is due to a dielectric tensor of the form [24]

$$\epsilon = \left\| \begin{matrix} \epsilon_0 & \epsilon_1 \\ -\epsilon_1 & \epsilon_0 \end{matrix} \right\|$$

where ϵ_1 is proportional to the magnetization M. A linearly polarized field E_p, incident on such material will, after the interaction, have a component E_n, in the orthogonal direction. This component is proportional to M. For complex structure such as sandwiches between various dielectric or metallic films, ϵ_0 and ϵ_1 are complex, so that E_n is also complex. A variable-phase plate with its principal axis parallel to the E_n, E_p axes can simultaneously convert the two elliptically polarized fields to linear ones (Fig. 7–9). The light is then split into two halves by a polarizing beam splitter BS2, such that each half may be detected by a separate avalance photodiode. The signal at the output of the differential amplifier is directly proportional to the intensity of the polar Kerr effect and may be processed by a filter before reaching the decision point without common-mode interference [25]. The advantage of this differential method is that it is much less sensitive to laser and reflectivity fluctuation.

Thermoplastic.   Thermoplastic material is a multilayer structure [18] consisting of a substrate, glass or Mylar; a thin conductive layer, Au, Ag, and so on; a photoconductor, polyvinyl carbonate sensitized with trinitro-9-fluorenone; and a thermoplastic, stoybelite Ester 10. Sensitivity is about 10^{-4} J/cm², and resolution over 2000 cycles/mm. Replicas can be made by the embossing technique.

The recording includes the following steps:

- A uniform charge is established on the surface of the thermoplastic, usually with a corona-charging device. The voltage is capacitively divided between the photoconductor and thermoplastic layers.

- Optical exposure of the material causes the photoconductor to conduct in the illuminated regions and thus discharges the voltage across it. The surface charge density on the thermoplastic surface is unchanged. The surface is recharged uniformly, the charge is added to the illuminated regions, and a force variation on the thermoplastic is established.

- The thermoplastic is heated to its softening temperature and the electrostatic forces deform the source into a relief pattern which corresponds to the optical recording information. When the thermoplastic is cooled, the relief pattern is permanently stored in the surface of the thermoplastic.

Information can be erased by heating the thermoplastic so that surface tension smooths out the relief variations.

Chalcogenide: Phase-Change Optical Media.   This erasable recording technique is based on a crystal-to-amorphous transition in tellurium or a tellurium alloy (Te-As-Ge, Te-Se-sulphur, for example). Both the crystal and amorphous forms of the alloy are stable, because they are separated by an energy barrier. Thus data recording sensitivity is 10^{-2} to 10^{-1} J/cm², and resolution approximately 1000 cycles/mm. No major degradation of system performance results from its erasability.

The recording mechanism is based on reversible switching between the amorphous and crystalline status of the materials by heating with a laser. The crystalline/

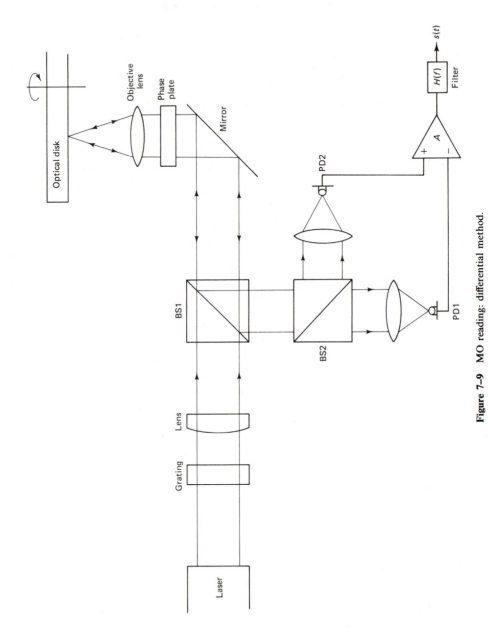

Figure 7-9 MO reading: differential method.

amorphous transition can be reversed simply by applying energy; data recording is then completely erasable. It takes less energy to initiate the amorphous-to-crystalline transition than it does to go the other way; erase power needs are only about half of writing power needs. Spot size has not reached the limit of the material.

Readout is provided by the difference in the optical properties between the two states. Because the crystalline form is both more reflective and more opaque than the amorphous material, such a disk could be used in either reflective or transmission playback mode.

The major disadvantage is the high temperature dependence of the material. There is a substantially lower margin for the reading and writing rates. Many phase-change media suffer from a degradation in performance during the reading process, which places a limit on the number of rereads possible before the material must be refreshed.

7.3.2.2 Limited storage.

Typical materials include photochromic, photoferroelectric, photoconductive/electro-optic, and electro-optic. Resolution in this group is 100 to 2000 cycles/mm, and the sensitivity 10^{-5} to 1 J/cm^2. Storage time is between several minutes and infinity.

The photochromic medium is a film containing both silver chloride and metallic silver, deposited on a transparent substrate. Material may exist in two or more relatively stable states having different optical constants and may be switched from one state to the other by photoirradiation [18]. Before exposure to light, the film transmits light equally without regard to polarization. However, exposure to polarized visible light impresses polarization sensitivity onto the material. Recording is accomplished entirely by the energy of the optical beam.

Reading is typically with an infrared beam whose polarization direction is rotated 45° from that of the write beam; it cannot write on the material, which is transparent at wavelengths longer than about 800 nm. The infrared beam passes through the rotating medium and a polarization analyzer, producing a modulated signal that contains the recorded information.

Erasure is by irradiating the material with a defocused visible beam whose polarization is rotated with respect to that of the original beam. Such erasing can be done ad infinitum, although for best results, writing energy should be kept low.

Photochromics have two severe drawbacks. Because they are sensitive to visible light, a couple of minutes in a well-lit room can bleach the data away. In addition, the material is relatively insensitive, requiring at least 200 mJ/cm^2 for adequate contrast ratios.

The photochromic can be classified as inorganic (CaF_2, etc.), silver halide-doped glass (borosilicates), and organic (thioindigo in epoxy, stilbene, etc.).

For the photoferroelectrics (Bi_4, Ti_3, O_{12}, etc.), particular orientations of the optical indicator and external polarizers must be used to obtain readout. The photoconductor absorbs the light and the resulting charge pattern induces switching between the stable remnant states of the ferroelectric. The erase process is typically the same as the record process. The sensitivity is 10^{-3} to 10^{-2} J/cm², and the resolution is 125 to 800 cycles/mm.

Photoconductive electro-optic materials (Bi_{12}, SiO_{20}, etc.) are both photoconductive and exhibit a linear electro-optic effect. The storage is several layers and the material can be reused indefinitely. Photo-generated carriers in the material drift to the insulating layers and the field of this charge distribution produces a spatially varying retardation in the crystal via the linear electro-optic effect. The retardation is detected between crossed polarizers at a wavelength chosen to minimize photoconductivity. Resolution is 1000 cycles/mm, and sensitivity is 10^{-5} to 10^{-4} J/cm². The erase process is similar to the recording process.

Electro-optic materials ($LiNbO_3$, SNB, etc.) have a storage time from a few hours to indefinite. Resolution is approximately 1500 cycles/mm, and the sensitivity is 10^{-1} to 1 J/cm². Recording is based on the generation of fields through photoexcitation and spatial rearrangement of electrons to cause changes in the refractive index of the material. The information is stored through refractive index modulation and readout is based on the index change. The electronic charge information can be converted by a thermal fixing process into optically stable ionic charge information which is permanently fixed. The ionic charge information can be erased by later thermal treatment of crystal [18].

7.4 STRUCTURES FOR OPTICAL RECORDING

The media, and overall system characteristics, can be improved by enhancement techniques and are greatly influenced by disk structure. If recording media oxidize or interact in any way in contact with some materials, a protective layer is required. The typical graph that represents size distribution of the defects for an unprotected coated disk reveals two broad peaks [26]. The first is centered between 1 and 2 μm; the second pair is centered between 5 and 10 μm and is strongly correlated with the distributions associated with airborne dust. For the first class, a clean-room environment, the second pair does not exist. Unfortunately, dirt is unavoidable in every optical disk system. The most widely used approach to a solution of the problem is to manufacture the active disk surface under the cleanest possible conditions, and then immediately cover this sensitive surface with a transparent layer. Recording and play take place through the protective layer, so dirt on the disk exterior is always out of focus. Two structures that provide this protective layer are commonly utilized. In the Philips DRAW system, two recording surfaces are assembled into a sandwich, with their substrates facing out and an air space in between (Fig. 7–10a). Thus the substrate itself becomes the protective layer. But in the case of metallic or metallic-like thin-film recording materials, the oxidation in an air sandwich proceeds from the cavity side of the sandwich structure and the rate of oxidation is determined by the equilibrium vapor pressure within the cavity of the air sandwich [27]. RCA has introduced coating of the recording surface with a transparent material (Fig. 7–10b). In both cases it is necessary to use optics designed for the appropriate transparent

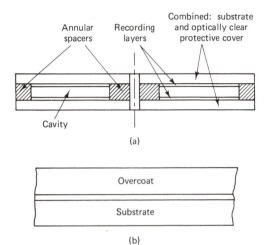

Annular spacers   Recording layers   Combined: substrate and optically clear protective cover

Cavity

(a)

Overcoat

Substrate

(b)

Figure 7–10   Simple optical disk structures: (a) air sandwich; (b) coating with transparent material.

layer. The protection in the air sandwich is afforded without loss of sensitivity. In the overcoated (encapsulated) structures only a few materials can be utilized without suffering a significant loss in working sensitivity. More complex variants of these structures are considered, such as those employing multiple layers of different polymers and antireflective structures employing a multiplicity of polymer and metallic layer.

The recording layer can be placed on the flat substrate, or on the pregrooved layer (e.g., a groove depth of about 70 mm). The pregrooved layer can be included in the substrate or added on. Pregrooved tracks are used for guidance and pregrooved data for indicating the position of the spot on the optical disk [28]. Pregrooved data corresponding to interruption of the track itself give after decoding of track and sector address a so-called hard preformat (Fig. 7–11a). Usually, a black-and-white track is defined depending on whether low- or high-level-output (read) light appears against the surrounding foreground (Fig. 7–11b and c). A tracking technique is adjusted for the particular class of disks, pregrooved or flat, to avoid possible interaction between pregrooved and postrecorded information [28].

7.4.1 Multilayer (Antireflection) Structures

The contrast between the disturbed (pit) and undisturbed portions of the track, and thus also the SNR, can be enhanced by optimization of the recording structure. Also, the noise associated with edge variation is influenced by the interference properties of the recording structure. Since the early 1940s [29] it had been noticed that Kerr rotation from the surface of a MO material could be significantly enhanced by coating that surface with dielectric layers of the appropriate thickness. In the early 1960s [30] it was demonstrated that maximum enhancement of the Kerr effect occurs when the dielectric layer thickness is such as to result in an antireflection condition. For thermal recording the melting or vaporization temperature and the thermal diffusivity will largely determine the minimum laser power required to record

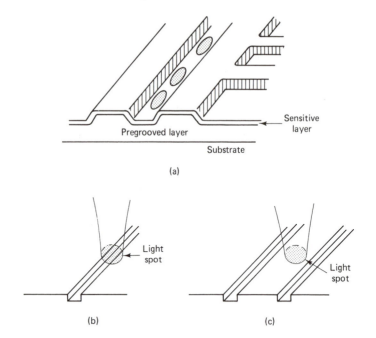

Figure 7–11 (a) Cross section of the pregrooved disk; (b) black track; (c) white track.

in ablative thin-film materials. However, the threshold for recording is also significantly affected by the combined optical and thermal efficiencies of the surrounding structure that incorporates the recording material.

The total absorption, A, in the multilayer structure is

$$A = \sum_j A_j$$

where the individual absorption in each of the layers, A_j, can be computed, for example using a matrix method [31]. Also:

$$A = 1 - R - T$$

where R and T are the energy reflectance and transmission coefficients, respectively.

Before discussing multilayer construction, it is worthwhile and convenient to analyze a monolayer structure. Consider the simplest configuration, a thin metallic layer deposited directly onto a substrate. Only a portion of the incident beam is absorbed by the layer, with the remainder being reflected or transmitted into the substrate. The fraction of the incident beam absorbed in the metallic layer represents the optical efficiency of the layer. The typical absorption curve is shown in Fig. 7–12. The critical thickness, d_c, for various materials on a glass substrate is given in Table 7–1 for $\lambda = 488$ nm.

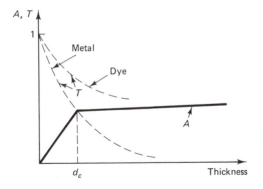

Figure 7–12 Typical absorption curve for the metal layer on a glass ($n = 1.50$) substrate, and typical transmission for metal and dye (dashed lines).

TABLE 7–1

Metal	n	d_c (nm)	A_c
Bismuth	1.6–12.7	20–25	0.4
Titanium	2.4–13.3	15	0.4
Rhodium	1.6–14.6	10	0.3

The thermal efficiency is a measure of how much of the heat generated in the recording layer remains localized during the pit ablation process. The thermal diffusion length l is

$$l = \sqrt{k\tau}$$

where k is the thermal diffusivity and τ is the duration of the pit formation process. Normally, $d_c < l$. This means that the heat loss from the absorption region is dominated by the diffusion of heat into the substrate disk. Thus it is important that the substrate material immediately adjacent to the recording material layer absorber have low thermal diffusivity [13].

Radial heat loss in the metal film will further reduce thermal efficiency. This can be significant if the thermal diffusivity of the metal is moderate to high. Indium, lead, bismuth, and metal-like tellurium have low melting points and low thermal diffusivity. They are less absorptive than metal but more difficult to work with since they tend to decompose rather than evaporate during vacuum deposition [13].

The playback signal contrast, S, is

$$S = \frac{I_{max} - I_{min}}{I_{max} + I_{min}}$$

where $I_{max} = \max \{I_1, I_2\}$ and $I_{min} = \min \{I_1, I_2\}$; I_1 and I_2 are the output light intensities from the track and the pit area, respectively. The signal contrast is, typically, proportional to the incident laser power, up to some threshold power; and then contrast can even be reduced for some range of power. But the playback SNR is limited by the material's noise component. Irregular pit geometry results from material impurities and imperfect redistribution of molten material during the recording process.

7.4.2 Bilayer Antireflection Structure

Higher optical efficiency and high SNR can be obtained using multilayer structures. The bilayer approach is best suited for use with moderately light-absorbing organic dye materials, and the trilayer structure is used most effectively with strongly absorbing materials such as metals [13]. Structures should be optimized for the desired wavelength.

In the bilayer antireflection structure, two layers are added on the substrate: reflector and absorptive recording material in quarter-wave thickness ($\lambda/4n$). The reflector can be a layer of aluminum (e.g., 30 nm thick, $R \simeq 0.9$) or a dielectric in quarter-wave thickness. During the recording, the light absorbed in the recording film (organic dye) causes its temperature to rise until a pit is formed while the reflector (aluminum) layer remains intact (Fig. 7–13a).

It is now convenient to introduce the following notation:

- B_1 boundary air/dye, B_2 boundary dye/reflector (aluminum surface)
- Light intensities:

  $I_{r1} = I_r^{(1)}$ reflected from B_1

  $I_r^{(2)} =$ reflected from B_2

  $I_{r2} =$ part of $I_r^{(2)}$ that passes B_1

  $I_1 =$ total (at photodiode) from the pit (aluminum)

  $I_2 =$ total from the (undistracted) dye

In Fig. 7–13a, a time axis is added to illustrate a light path when the light beam falls on the dye surface. Destructive interference will occur, producing lower reflectivity to the dye surface. The reflected light from B_2 is out of phase with the reflected light from B_1 when they reach the photodiode (at B_1 also) because there is a difference of $\lambda/2n$ in the traveled distance: an optical path length of one half-

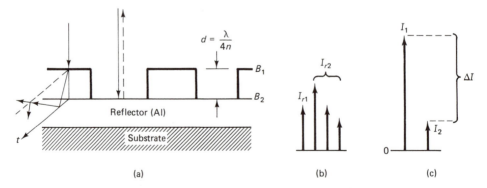

Figure 7–13   Bilayer antireflection: (a) structure (cross section); (b) antireflection—intensities and photodiode; (c) contrast—reflection and antireflection signals.

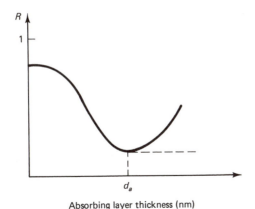

Figure 7–14 Reflectivity of the bilayer structure as a function of the thickness of the recording film (absorber).

wave. This is a basis for the antireflection phenomenon. Complete cancellation will occur (Fig. 7–13b) only if $I_{r1} = I_{r2}$. Otherwise, $I_2 \neq 0$ (Fig. 7–13c).

Theoretically, the reflectivity is a periodical function of the thickness of the recording fiber: antireflection should occur if

$$d = (2k + 1)(\lambda/4n); \qquad k = 0, 1, 2, \ldots$$

Figure 7–14 shows a first minimum of the typical reflectivity versus thickness curve. The antireflection thickness, d_a, is on the order of 60 nm. The SNR may be improved if the volume of the (melted) material that is redistributed around the pit rim is reduced. This implies metal layer (thickness) reduction.

Because metal is a highly absorbing material, due to imbalance in the amplitude ($I_{r1} \neq I_{r2}$), a strong antireflection minimum cannot be obtained. Also, the organic dye material has low red absorption; thus the bilayer structure cannot be optimized for the He-Ne or Ga-As laser, even though the intrinsic sensitivity of the dye theoretically is high enough to allow recording with these sources.

7.4.3 Trilayer Antireflection Structure

In the trilayer antireflection structure (Fig. 7–15a), a transparent dielectric (SiO_2, a polymer dielectric material, etc.) is introduced between the recording metal film and the reflector (aluminum): a boundary absorber/dielectric (B_3) as well as air/dielectric (B_3') are introduced. Any reflection at B_3' is a loss for the system, and the optical path length between B_1 and B_2 is a quarter-wave. The quarter-wave thickness of the dielectric phase layer results in destructive interference of the two beams, causing the reflectivity R for (unrotated polarization) the recording media/air or overcoat boundary to be very small. Practical trilayer structures include an overcoat layer over the absorber. The thickness of the dielectric layer at $\lambda = 488$ nm is on the order of 70 nm, and the absorber layer is 3 to 10 nm thick. The optical efficiency of the structure is extremely high because the maximized absorption (minimized R)

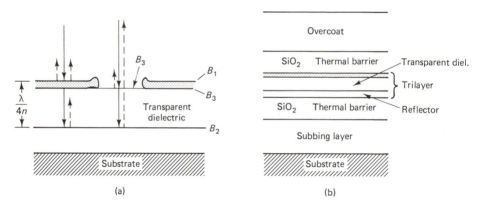

Figure 7–15 Trilayer antireflection structure: (a) minimum; (b) fully encapsulated.

may be over 95%. The optical properties of the trilayer are periodical with the value of the dielectric thickness.

The reduced thickness of the recording (metal) layer reduces the volume of the material melted and redistributed during the process of pit formation. Thus the SNR is improved. The threshold for recording is reduced in the trilayer configuration compared to that of the monolayer structure, as a result of the improved optical efficiency.

The thermal efficiency depends on the thermal properties of the recording medium and the material in contact with this layer, but also on the duration of the pit-formation process, which is determined by system-related factors such as recording signal bandwidth, duty-cycle (signal and layer output), turntable speed, and laser spot size. The presence of the aluminum reflector layer represents a potential heat sink, which may reduce the thermal efficiency of the recording process. The parameters of the dielectric layer must be specified such that the thermal time constant for heat to diffuse from the layer of the recording medium to the aluminum layer exceeds the duration of the pit-formation process [16].

An overcoat protection is normally required. But the melting point of some recording materials (e.g., 1668°C for titanium) far exceeds the temperature at which organic materials will melt or decompose [32]. To prevent thermal degradation of the plastic material adjacent to the metal trilayer structure, it is necessary to isolate the plastic thermally by using a thermal barrier layer to prevent heat generated during recording from reaching the plastic materials (Fig. 7–15b). Any added layer, and also the overcoat and substrate layers, increase total material noise and dropouts due to localized disruption. With careful structure optimization, the thermal barrier between the aluminum reflecting layer and the substrate layer can be eliminated. For example, a second minimum thickness of SiO_2 can be enough. The subing (surfacing) layer is added, in general, to hide the substrate surface defects. A structure with overcoat, recording film, dielectric layer, and reflector is sometimes called quadrilayer configuration [25].

7.4.4 Recording Modes

Here the term "mode" refers to the (recording) process, not to the signal-element change with radii: constant angular or linear velocity (CAV, CLV).

In the previous structure discussion, attention was directed primarily to the ablative pit-forming mode of optical recording. Low-melting-point (metal) material was a primary choice. But nondestructive recording modes are also possible for all the structures discussed, not only for the trilayer configuration. Also, all structures can be used for both flat and pregrooved disks.

In the trilayer construction, instead of a low-melting-point absorber, an erasable film can be used. High reflectivity spots are formed during recording by changing the optical properties of the recorded region; the material switches between the crystalline and the amorphous phase. There is no topographical disruption in the recording media. The incident beam is plane polarized. The MO film rotates or not with polarized light, depending on the direction of the magnetization of the medium at the reading point. The trilayer structure suppresses the unrotated component of the reflected beam, while the polarization-rotated component is reinforced. The operation of the encapsulated trilayer structure is essentially the same; the dielectric overlayer increases the flexibility in the choice of layer thickness in the optimization of the design for maximum SNR.

In the bubble-forming or blister mode [33,34] the absorbing layer is a high-melting-point material (metal) such as gold, platinum, or tellurium, and the dielectric layer material has a vaporization temperature well below the melting temperature of the absorbing layer. Upon exposure to the laser beam, bubbles are formed as the top absorber layer bulges because of the gas pressure from the vaporized dielectric layer beneath. In Fig. 7–16, schematic representations of the conditions prior to and after writing are shown. The interference conditions at any wavelength can be adjusted in the design so that initially either a maximum or a minimum reflectivity is obtained [35]. When a bubble is formed in the metal layer, these conditions must clearly be destroyed. In the antireflection trilayer stack, the formation of the surface protuberance destroys the antireflection condition [26]. Thus a high reflectivity is obtained from the bubble. Obviously, a scattering effect will reduce the readout contrast.

In the reading process, the bubbles are easily observed using a lower-power laser. In the trilayer antireflective structure, bubbles are a highly reflective surface. In the bilayer structure without the antireflection feature, the bubble, having a high

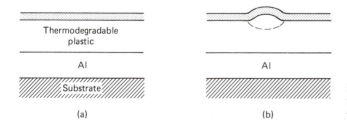

(a)   (b)

Figure 7–16 Bubble recording model: (a) trilayer structure prior to writing; (b) after writing a bubble in the metal layer.

reflective surface before recording, tends to scatter the beam, thus decreasing the reflectivity.

An important advantage of bubble media over hole-burning materials is amenability to replication. The bubbles are much sturdier than pits, and a single blistered disk can serve as the template for a large number of copies without suffering damage.

New materials being tried for optical recording include amorphous hydrogenated semiconductors [36,37]. The recording mechanism is based on the hydrogen evolution or effusion within the recording layer. In one of three processes without melting, hydrogen evolution on amorphous hydrogenated silicon films (α-Si:H) deforms the film into microbulges or bubbles. In two other processes: (1) ablation in thick α-Si:H films reveals holes with sharp edges, and (2) the effusion of hydrogen in α-Ge:H can be described as microswelling, where the imprint resembles a sponge with holes a few hundred angstroms in diameter.

A texture recording technique depends on a change in the surface morphology [38]. Films of Ge or Si, textured by reactive ion etching, have random arrays of decoupled columns or cores (Fig. 7–17) with cross-sectional dimensions of less than 100 nm. This structure substantially reduces the reflectance of the surface. The recording light beam melts the columns to produce a smooth reflective surface. This writing process tends to be self-limiting, in that less laser power is absorbed after the surface has been smoothed. Because of the threshold effect (i.e., the threshold is larger for the Si than for the Ge surface) and because of the low lateral thermal conductivity inherent in the columnar structure, the edge definition of the recorded spots is sharper than the Gaussian laser beam profile. Basically, this is a texture-destructive recording mode. Low-power laser beams have been used to record in this permanent optical storage technique.

If very small, high-electrical-conductivity metal spheres or spherical particles are distributed throughout a dielectric medium, the effective dielectric constant or reflective index will rise, owing to the added induced dipoles of the metal particles [39]. Based on this characteristic, a nontoxic reflective laser recording storage medium (Drexon), which may be manufactured without the use of a vacuum system and on a continuous basis, and which may be used to record low-reflective spots in a reflective

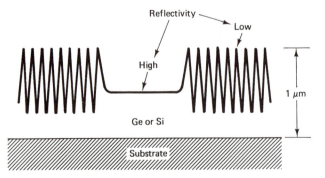

Figure 7–17 Cross section of the texture recording structure.

field with relatively low energy laser pulses, has been developed (U.S. Patent No. 4,269,917). The reflective surface layer is typically less than 1 μm thick, has a reflectivity of 15 to 50%, and is a nonconductor electrically and a poor conductor thermally because the matrix is typically gelatin (the high concentration of tiny particles and agglomerates of silver particles are separated and isolated from each other by the gelatin matrix). The recording sensitivity is in the class of bismuth and tellurium.

In the modified version, Drexon II, the medium is a double-layer configuration of a crust containing silver halide particles and an insulating underlayer devoid of the metal. A diode laser heats the medium so that the silver halide particles absorb the laser energy. As the temperature rises to about 200°C, the polymer film melts and creates spots of low reflectivity in a field of high reflectivity. In this recording mode, the laser shrinks the gelatin in the medium, rather than throwing up craters around the pit as is typical in ablative techniques.

7.5 OPTICAL DISK EVALUATION: TESTINGS AND MEASUREMENTS

An important part of the successful development of media technology involves the various techniques for characterization of the media [40]. In general, the quality control problems and techniques discussed for the videodisc in Chapter 6 and in Sections 2.3.1.5 and 2.3.1.6 are also valid for optical disks (PROD and EPROD). When used for video signal recording, the SNR, intermodulation products, dropout statistics, duty-cycle variation, track-to-track crosstrack, "track kissing," and so on, are as important as they are for videodisc systems. Also, for digital data recording the SNR and raw BER are the most important parameters.

Two evaluations are possible: before and after recording. For the erasable disk, evaluation is required after each recording. For disks with reading and nonreading areas, evaluation of the reading area is a requisite. The reading-before-writing feature may be incorporated. In any step, a polar figure for the dropouts and noise, for example, would be helpful.

Archival stability must be evaluated for optical disks. Optimum recording power can be determined by evaluating the distortion of the playback waveform. The "eye pattern" change with recording power has been studied for the delay modulation recording code [41]. The values commonly obtained today are: SNR, 40 to 70 dB (measured at the 30-kHz slot); raw BER, 10^{-3} to 10^{-6}; and lifetime, 10 years (estimated). The typical distributions are similar to those for optical videodiscs [42]. Current magnetic disk storage media have a lower BER than that of optical data storage prototypes. But there is no intrinsic advantage associated with magnetic media; it is simply that optical media are operating on a much higher data storage density. When we compare BER per square centimeter, the figures are not as favorable for the magnetics.

7.6 MODIFICATIONS AND VARIATIONS

Practically, modifications in videodisc systems can be applied to optical disks, both PROD and EPROD. Programmability and erasability features permit further modifications to be introduced.

An optical coding is used in videodisc systems such that two signal components scarcely influence one another [43]. The spacing of the pits in the track direction carries the luminance information, while the undulation of the track carries the color or sound information. In the same way, the video signals, or high-bit-rate and somewhat lower-bit-rate signals, can be recorded and played back in optical disk systems. Of course, the same limitations and problems exist: for example, crosstalk, intermodulation products, and so on.

CAV and CLV optical disks can be recorded. For proper optical disk structure, the standard TV format can be recorded and the disk can be played as an optical videodisc on a standard (laser) player. A hybrid mode, the modified CLV (MCLV), is also possible [44]. In this method, the optical disk recording area is divided into small areas, A_1 to A_m, m zones, in the radial direction (similar to Fig. 7–3a). Every zone may contain an equal number of tracks. The angular velocity varies discretely from ω_1 to ω_m, but is constant in each zone. The mean linear velocity value for each zone is equal. The zones are divided into sectors. Thus the MCLV combines the properties of the CLV—the physical lengths of segments are practically equal all over the disk—and in the CAV—the track addresses, the number of segments in a track revolution, the disk angular speed, and the start position for each segment are all determined physically. Simultaneous multichannel parallel readout is possible, although this increases the technical problems. Up to nine channels are reported [45].

Optical memories can also be produced as optical tapes and optical cards (as debit cards, credit cards, ID cards, personal history cards, etc.) [46]. Up to 5 megabits of data can be stored on a 1.25 cm × 7.5 cm optical card [47] compared to the 1700 bits now used on credit cards. Data are recorded on a Drexan recording stripe through a clear plastic encapsulation by use of a low-power laser. Both machine-readable and human(eye)-readable data can be recorded. Of course, both sides of the card can be used. The application areas also include software distribution for personal computers, microprocessors, word processors, computer game software distribution, and security. In this kind of application, the emphasis is on very high storage capacity, storage performance, and low cost per megabyte of storage.

7.7 LOOKING AHEAD

The technology obviously has a long way to go. For example, magnetic disk technology is much farther along its learning curve than is optical technology.

Future improvements in the media and (semiconductor) laser development will be a trade-off between writing sensitivity and readout power. For the erasable materials, improvements in media stability and reproducibility of fabrication can be expected.

The consumer videodisc system maps roughly 6 bits of information on each pit in a disc. Used as an analog video channel for digital data recording [2,3], approximately 400 bits per (NTSC) line has already been accomplished (DVA, 1980). Programmable optical disk technology, at least in the laboratories, can now offer similar results, for example in disks structured to be played on standard consumer players. Also, the applications for training manuals and encyclopedias, based on an optical disk containing pictures, sound, and digital data, similar to a videodisc [48,49]—a combination "electronic book" and "visual personal computer"—are attractive.

Digital memory optical disk systems are expected to store at least 10^{11} bits per one side of disk, to have a data rate of at least 50 Mbits/s, and to have an access time on the order of tens of milliseconds. A "jukebox" approach could drastically increase storage capacity, with a few seconds of access time to the data on a random disk.

In principle, an optically focused laser beam can be combined with a selective field-optical or nonoptical (Fig. 7–18) to form a three-dimensional optical memory. The total SNR could be improved by quantization of the field-plane positions. Multi-layer disks and the jukebox are special cases of three-dimensional memories.

Work is reported on the idea of increasing optical disk packing density drastically by recording in various colors and using filters to read only the data desired [9]. It is expected that a photochemical medium can store many bits at each location using certain substances with wavelength-dependent absorption [50,51]. During recording a beam is scanned spectrally. To write, a laser is turned to any of up to 1000 discrete frequencies. At each frequency, a small portion of the molecules of the irradiated site undergo a chemical transformation which robs them of their ability to absorb again at that frequency. In effect, "holes" are bleached at selected spectral positions on the medium's absorption bond. The presence or absence of an absorption hole, when the reading laser is scanned across the same frequency range, indicates a binary 1 or 0. This is thus a frequency-spatial memory.

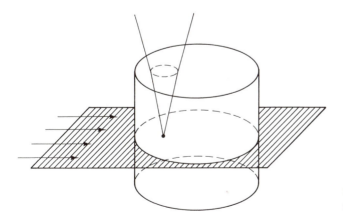

Figure 7–18 Three-dimensional optical memory.

REFERENCES

1. J. Isailović, "A new method for encoding and decoding digital data for optical memories," Proc. 11th Congr. Int. Comm. Opt., Madrid, 1978, pp. 363–366.

2. J. Isailović, "Binary data transmission through TV channels by equivalent luminance signal," *Electron. Lett.*, Vol. 17, No. 6, Mar. 19, 1981, pp. 233–234.

3. J. Isailović, "Binary data transmission through TV channels by equivalent chrominance signal," Proc. SPIE 329, Los Angeles, Feb. 18–22, 1982.

4. D. Lou, F. Zernike, G. Kenney, A. Chan, P. Janssen, R. McFarlane, and J. Wagner, "A prototype optical disc recorder," Conf. Rec. 1977 IEEE/OSA Conf. Laser Eng. Appl., Washington, D.C.

5. R. Pepper, and I. Sander, "High density direct read after write (DRAW) recording," *Opt. Acta*, Vol. 24, 1977, pp. 427–431.

6. E. E. Gray, "Laser mass memory system," *IEEE Trans. Magn.*, Vol. MAG-8, 1972, pp. 416–420.

7. H. R. Dell, "Design of a high density optical mass memory system," *Comput. Des.*, Aug. 1971, pp. 49–53.

8. K. McFarland and H. Hashiguchi, "Laser recording unit for high density permanent digital data storage," Conf. Rec., Fall Joint Comput. Conf., Vol. 33, 1968, pp. 1369–1380.

9. E. Rothchild, "Optical-memory media," *Byte*, Mar. 1983.

10. *Optical Memory Newsletter*, No. 4, July–Aug. 1982, p. 4.

11. *Optical Memory Newsletter*, No. 9, May–June 1983, pp. 12–13.

12. G. C. Kenney et al., "An optical disk replaces 25 mag tapes," *IEEE Spectrum*, Feb. 1979, pp. 33–38.

13. R. A. Bartolini et al., "Optical disk systems emerge," *IEEE Spectrum*, Aug. 1978, pp. 20–28.

14. G. J. Ammon, C. W. Reno, and R. J. Tarzaiski, "Digital optical video disc," Final Rep. Govt. Contract F30602–78–C–0050, Rome Air Development Center, Apr. 1979.

15. R. A. Bartolini, "Media for high-density optical recording," *Opt. Eng.*, Vol. 20, No. 3, May–June 1981, pp. 382–386.

16. R. A. Bartolini, "Optical recording: high-density information storage and retrieval," *Proc. IEEE*, Vol. 70, No. 6, June 1982, pp. 589–597.

17. A. E. Bell, "Critical issues in high-density magnetic and optical data storage," Part 1, *Laser Focus*, Aug. 1983, pp. 61–66.

18. R. A. Bartolini, H. A. Weakliem, and B. F. Williams, "Review and analysis of optical recording media," *Opt. Eng.*, Vol. 15, No. 2, Mar.–Apr. 1976, pp. 99–108.

19. T. A. Allen and G. S. Ash, "Optical properties of tellurium films used for data recording," *Opt. Eng.*, Vol. 20, No. 3, May/June 1981, pp. 373–376.

20. M. Mashita and N. Yasuda, "Amorphous Te-C films for an optical disk," Optical Disk Technology, Proc. SPIE 329, 1982, pp. 140–144.

21. N. Akahira, T. Ohta, N. Yamada, M. Tokanoga, and T. Yamashita, "Sub-oxide thin films for an optical recording disk," Optical Disk Technology, Proc. SPIE 329, 1982, pp. 195–201.

22. R. M. White, "Magnetic disks: storage densities on the rise," *IEEE Spectrum*, Aug. 1983, pp. 32–38.

23. T. Togami et al., "Amorphous thin film disk for magneto-optical memory," *Optical Disk Technology*, Proc. SPIE 329, 1982, pp. 208–214.

24. D. Cheng, D. Treves, and T. Chen, "Static tests of TbFe films for magneto-optical recordings," Optical Disk Technology, Proc. SPIE 329, 1982, pp. 223–227.

25. M. Mansuripur, G. A. N. Connell, and J. W. Goodman, "Signal-to-noise in magnetoptic storage," Optical Disk Technology, Proc. SPIE 329, pp. 215–222.

26. R. P. Freese, R. F. Willson, L. D. Ward, W. B. Robbins, and T. L. Smith, "Characteristics of bubble-forming optical direct-read-after-write (DRAW) media," Optical Disk Technology, Proc. SPIE 329, 1982, pp. 174–180.

27. T. W. Smith, G. E. Johnson, A. T. Ward, and D. J. Luca, "Barrier coating for optical recording media," Optical Disk Technology, Proc. SPIE 329, 1982, pp. 228–235.

28. C. Bricot, J. Cornet, J. L. Gerard, and F. LeCarvennec, "Tracking techniques in pregrooved optical disc technology," Optical Disk Technology, Proc. SPIE 329, 1982, pp. 94–97.

29. M. M. Noskov, *Compt. Rend. Acad. Sci. USSR*, Vol. 31, 1941, p. 111.

30. P. H. Lissberger, *J. Opt. Soc. Am.*, Vol. 51, 1961, p. 957.

31. A. E. Bell and F. W. Spong, "Antireflection structures for optical recording," *IEEE J. Quant. Electron.*, Vol. QE-14, No. 7, July 1978.

32. A. E. Bell, R. A. Bartolini, and F. W. Spong, "Optical recording with the encapsulated titanium trilayer," *RCA Rev.*, Vol. 40, Sept. 1979, pp. 345–362.

33. J. Cornet, J. C. Lehureau, and F. LeCarvennec, "Characteristics and performance of a novel recording material," Conf. Lasers Electro-Optics (CLEO), Washington, D.C., June 1981.

34. W. Robbins, R. Freese, T. Smith, and R. Willson, "Bubble forming media for optical recording: a new approach," Conf. Lasers Electro-Optics (CLEO), Washington, D.C., June 1981.

35. G. A. N. Connell, R. I. Johnson, D. Kowalski, and C. De Puy, "Trilayer bubble forming optical recording media," Optical Disk Technology, Proc. SPIE 329, 1982, pp. 166–173.

36. M. A. Bosch, "Hydrogenated semiconductors as optical recording media," Optical Disk Technology, Proc. SPIE 329, 1982, pp. 181–185.

37. A. R. Tebo, "Advances in thermal and magneto-optic recording," *Electro-Optics*, Apr. 1983, pp. 40–47.

38. H. G. Graighead, R. E. Howard, R. W. Smith, and D. A. Snyder, "Textured optical storage media," *Electro-Optics*, Apr. 1983, pp. 202–205.

39. *Principles of Microwave Circuits*, ed. C. G. Montgomery, McGraw-Hill, New York, 1948, pp. 376–397.

40. D. Y. Lou, "Characterization of optical disks," Optical Disk Technology, Proc. SPIE 329, 1982, pp. 247–251.

41. G. J. Amon, R. F. Kenville, A. A. Litwak, M. S. Nigro, and C. W. Rene, "Performance measurements from digital optical disk systems," Optical Disk Technology, Proc. SPIE 329, 1982, pp. 33–39.

42. J. Isailović, "Channel characterization of the optical videodisc," *Int. J. Electron.*, Vol. 54, No. 1, 1983, pp. 1–20.

43. J. J. M. Braat and G. Bouwhuis, "Optical video disk with undulating tracks," *Appl. Opt.*, Vol. 17, No. 13, July 1, 1978, pp. 2022–2028.

44. T. Murakami, I. Hoshino, and M. Mori, "Optical disk memory system," Optical Disk Technology, Proc. SPIE 329, 1982, pp. 25–32.

45. D. C. Kovalski, D. J. Curry, L. T. Klinger, and G. Knight, "Multichannel digital optical disk memory systems," Optical Disk Technology, Proc. SPIE 329, 1982, pp. 8–20.

46. D. Wilson, "Optical Disks—new media promises high bit density," *Digit. Des.*, Mar. 1983, pp. 64–67.

47. J. Drexler, "Laser card for compact optical data storage systems," Optical Disk Technology, Proc. SPIE 329, 1982, pp. 61–67.

48. A. Lippman, "The computational videodisc," *IEEE Trans. Consum. Electron.*, Vol. CE-27, No. 3, Aug. 1981, pp. 315–319.

49. D. S. Backer, "Personalized electronic publications," Natl. Comput. Graphics Assoc. Conf., Chicago, June 1983.

50. G. C. Bjorklund, W. Lenth, C. Ortiz, "Cryogenic frequency domain optical memory" SPIE Vol. 298 Real-Time Signal Processing IV (1981) pp. 107–114.

51. P. Pokrowsty, W. E. Moerner, F. Chu, and G. C. Bjorklund, "Reading and writing of photochemical roles using GaAl As-diode lasers" *Optical Letters*, Vol. 8, No. 5, May 1983, pp. 280–282.

APPENDIX 7.1 RECORDING CODES

Many different wave shapes for encoding digital information are being used, each with its own application and limitations [1].

If clock and data are derived from the read waveform, transitions must occur frequently enough to provide synchronization pulses for the clock. On the other hand, consecutive transitions must be far enough apart to limit the interference to an acceptable level for reliable detection. Because of this, the binary data are coded in binary sequences that correspond to waveforms in which the maximum and minimum distances between consecutive transitions are constrained by prescribed coding rules. The most common waveforms are shown in Fig. A7.1–1. A digital word 00010101110 is used to compare the various codes.

A7.1–1 NON RETURN TO ZERO (NRZ)

Two main classes exist in non-return-to-zero coding: NRZ-level and NRZ-mark. In NRZ-level (NRZ-L) 1's and 0's are defined by the corresponding levels (Fig. A7.1–1a):

- $1 =$ level L_1
- $0 =$ level L_0

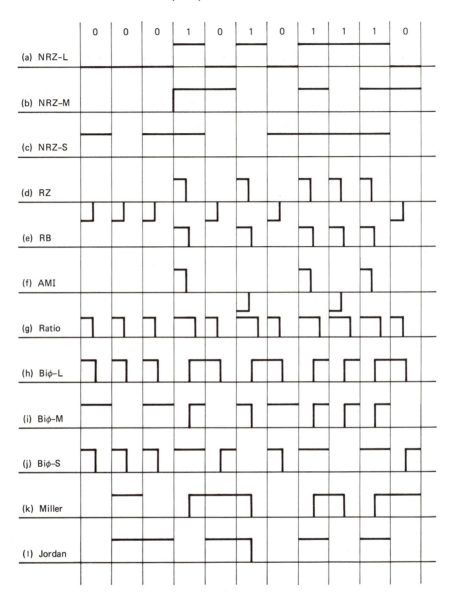

Figure A7.1–1   Code waveforms.

Transitions occur only when the data bits change. If levels L_1 and L_0 are chosen such that $L_1 = -L_0$, a bipolar code is obtained.

In NRZ-mark (NRZ-M), 1's in the data stream are recorded (i.e., marked) as transitions in the middle or at the beginning of the data bit interval, and 0's are ignored (Fig. A7.1–1b):

- 1 = a change in level
- 0 = no change in level

Basically, this is a differentially encoded waveform, and sometimes is referred to as "differentially coded 1's."

NRZ-space (NRZ-S) is the complement of NRZ-M (Fig. A7.1–1c):

- 1 is represented by no change in level
- 0 is represented by a change in level

NRZ-M and NRZ-S are the most commonly used systems with multitrack systems (e.g., in computers), because with these two codes a clock rate can be produced with the advantage of check or parity information. One method of producing a clock rate pulse is to record the complement on a second track, instead of recording a clock track.

NRZ is not a run-length-limited code. Large strings of 0's or 1's could result in no transitions, giving the signal a very low frequency component. In consequence, the "eye" pattern can become completely closed, rendering simple binary threshold detection impossible.

Enhanced NRZ (ENRZ), also referred to as S-NRZ, is simply NRZ-L recording with an extra bit every 8 bits to provide a lower bound on bandwidth requirements and clock tracking. Rather tricky electronics and buffering are required to squeeze an extra bit for every 8 in encoding and then clip it out again in decoding. The basic idea is to add extra transitions to NRZ-L to make it run-length limited. This assures clock extraction for any data input. Two variations of ENRZ have been described in the literature (not shown in Fig. A7.1–1). In both versions an extra interval is inserted after 8 (or 7) code bits. In ENRZ-parity, the added bit creates an odd parity with the previous code bits. In ENRZ-complement the added bit is the complement of the last code bit. A single bit memory is required to implement both codes.

Group coded recording (GCR), which is also called run-length-coded NRZ, is a technique of slicing the incoming data into blocks or groups and transforming these data groups into longer code words, which are then recorded as NRZ-M. The transformation may be performed according to algebraic rules or by using a look-up table or dictionary. The advantage of transforming the data words into large code words is that undesirable code words may be rejected and favorable ones retained. GCR avoids more than two 0's in succession by transforming 4 bits into 5. Of 32 combinations of 5 bits, those that begin or end with more than one 0 and those that have more than two 0's internally are not used. After eliminating these, the combination 11111 is reserved for synchronization. The remaining 16 combinations of 5 bits are correlated with the 16 possible combinations of 4 bits. After each group of 4 bits in a sequence of data is transformed into a corresponding group of 5 bits, NRZ-M is used (not shown in Fig. A7.1–1). A 4-data-bit memory is thus required. The 16 (out of 32) chosen code words are shown in Table A7.1–1.

TABLE A7.1–1   FOUR-FIVE GCR

Decimal value	Hexadecimal value	Data word	Code word
0	0	0000	11001
1	1	0001	11011
2	2	0010	10010
3	3	0011	10011
4	4	0100	11101
5	5	0101	10101
6	6	0110	10110
7	7	0111	10111
8	8	1000	11010
9	9	1001	01001
10	A	1010	01010
11	B	1011	01011
12	C	1100	11110
13	D	1101	01101
14	E	1110	01110
15	F	1111	01111

A7.1–2 RETURN TO ZERO (RZ)

In return to zero, 1's and 0's are defined as (Fig. A7.1–1d)

- 1 = positive pulse (a ½-bit wide)
- 0 = negative pulse (a ½-bit wide)

Between pulses, the signal is zero. In the return-to-bias (RB) system, 1's and 0's are represented as (Fig. A7.1–1e)

- 1 = a ½-bit-wide pulse
- 0 = no pulse condition

This system is more efficient than RZ but still requires high resolution. It also requires a reference at the bit rate because if a long series of 0's are being retrieved, the number of missing pulses must be determined.

In the bipolar or alternate-mark-inversion (AMI) code, a binary 0 is uncoded and is represented by a space or absence of a pulse. On the other hand, a binary 1 is represented by a mark, or pulse, which alternates in polarity (Fig. A7.1–1f).

Ratio code is similar to RZ in the sense that it is a self-clocking code. In ratio coding, a positive-going transition is always at the leading bit cell edge while the position of the negative-going transition determines the data (Fig. A7.1–1g).

- 1 = negative-going transition is in the second half of the bit cell
- 0 = negative-going transition is in the first half of the bit cell

Compared with RB, the price for self-clocking is lower SNR.

A7.1–3 DOUBLE-FREQUENCY CODES

In this class of (line) coding, redundancy is introduced by using a signaling rate equal to twice the input binary data rate. The coded (line) signal is dc free and contains a large number of transitions from which timing information can be recovered. Double-frequency coding (and its variants) has many names, such as "biphase," "split-phase," "Manchester codes," "phase encoding," "frequency-shift keying," "frequency modulation (FM)," "conditioned diphase," or "two-level AMI class II coding."

In biphase-level (or split-phase, phase encoding, Manchester II + 180°), there is the following correspondence (Fig. A7.1–1h):

- 1 = positive-going transition in the middle of bit cell
- 0 = negative-going transition in the middle of bit cell

A transition is made during a clock time only if it is required to give the correct phase transition for the following bit; this means that a transition always occurs in the middle of each bit interval and also between identical bits. This can also be expressed as

- 1 is represented by a 10
- 0 is represented by a 01

In biphase-mark (or Manchester I) codes, a transition occurs at the beginning of every bit period (Fig. A7.1–1i) and

- 1 = second transition, ½ bit period later
- 0 = no second transition

The direction of the transition is immaterial; therefore, the phase or polarity of the signal is not important. The decoding of the signal is more complex because the clock edges and the bit information edges have to be separated.

The biphase-Space (or Manchester I + 180°) is the complement to biphase-mark. A transition occurs at the beginning of every bit period (Fig. A7.1–1j) and

- 1 = no second transition
- 0 = second transition, ½ bit period later

In bi-phase-mark or space, a 1 and a 0 correspond to the presence or absence of a transition in the center of the corresponding bit cell.

A7.1–4 DOUBLE-DENSITY CODES

These codes are known by two other names: delay modulation and modified frequency modulation (MFM). They are basically phase-shift codes. The bandwidth requirements of these codes are slightly greater than for NRZ. The price of this is the double-frequency clock, a 3-dB loss in the signal-to-noise ratio, and the requirements for the data pattern to establish a correct phase of the clock. The encoders for double-density codes are relatively simple, while the decoder requires a slightly more sophisticated system.

The algorithms for Miller and Jordan codes will be adduced here. The Miller code can be defined as follows (Fig. A7.1–1k). A 1 is represented by a transition in the middle of a bit cell. A 0 has no transition unless it is followed by another 0, in which case there is a transition at the end of the bit cell of the first 0. The synchronizing data pattern is 101.

The coding rules for the Jordan code are as follows (Fig. A7.1–1l). A 0 is represented by a transition either at the beginning or at the end of a bit interval, and a 1 is represented either by a transition at the beginning and end of the bit interval or by a transition in the middle of the bit interval, such that the least distance between two adjacent transitions is one bit interval in length. The synchronizing data pattern is 000. The definitions of a data 1 and 0 can be interchanged for both codes.

The coding rules for these two codes may be implemented by using an auxiliary sequence of transitions. The signal in the desired code is obtained by dividing the auxiliary sequence by 2 (e.g., one flip-flop).

The auxiliary sequence for the Miller code is such that a transition always occurs in the middle of each bit interval and between identical bits. The auxiliary sequence for the Jordan code is that a transition always occurs between bits and in the middle of the bit interval of a data 1. Obviously, the auxiliary sequences for the Miller and Jordan codes are biphase-level and biphase-mark, respectively.

In Fig. A7.1–2 the spectral power density curves for NRZ-L, biphase-L, and delay codes are compared. The normalized frequency scale, $1/T$, where T equals bit period in seconds, is used. Note that the curves differ significantly.

A7.1–5 OTHER CODES

There are a number of codes that can be used for digital recording. One author reported over 100 self-clocking codes, but most of them can be reduced to the codes mentioned previously. Some algorithms have been developed to match the characteristics of the magnetic recording channel. Among these are zero modulation (ZM),

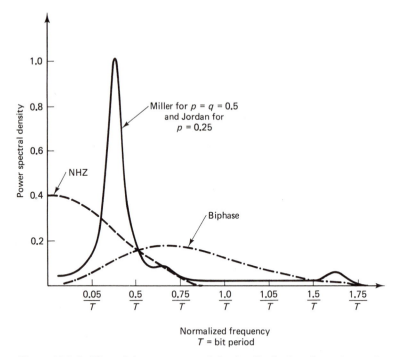

Figure A7.1–2 Plots of the power spectral density distributions for various codes.

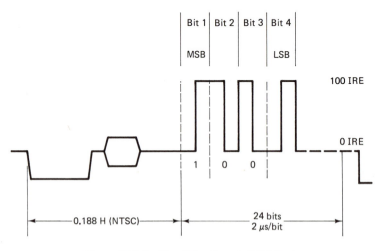

Figure A7.1–3 Bit cell length and digital level.

M², variable cell width (VCW), and three-position modulation (3PM). These codes are not being discussed here.

In Fig. A7.1–3 an idealized form of the (video) signal containing binary data is shown. Four bits (1001) are shown in the biphase-L code. Each bit cell is 2 μs long with a digital level between 0 and 100 IRE.

To select the most promising codes for optical recording, the channel parameters must be incorporated [2]. Complexity and cost, code efficiency, code overhead, SNR required for given BER, sync problem, error propagation, etc. should be compared for different codes. For example, enhanced NRZ, double density, and 3PM[3] codes can be taken as reference codes for optical recording. The (2,7) code (US patent No. 3 689 899) is similar to 3PM code in properties. Any other promising codes should then be compared with these codes.

REFERENCES

1. J. Isailović, *Videodisc Systems,* Ch. 6.

2. J. Isailović, "Background study of various codes for optical recording," (June 1984), Anaheim Hills, CA.

3. G. V. Jacoby, "A new look ahead: code for increased data density," *IEEE Transactions on Magnetics,* Vol. MAG-13, No. 5 (September 1977), pp. 1202–1204.

APPENDIX 7.2 ODC Laser Videodisc Recording System LVDR 610

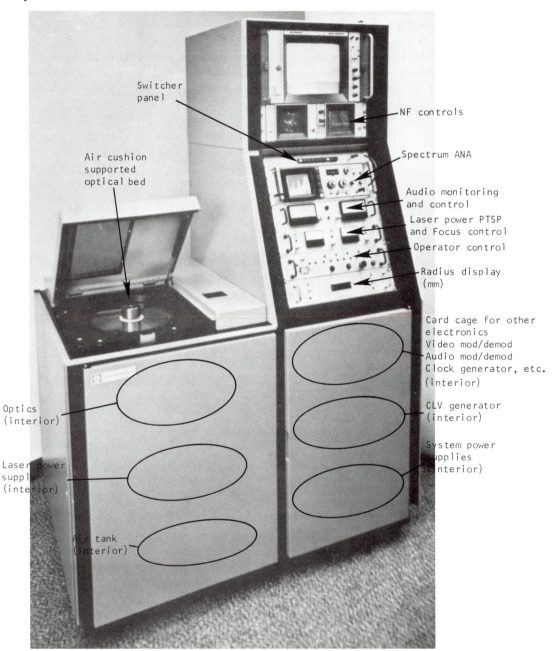

Switcher panel

NF controls

Air cushion supported optical bed

Spectrum ANA

Audio monitoring and control

Laser power PTSP and Focus control

Operator control

Radius display (mm)

Card cage for other electronics
Video mod/demod
Audio mod/demod
Clock generator, etc.
(interior)

Optics (interior)

CLV generator (interior)

System power supplies (interior)

Laser power supply (interior)

Air tank (interior)

INDEX

Abbe's law, 85
aberration-free system, 223
aberrations, 87, 108, 185, 190, 191, 192, 193,
    194, 218
ablation, 41, 302
ablative:
    materials, 38
    recording media, 38
    techniques, 317
    thin films, 301, 302, 310
abrasive wear, 180
absolute amount of defocus, 218
absolute track drop (TD), 58
absorption, 226
    curve, 310
    modulation, 54
    plate, 226
absorptivity, 88
access time, 297
accordian surface, 160
acoustic beam, 235
acoustic subcarrier, 237
acoustic wave, 235, 236
acousto-optical devices, 235,
acousto-optical modulator, 31, 36, 55 233, 238,
    257, 299
acrylic, 48, 66
active aerodynamic stabilizer, 125
adaptive aperture, 62
additive noise, 250
additives, 33, 65
admittance, 215
advanced optics, 185
aerodynamic stabilization, 124

AGC circuit, 182
air bearing, 36, 61, 113
air cushion, 36
air sled, 299
airborne debirs, 51
airy disc, 210, 224
aluminum, 34, 50
    coating, 50, 51, 52
    cover, 50
    layer, 51
AM, 55
AMI code, 325
amorphous hydrogenated, 316
amorphous hydrogenated silicon films, 316
amorphous magnetic, 303
amorphous Te-C films, 302
amorphous transition, 307
amplifier, 131
amplitude, 33, 60, 95, 96, 108, 114, 119, 235
    characteristic, 98, 99
    detector, 173
    distribution, 212
    function, 132
    grating, 210
    imbalance, 133
    modulation, 124, 160, 165
    modulations, 302
    or phase gratings, 210
    probability density (APD), 260
analog, 233, 245, 294
analog read-only memory (ROM), 293
analog-video channel, 298
analyzer, 234
angle of incidence, 186, 187, 227, 237

angle of reflection, 186
angular frequency, 108
angular modulation, 108
angular sectors, 296
angular velocity, 105
anhydrous isoproponal, 48
anisotrophy, 227
antimony, 42
antireflection, 313, 315
    phenomenon, 313
    thickness, 313
antireflective-coated beam splitter, 43
APD, 260, 261
aperture, 69, 77, 163, 191, 192, 203, 207, 210,
        212, 217, 218, 226
    compensation circuit, 97
    compensator, 107
    corrector, 73
    function, 231
    stop, 190, 192, 194
apodization, 212, 223, 226
archivability, 295
archival stability, 317
archival storage, 38
argon laser beam, 36, 42
argon-ion laser beam, 41, 53
arm-advance motor, 173
arm dynamics, 174
arm stretcher, 172, 177
    response, 174
    system, 168
    transducer system, 167, 176
aspheric condenser system, 128
astigmatic focii, 192
astigmatic lens, 115
astigmatic sensor, 124
astigmatism, 191, 192, 193, 194
asymmetrical aberration, 223, 224
asymmetrical duty cycle, 46
attenuation, 259
    slope, 73
attenuator, 299
audio, 172
audio carrier, 163
    amplitude, 163
    signal, 46
audio discs, 33, 52
    processing, 43
    replication, 53
audio disturbance, 170
audio FM signal, 170
audio LP records, 33, 159
audio players, 176
audio record, 33
audio signal 31, 159, 169
audio signal-to-noise ratio, 72
autocorrelation function, 278, 279, 280, 281, 284,
        285
automatic focusing, 128
automatic gain control (AGC) circuit, 180, 182
auxiliary, 116
auxiliary beam, 119

avalanche mode, 128
avalanche photodiode, 259, 305
average duty cycle, 46
averaging, 275
axial object, 189
axial symmetry, 187
axis, 194

background noise level, 30
ball screw, 299
ballistic galvanometer, 123
banding, 66
bandpass, 107
    filters, 182
bandwidth, 97, 112, 123, 124, 127, 165, 167,
        250
bar charts, 225
barrel distortion, 194
bars, 210
base error, 179
base material, 33
baseband noise, 247
baseboard frequency, 247
basic field curvature, 192
basic operations, 31
beam, 31, 39, 52, 53, 85, 87, 109, 116
    polarization, 94
    power, 39
    splitter, 31, 43, 76, 94, 229
BER, 297, 317, 329
Bernouilli effect, 125
Bernouilli's formula, 126
Bessel function, 1, 209, 241
bias current, 253
biconcave, 187
biconvex, 187
bilayer, 70, 312
binary, 245
binary data, 329
binomial distribution, 274
binominal pdf, 275
biphase-L code, 329
birefringence, 50, 52, 227, 228
bismuth, 42
black body, 226, 227
black level, 169
blanking interval, 173
bleed-out, 65
blistered, 316
blurred edges, 256
blurring, 242, 243
Boltzmann's constant, 252
borosilicates, 307
Bose-Einstein formula, 252
Bragg angle, 237
Bragg angle modulation type, 237
Bragg cell, 237
Bragg regime, 236
breakup, 173
bridge circuit, 112
bumps, 267
buried-subcarrier, 169

burst, 123, 134, 157, 168
  correction, 151, 157
  signal, 112, 134, 155

caddy, 35, 57
camera, 31
capacitance, 162
  pickup, 158
  stylus, 158
  variations, 160
capacitive:
  disc, 33, 34, 67, 129, 130, 179
  pickup, 162
  sensor, 61, 160
  systems, 161, 169, 179
  variations, 160
  videodisc, 56, 61, 70
  videodisc recording, 74
  videodisc systems, 179, 184
carbon, 33, 66
carbon loading level, 251
cardinal points, 188
carriage drive, 78
carriage servo, 123
carriage servo-drive motor, 123
carriage velocity, 123
carrier, 56, 108, 160, 250, 260, 261, 262, 264,
    308
  frequency, 55, 152, 157, 160, 168
  level, 35
  lifetime, 253
  light beam, 87
  mobility, 253
  output, 251
  output port, 165
  to-noise ratio, 46, 251, 289
  train, 211
cartridge, 172
  housing, 172
  pickup arm, 170
casting, 34, 39
  method, 49
  process, 34
catastrophic defect, 266
catastrophic failure, 266
Cauchy pdf, 273
CAV, 73, 107, 267, 315, 318
CAV mode, 78, 133
CCD delay line, 172
CED system, 166
cellulose triacetate, 249
centering error, 167
central limit theorem, 270
central moments, 276
central obscuration, 223
chalcogenide films, 301, 303
channel impairment, 247
channel radii, 94
channeltron, 128
characteristic impedance, 12, 132
characteristic transfer function, 132

charge-couple device, 122
chief ray, 192
chroma, 31, 133, 167
chroma burst, 151
chroma phase errors, 167, 169
chroma phase-locked loop, 169
chroma signal, 133
chromatic aberrations, 191, 194
chrominance information, 172
circular aperture, 216, 226
circular function, 241
circularly polarized, 228
circumferential velocity, 105
clean-room environment, 49
closed loop, 152, 156
  configuration, 128
  correction systems, 167
  response, 168
  stability, 176
closed servo loops, 37
CLV, 73, 267, 315, 318
CLV mode, 78
coated capacitive discs, 67, 68
coated surface, 39
coating, 34, 35, 38, 39, 40, 53, 67, 68
code efficiency, 329
code overhead, 2, 329
coding rules, 322
coherent and incoherent imaging, 185
coherent laser source, 127
coherent light, 203
coherent noise, 127
cold flow, 50, 51
  birefringence, 50
color, 167
  burst, 172
  carrier frequency, 152
  comb filter, 172
  noise, 285
  reproduction, 167
  subcarrier, 77, 167, 168, 169, 172
  tv signal, 34
colorant, 66
colored noise, 285
coma, 191, 192, 194, 224
coma filter, 170, 172
comb function, 231, 240
combed chroma signal, 172
combed color signals, 169
combed luminance signal, 169, 172
communication system, 244
compact disc (CD), 293
compensation, 73, 105
  circuit, 104, 105, 122
compensator, 107, 121
compensator systems, 96
complex amplitude, 196
complex frequency response of an optical system,
    213
component amplitudes, 203
component intensities, 203
composite video, 31, 169

compound, 33, 48, 49
  contaminants, 72
  matrix, 65
  melt flow, 49
  moldings, 52
  processing, 64, 65
  stability, 33
compounding:
  process, 33
  stage, 64
compressed-bandwidth, 169
compression, 66, 67, 68
  moldings, 33, 49, 65, 67, 68
  process, 67
compression-molded discs, 68
computers, 295
concave, 187
concept of transformation, 185
concepts of optics, 184
conductive discs, 68, 161
conductive plane, 69
cone of light, 190
confidence level, 71
constant angular velocity, 39
  discs, 247
constant rotation, 30
constant valley level, 61
contact-free playback, 129
contact-free readout, 83
contact-printing, 303
  methods, 55
contact transfer, 220
contamination, 71
contrast:
  (modulation), 220
  power, 89
  resolution, 34
control, 109, 170,
  circuit, 108
  with pressure sensors, 49
  servo system, 169
  signal, 74, 112
  system, 87, 119
  test, 35
convex, 187
convex lens, 188
convolution, 212, 214, 215, 220, 231, 241
  integral, 281
  theorem, 242
copies, 31, 34
copolymers, 67
copper, 70
corona-charging device, 305
correction peak, 168
correlation, 241
  function, 279
  theorem, 242
corrosion, 70
cosine compensator, 96
cosine equalizer, 96, 107
costs yields, 35
credit cards, 318

critical angle, 187
critical shear rate, 67
critical thickness, 310
critically damped, 113
cross-sectional, 33
cross-talk, 53, 233, 318
cross-track, 73
CRT, 265
crystal oscillator, 168
crystal-controlled, 108
crystalline status, 305
Curie point, 304
Curie temperature, 303, 304
curling, 249
curvature, 192
curvature of field, 191
cutoff frequency, 105, 119, 217, 219, 230
cutoff period, 129
cutoff spatial frequency, 219
cutterhead, 59, 60
cutting, 31, 41, 43
  beam, 31
  head, 31, 41
  objective, 43
  process, 41
  spot, 43
cycle time, 50, 67
cylindric lens, 109, 114
  astigmatic, 114

damage, 34, 50
damping, 177
data, 31, 38
  integrity, 295
  storage, 301
dc bias point, 62
dc error, 169
dc motor, 116
De Moivre, 271
debit cards, 318
debris, 48
decoder, 245
defects, 33, 38, 46, 301, 314
  in the master, 30
deflection, 113
defocus, 218, 223
degradation, 174
delamination, 50
delay, 102
  line, 97, 98, 132, 170, 172
delta function, 239, 281
demodulation, 180, 245, 260
dense medium, 187
density domain, 250
deposition processes, 69
depth of focus, 41, 85, 113, 184, 219, 229
  of the lens, 85
  of the system, 219
descent source, 128
design, 244
destructive interference, 112, 204

detection, 159
  of burst, 134
  of carrier, 164
  of response, 164
detector, 46, 92, 112, 124, 127, 128, 129, 259,
    266
  aperture, 90
  plane, 233
  surface, 115
developed disc, 39
developer, 40
development, 35, 39, 40, 41, 249, 250
  time, 40
diamond stylus, 175
diazo copy, 55
diazo process, 82
dichroic mirror, 43
dielectric, 69, 175, 227, 258, 259, 315
  coatings, 70
  constant, 162, 163
  layer, 69, 70
  stack, 229
  tensor, 304
  thickness, 314
differential:
  detector, 91, 129
  mode, 77
  track drop, 59
differentiation, 241
diffracted, 235
diffracted light, 90, 91
diffracting:
  aperture, 204
  body, 206
diffraction, 83, 89, 90, 129, 185, 203, 204, 213,
    221, 222, 235
  beam, 237
  by a circular source, 208
  by a single slit, 207
  grating, 129, 210
  integral, 204
  orders, 45, 89
  limited circular lens, 217
  limited system, 43, 216, 219
  pattern, 89, 204, 212, 224, 226
  problems, 203
  spectrograph, 46
diffractive scattering, 129
digital, 233, 293
  audio, 293
  channel, 298
  data, 109
  data recording, 301
  disc, 294
  errors, 275
  information, 322
  memories, 294
  optical memory, 294
  ROM, 293
  signal, 31
  storage, 294
  systems, 43

diode, 114, 119, 301
dip-coating, 39, 40
Dirac delta function, 239, 284
direct analog recording, 55
direct-read-after write (DRAW) discs, 31, 36, 38,
    43, 294
direction, 233
  of polarization, 227
  of propagation, 203
  dirt, 87
disc, 30, 31, 33, 34, 35, 36, 39, 40, 41, 43, 45,
    46, 48, 49, 50, 51, 52, 53, 61, 64, 73, 85,
    87, 109, 112, 113, 114, 116, 124, 126,
    128, 129, 134, 138, 158, 159, 160, 162,
    167, 168, 170, 172, 174, 175, 176, 180,
    229, 245, 247, 249, 251, 269
disc angular velocity, 79
  capacitance, 159
  coating, 50, 56, 64
  compounds, 49
  conductivity, 33
  construction, 87
  eccentricity, 108, 135, 139, 143, 145, 148,
    151, 168, 251
  flatness, 51
  linear velocity, 246
  material, 53
  mold cavity, 49
  molding, 64, 65
  noise, 250, 260, 263
  playback system, 167
  PVC compounds, 67
  quality, 43
  readout, 102
  reflectivity, 77
  resistivity, 251
  rotation speed, 158
  runout, 176
  stabilization, 113
  stampers, 64
  surface, 34, 41, 94, 95, 128, 180
  systems, 129
  vibration, 229
  warp, 68, 166, 176
  warpage, 72
    rotation frequency, 123
    molding, 59
    stress, 67
    wear, 67, 68
discrete defects, 71
discrete noise component, 155
discrete pdf, 274
disk, 294, 298, 301, 308, 309
disk structure 308
dispersion, 72, 276
distance between, 31
distortion, 43, 62, 73, 108, 124, 144, 177, 191,
    193, 194
disturbance, 168, 172
diverging lens, 43
document storage, 297
double convex lens, 212

double density, 329
  codes, 327
double-frequency codes, 326
double refracting, 227
double-sided, 52
  disc, 50
  transmissive discs, 53
DRAW, 43, 53, 55, 294, 296, 299, 302
draw bits, 43
draw function, 43
draw optical system, 43
DRAW signal, 299
DRAW signal detector, 299
Drexan recording stripe, 318
Drexan II, 317
drive motor, 116
driver, 62
driving voltage, 60
drop tests, 67, 68
dropout, 35, 43, 46, 51, 109, 172, 246, 247, 264,
    265, 266, 267, 317
  compensation, 169
  density, 267
  detector, 265, 266
  distribution, 265
  pulse, 265
  statistics, 317
dry molding compound, 49
durability, 87
dust, 34, 36, 48, 50, 51, 69, 174, 180, 308
dust particles, 125
duty cycle, 41, 46, 73, 256, 314
  variation, 317
dye, 312
dynamic stress, 60

eccentricity, 50, 107, 108, 116, 119, 152, 172,
    251
edge blurriness, 251
edge detection, 242, 243
effective frequency response characteristic, 96
elastomer, 34
elastomeric material, 60
electrical birefringence, 234
electrical circuits, 215
electrical (domain), 244
electrical signal, 35, 36, 94
electrical transfer function(s), 230
electric and magnetic fields, 185
electric field, 227, 234
  vectors, 227, 228, 234
electrode, 159, 161
  lifting, 166
electroforming, 33, 47, 48, 56, 245
  of metal, 33
  of nickel, 33, 67
  process, 48
  stamper replica, 35
electromagnet, 234
electromagnetic bender motors, 94
electromagnetic spectrum, 184

electromechanical:
  master, 60, 129
  mastering, 56, 60
  recorder (EMR), 30, 56, 59
  systems, 31, 168
electromodulation, 244
electromotive force (EMF), 112
electromotor, 112
electron beam, 129
electron-beam:
  masters, 30, 62
  radiation, 34
  recordings, 30
electronic compensation, 122
electronic servo, 127
electronic signal recovery (playback), 68
electronics, 94
electron microscopy, 46
electrons, 184
electro-optical, 31, 233, 299, 301, 307
  modulation, 228
  modulator, 36, 41, 62, 234
  transducer (optomodulator and photodiode),
      244, 252
electroplating, 56, 63, 64
  of the master, 64
electrostatic forces, 69
elementary imaging system, 189
elliptical deformation, 114
elliptically polarized beam, 227, 228, 234
embossing, 33, 53, 303
  or web press techniques, 53
emission wavelengths, 128
emulsion layer, 249
encoded mold surface, 34
encoded signal, 31, 40
encoder, 245
encoding, 31, 33, 35, 43, 169
  electronics, 31
energy, 184
  reflectance, 310
enhanced NRZ, 324
ENRZ-parity, 324
entrance pupil, 190
envelope, 97, 98, 202
  delay, 259
  of the wavefronts, 203
environmental conditions, 72
environmental testing, 72
EPROD, 294, 317, 318
equalization, 60
equalizing circuit, 60
erasable, 297
  optical disks, 297
  PROD (EPROD), 295
erasure, 307
ergodic, 269, 275
erodic random process, 278
error, 167
  correction, 109
  propagation, 329

signal, 111, 112, 116, 119, 124, 129, 134, 151, 152, 168, 180
excess resist, 40
exit pupil, 190, 218
exposed film density, 250
exposed glass, 43
exposure, 35, 39, 40, 41, 245, 249, 250
    energy, 40
extended play time, 31, 169
extruder hopper, 66
eye, 221
    pattern, 317

fabrication, 174
    errors, 194
fan-out process, 47, 48
Faraday effect, 234, 304
fast flow, 126
fast forward, 177
feedback, 74, 107, 122
    loop, 122
Fermat's principle, 186
ferroelectric, 307
fiber optics, 128
field curvature, 194
field numbers, 173
field stop, 190
filaments, 128
film, 31, 55, 221, 303
    -based videodiscs, 54
    disc, 127
    -grain noise, 249, 250
    master, 55
    of silicone elastomeric rubber monomer, 34
    structure, 54
filtering, 213
final disc, 48
fingerprints, 249
first bright ring, 210
first dark ring, 210
first law of reflection, 186
flags, 31
flat capacitive disc, 70, 179
flat capacitive videodisc, 70
flat disc, 180
    grooveless, 177
flatness, 50
flexible discs, 50, 51, 124
floppy, 124
flow modifiers, 66
flux, 227
flylead, 172, 173, 176
flywheel synchronization circuit, 109
FM, 55, 250
    demodulator, 73
    detection, 155
    -encoded video, 41
    limiting process, 259
    periods, 249
    recording signal, 41
    signal, 109, 133, 169, 264

FM carrier, 170
    frequency, 42
focal depth, 189, 229
focal intensity distribution, 95
focal length, 94, 188, 192, 194
focal plane, 212, 213
focal point, 124, 188, 189
focus, 111, 188, 191, 219
    signal generation, 109
focus detector, 76
focused laser beam, 42
focused spot size, 95
focus error, 116
focus offset, 115
focus servo, 115, 124
focusing, 45, 109, 130, 218
    depth, 249
    lens, 36, 94
    signals, 45
format, 31
formation, 185
Fourier, 278, 279, 287
    analysis, 185
    optics, 185, 210, 225
    plane, 213
    series, 98, 141
    transform, 212, 213, 215, 231, 232, 233, 239, 240, 280, 281
        detector plane, 231
        FFT, 206
    transition, 213
frame search, 299
frames per revolution, 166
Fraunhofer diffraction, 212
    equation, 207
    pattern, 204, 212
freeze frame, 70, 177
frequency, 56, 60, 61, 73, 94, 102, 107, 109, 112, 120, 233, 235
    band, 96
    change, 111
    characteristic, 73
    -dependent amplitude, 214, 215
    -modulated carrier, 138
    modulator, 169, 170
    offset and nonlinearities, 245
    response, 71, 107, 150, 184
        function, 223
    -selective circuit, 134
    shift, 242
    -spatial memory, 319
    spectrum, 74
Fresnal:
    approximation, 205
    diffraction, 206
    diffraction pattern, 204
    integrals, 204
    transform, 206
    -type diffraction, 212
frictional, 66
fringes, 203

fundamental frequency, 62
fundamental resolution, 94
fusion, 65

gain, 123, 132
galvanic processing, 39
galvanic technique, 43
gate switching circuit, 182
Gauss's lens equation, 189
Gaussian, 226, 231
  amplitude distribution, 266
  curves, 233, 261, 262
  distribution, 247, 263, 271
  distribution noise, 283
  function, 231, 239, 240
  noise, 272
  opening, 226
  optics, 185
  pdf, 271, 276
  probability density function (pdf), 270
  random processes, 287
  random variables, 288
  type of distribution, 225
  white noise, 266, 285
gelatin, 317
gelatine, 249
general moments, 276
generation-recombination noise, 252, 253
geometrical distortion, 34
geometrical optics, 184, 185
geometric considerations, 185
geometric optics, 184, 185
geometric path length, 187
geometry of diffraction, 204
glass, 39
  disc, 39
  master disc, 53
  masters, 39
  substrate, 31, 39
gold/platinum alloys, 302
grating, 45, 109, 211, 230
  length, 236
  phase modulation, 88
grazing high-intensity light, 71
grids, 210
grime, 180
groove, 34, 159, 165, 174
  size, 94
  velocity, 172
grooveless capacitance systems, 177
grooveless discs, 160
group coded recording (GCR), 324
guidance systems, 116
guides, 125

half-cycle period, 109
half-cycles, 109
half-wave plate, 43
halogen lamp, 128
handling, 87
  damage, 35
hard limiter, 146

harmonic distortion, 73, 74
harmonics, 62, 143
He-Ne laser, 43, 76, 129, 256
  reading beam, 43
heat stabilizers, 65
heterodyne technique, 168
high-density, 37
high density of energy, 30
high-frequency, 96
high information densities, 54
high resolution, 39
high-spatial frequency, 213
hole, 34, 36, 432
holography, 185
horizontal phasing circuits, 167
horizontal sync, 167
  variations, 167
hue shifts, 167
hue, 151
Huygen's principle, 203, 204, 207

ID cards, 318
idealized optical system, 216
ideal lens, 187
ideal optical disc, 87, 88
ideal optical systems, 187
ideal reference sphere, 218
ideal spherical wave, 218
identification number codes, 31
illuminating spectrum, 229
illumination, 209, 212
  pattern, 210
illuminator, 229
IM products, 46
image, 128, 189
  defects, 191
  energy distribution, 222, 223
  formation, 185, 214, 215
  -forming, 221
    process, 185
    quality, 213
  modulation, 220, 223
  offset, 219
  plane, 190, 212
  point, 212, 213
  space, 190
  tubes, 221
imaging system, 191, 212
imbalance, 124
immunity, 301
impairment, 244, 245
impedance, 60, 132
imperfections, 108, 109
impulse noise, 247, 258
impulse response, 212, 222
incandescence, 128
incandescent light bulb, 128
incandescent light source, 127, 128
incident and reflected rays, 227
incident beam, 229, 235
incident flux, 226, 227
incident light, 234

incident plane-polarized, 234
incident ray, 111, 186
incident wave, 202
incoherent illumination, 207, 214, 215
incoherent light, 203
incoherent optical transfer function, 216
incremental noise element, 155
index of reflection, 302
index of refraction, 186, 190, 192, 199, 227
indirect addressing, 297
induced modulation, 234
information, 36, 38, 43, 45
    channels, 177
    elements, 34
    holes, 42
    pits, 177
    reading, 45
    sinc, 257
    surface, 48
    track, 158
infrared absorbtion, 42
inhomogeneities, 246
injection, 49, 65
    cycle, 49, 67
    -molded discs, 51, 68
    molding, 33, 47, 49, 65, 67, 68
    -molding, 48, 50, 65, 66
    -molding machine, 48
    -molding method and process, 50, 52
    nozzle, 65
    pressure, 67
    process, 49
inner radius, 102, 133
input intensity, 234
input power, 237
inside radius, 105, 230
inspection, 35
instant playback, 38
instantaneous angular frequency deviation, 155
instantaneous frequency, 109, 165
instantaneous intensity, 237
instantaneous RF envelope voltage, 237
insulator, 163
integration, 241
integrity, 297
intensity, 73, 128, 206, 233
intensity domain, 250
intensity modulation, 234, 235
intensity of a laser beam, 35
intensity of the radiation, 220
intensity-modulated laser beam, 39
interfaces, 87
interfacial frictional effects, 65
interference, 201, 203, 244, 251
    diffraction, 226
    patterns, 210
interferometric reference mirror, 129
intermediate parts, 129
intermittent motion, 123
intermodulation (IM), 46, 262
    products, 35, 46, 317, 318
    terms, 247

intersymbol interference (ISI), 259
intervals, 210
ion sputter etching, 64
irradiance distribution, 197

jitter, 172, 251
Jordan code, 27, 293, 327
jump back, 121
jumping, 109, 122
junction diode lasers, 43

keel, 175
Kerr:
    effect, 234, 304, 309
    magneto-optic, 234
    rotation angles, 304
    rotation, 309
kickback pulse, 182
    -forming circuit, 182
kicker coil, 173
knife-edge scanning techniques, 95

lacquer, 51
laminar flow of velocity, 126
laminated, 87
Laplace pdf, 273
laser, 30, 31, 42, 43, 94, 301, 317
    (optical) recording, 179
    beam, 36, 55, 73, 84, 93, 126, 231, 315
    cavity, 94
    control, 73
    current, 73
    defect detector, 71
    diodes, 244
    encoding, 50
    light, 31
    power, 36, 39
    pulse, 42
    recorder, 127
    spot size, 314
    tracking control, 47
    tube, 93
    wavelength, 237
lateral chromatic aberration, 191
lateral magnification, 194
lateral thermal conductivity, 316
laws of reflection and refraction, 185, 194
lead-screw, 37, 41, 47
LED displays, 173
length of pits, 31
lens, 36, 41, 43, 45, 95, 185, 187, 188, 189,
        191, 192, 193, 194, 212, 213, 221, 229,
        231, 244, 256
    characteristic, 105
    focus control, 37
    magnification, 190
    mount, 61
    surface, 114
light, 36, 95, 126, 185, 213
    amplitude modulation, 302
    beam, 36, 83, 121, 122, 124
        intensity, 36

distribution, 124
flux, 226, 256
modulator, 35
patches, 201
propagation, 234
rays, 185, 186
signal, 107
source, 93, 124
spot, 41, 83, 84, 114
waves, 201
limitations, 167
limiter, 61, 74
line frequency, 123
line-spread frequency, 223
linear detector, 129
linear distortion, 257, 260
linear electro-optical (Pockels) effect, 234
linear electronic circuits, 215
linear filtering operation, 185
linear motor, 299
linear phase, 73, 99
linear polarizer, 228
linear recording, 61
linear response, 61
linear speed, 108, 116
linear systems, 203, 214, 215
theory, 185
linearity, 241
theorem, 242
linearly polarized beam, 227
linearly polarized incident field, 228
liquid resin, 34
monomer, 34
locked-groove defects, 173
longitudinal, 192
chromatic aberration, 191
modulator, 234
loop gain, 115
loss-less delay line, 131
low chloride nickel sulfamate, 48
low-frequency noise, 252, 254
low-intensity laser, 87
low-pass, 96, 107, 206
filter, 119, 133, 168, 180, 213, 251
low-power He-Ne laser tube, 93
lubricant, 34, 65, 68, 69, 70, 159
layers (boundary lubrication), 70
lubricity, 66
luma signal, 133
luminance, 151, 152, 172
band, 169
comb filter, 172
information, 172, 318
signal, 172
luminous energy, 199
luminous flux, 199
lumped-element circuit, 164

magnetic field, 234
magnetic interaction, 227
magnetic media, 317

magnetic tape, 31
magneto-optic readout, 304
magneto-optical (MO), 233, 301, 303
magnification, 192, 194
main carrier, 46
unmodulated, 157
Manchester codes, 326
manufacturing, 64
processes, 73
Marcum Q function, 271
masked photomultiplier tube, 128
masking, 301
master, 30, 33, 42, 43, 47, 53, 63
negative, 34
record, 42
recorder, 43
recording, 31, 37, 55, 64
process, 31, 43
videodisc, 83
master disc, 31, 34, 35, 36, 39, 40, 41, 42, 47,
56, 129, 247, 266
substrate, 36
mastering, 31, 35, 36, 41, 46, 52, 53, 56, 57,
109, 112, 229, 244, 245, 247
machine, 41
methods, 35, 39
process, 43, 46, 47, 73
recorder, 43
system, 31
material, 38, 51
inhomogeneity, 194
matrix, 33, 41, 56
processing, 33, 56
Maxwell pdf, 274, 278
MCLV, 318
mean, 276
mean noise power, 287
mean-square noise voltage, 258
mechanical, 69, 244
resonance, 168
tracking, 158
vibrations, 94
media stability, 319
medium's sensitivity, 42
melt flow, 64, 67
variations, 72
melt holes in metal films, 31
melt viscosity, 251
melting, 41, 43, 302
point, 42
threshold, 41
meniscus:
concave, 187
convex, 187
metal, 39, 43, 50, 68
-coated master, 47
coating, 52
composition, 70
film, 34, 35, 36, 38, 39, 41, 42, 43, 46
film mastering, 41, 46, 53
master, 33, 47, 64

stampers, 33
-to-metal duplication, 48
-to-metal replication, 48
metalization, 35, 50, 246, 251
metalized discs, 50
metallic film, 36
MFM, 327
microcrystalline, 303
microscope, 211
  lens, 36, 37
    objective, 41, 43, 55
microswelling, 316
microwrinkles, 126
Miller code, 327
minimum recording wavelength, 165
mirror, 36, 43, 184, 185, 187, 244
mirror transducer, 94
mistracking, 174
MO, 309, 315
modulate, 237
modulated focused light beam, 38
modulated radius, 107, 137
modulated signal, 155
modulated video information, 114
modulating signal, 74
modulating signal information, 34
modulation, 107, 128, 144, 160, 219, 220, 233,
    235, 241, 247, 250
  depth, 80
  plots, 220
  theory, 287
  transfer function (MTF), 94, 95, 219, 225
  voltage, 234
modulator, 61, 62, 160, 234, 235, 237, 245, 353
  signal, 39
moire fringes, 210, 225
mold, 33, 34, 47, 49, 50, 52, 64
  cavity, 33, 48
  pressures, 52
  surface, 49
  temperature, 67
molded disc, 34, 49
molded signal track, 52
molding, 33, 35, 48, 49, 50, 56, 64, 65, 66
  stampers, 33
monitors, 60
monochromatic, 227
  aberrations, 191
  light, 227
  waves, 203
monomers, 70
mother, 47, 48, 64
MTF, 99, 107, 128, 133, 150, 162, 219, 220,
    223, 225, 229, 230, 260, 261, 266
  compensation, 95, 96, 99
    circuit, 99
  curve, 96, 98
multilayer structures, 312
multilevel coding, 31
multilevel continuous or pulse signals, 55
multiple-line dropouts, 267

multiplication, 241
multiplicative noise, 245
  process, 250
Mylar, 305

NA (numerical aperture), 116, 189, 229
narrow-band:
  Gaussian noise, 287
  noise, 286, 287
  random process, 285
negative meniscus, 187
negative photoresist material, 42
negative replicas, 33
negative resist, 42
Newton's form, 189
nickel, 33, 43, 48
  -coated photoresist master discs, 46
  plating, 47, 294
  replicas, 31
  stamper, 48
noise, 38, 43, 154, 156, 157, 163, 182, 244, 245,
    246, 247, 253, 254, 256, 257, 258, 259,
    260, 261, 265, 269, 270, 275, 288, 303,
    309, 311, 317
  -free, 128
  immunity, 88
  -reduction, 74
  signal, 260
  source, 266
  spectrum, 72
  voltages, 253
nominal frequency, 168
nonconductive discs, 68
noncontact optical read method, 128
nonhomogeneities, 247
nonlinear phase, 73
nonlinear recording, 61
nonlinear response, 163
nonlinearities, 62, 251
non-normal-incidence interference, 229
nonuniformity, 72
normal, 186
normalized energy distribution, 81
normalized envelope, 131
normalized Fourier transform, 216
normalized transfer function, 162
nozzle, 67
NRZ, 322, 329
NTSC, 31, 112, 151, 295, 296, 299
  (PAL or SECAM) bandwidth, 169
  signal, 170, 172
  system, 83
  television receiver, 109
null of response, 218, 219
numerical aperture (NA), 36, 43, 84, 94, 95, 113,
    128, 129, 189, 224, 229, 255, 256
numerical objective, 55

object, 188
  luminance, 212
  modulation, 221

plane, 194
point, 212, 213, 215
position, 192
space, 190
spatial frequencies, 229
objective, 45, 94, 108, 113, 128
 focusing servo, 45
 lens, 45, 84, 93, 94, 113, 229
obliquity factor, 205
observation plane, 208
observation point (screen), 206
ODC laser videodisc recording system, 329
off-axis, 194
 distance, 194
OMC/252, 256, 257
OMTF, 220, 223, 224
on-off, 55
one-shot, 109
open-circuit noise voltage, 253
open loop, 128
open-loop gain, 120, 122
open-loop response curve, 116
optical:
 admittance, 215
 axis, 185, 188, 191, 192
 beam, 62, 234, 303
  deflection, 23-53
 and capacitive discs, 130
 cards, 318
 channel, 244, 245
 contrast, 43
 density, 302
 digital disc, 294
 discs, 38, 124, 129
  system, 37
 disk, 297, 301, 317, 318
 effect, 42
 efficiency, 302, 310, 312, 314
 elements, 190
 energy, 235
 exposure, 39, 303
 feedback, 96
 Fourier transforms, 212
 intensity transform function, 216
 interference, 42
 laser mastering, 36
 lens, 94
 master recorder, 61
 mastering, 61
 media, 317
 memory, 318
  system, 294
 modulation channel (OMC), 252
 modulation transfer function (OMTF), 220, 255
 modulators, 185
 part, 184
 path, 128, 212
 path length, 187
 pickup, 229, 260
  system, 90

power, 42
reader, 129
reading, 83
readout, 94, 129, 130, 255, 262
 systems, 266
recorder, 36
 system, 61
recording, 30, 73, 297, 315
 media, 38
 system, 36, 37
reflective disc, 34
resolution, 303
rideout, 185
signal 94
storage, 316
systems, 31, 43, 83, 87, 94, 185, 186, 190,
  214, 215, 220, 224
tapes, 318
testing, 47
 and thermal efficiencies, 310
 transfer function (OTF), 213, 215, 216, 252
 transmissive disc, 52, 124
 videodisc system, 184, 190, 244, 259
 videodiscs, 129, 159
 wave, 23, 233
 wavelength, 236
optics, 184, 185
optimum depth of the pits, 40
optimum detection, 41
optimum exposure energy, 40
optimum modulation depth, 40
optomechanical, 127
 system, 36
optomodulation, 244
optomodulator, 36, 256, 299
organic dielectric filter, 34
original master, 33, 47
oscillator, 87, 108, 159, 160, 164, 172, 259
 frequency, 164
 port, 165
OTF, 213, 214, 215, 216, 218, 219
out-of-flatness, 51
out of focus, 53
outer diameter, 158
outer radius, 133
output, 142, 245
 burst, 151
 error, 152
 impedance, 131
 intensity, 234
 mean-square value, 155
 modulated video signal, 151
 noise, 269
 noise power, 154, 155
 pilot, 151
 power, 91
 signal, 89, 91, 92, 96, 97, 98, 131, 136, 137,
  139
 tracking error signal, 180
outside radius, 102, 105, 165, 230

overall frequency response, 102
overcoated disc, 50

packaging, 56
PAL, 295, 296
    system, 82, 83
paraxial:
    focus, 191
    image, 193
    rays, 186
particulate, 249
passband filter, 112
passive aerosynamic stabilizer, 125
path-length error, 218
PCM, 275
pdf, 270, 276, 281, 287, 288, 289
peak-to-peak exposure, 73
pelletized, 66
perfect focus, 219
perfect polarizer, 228
permanent recording, 38
permanent storage, 303
permeability, 227
permittivity, 227
personal history cards, 318
Petzval:
    curvature, 192
    surfaces, 193
phase, 60, 96, 111, 233, 235
    amplitude, 246
    change, 213
        optical media, 305
    characteristic, 259
    contrast, 129
    curvatures, 218
    detector, 78, 112, 155
    deviation, 154
    difference, 89, 90
    disturbance, 163
    encoding, 326
    errors, 108
    function, 132
    grating, 210
    inversion, 128
    jitter, 245, 260
    -locked loop (PLL) circuit, 111, 167
    margin, 168, 177
    modulated, 143
    objects, 129
    -relief modulation, 52
    rotation, 218
    shift 95, 107, 203, 214, 215, 223
        plate, 218
    transfer function (PTF), 219
    transmission, 213
    velocity, 227
Philips DRAW system, 308
phosphors, 221
photo diode (detector), 230

photocells, 124
photochemical, 302
photochromic, 301, 307
    medium, 307
photoconductive electro-optic, 301, 307
    materials, 308
photoconductor, 305, 307
    noise, 254
photocurrent, 94
photodetectors, 45, 46, 73, 74, 75, 87, 90, 252,
    254
photodiodes, 45, 55, 84, 87, 94, 114, 119, 258,
    312
    noise, 94
photoelastic effect, 235
photoelectric detectors, 252
photoelectrical transducer, 74
photoexcitation, 308
photoferroelectric, 301, 307
photographic contact-printing techniques, 34
photographic films, 301, 302
photographic lenses, 211
photographic methods, 34
photographic negative, 34
photographs, 34
photoirradiation, 307
photometry, 199
photomultiplier, 259
photon detectors, 252
photon noise, 252, 261
photopolymerization techniques, 34
photopolymers, 301, 302
photoresist, 31, 34, 35, 38, 39, 40, 42, 43, 46,
    47, 48, 61, 62, 73, 245, 251, 266, 301,
    302, 303
    disc, 47
    exposure, 46
    film, 40
    -like organic material, 33
    master, 264
    mastering, 39, 46, 53
    surface, 31, 36
photosensitive material, 35, 39
phototransistor, 111
photovoltaic detector, 254
physical optics, 184, 185, 201, 226
pickup, 49, 165
    circuitry, 259
    loop, 164
picture quality, 46
piezoelectric, 94
    transducer, 60
pilot, 152, 157
    and/or burst, 156
    frequency, 152
    signal, 151, 153, 154, 157, 179, 180, 182
PIN photodiode, 43
    detector, 94
pincushion distortion, 194
    results, 194

pinholes, 50, 52
  dust, 38
pit, 39, 40, 41, 89, 90, 91, 311, 312
  capacity, 299
  density, 39
  depths, 40, 89, 129, 247
  dimensions, 40
  edge, 91
  geometry, 40
  pattern, 39, 87, 89, 91, 160
  profiles, 39, 40
  rim, 313
  size, 53, 299
  walls, 40
  width, 89
  pattern, 160
pitch, 116
  measurements, 47
pits, 31, 39, 41, 47, 52, 129, 229, 233, 302, 318
pivoting mirror, 122
Planck's constant, 252
plane of incidence, 186, 234
plane-polarized, 229, 234
  beam, 228
  laser tube, 94
  light, 94, 234
plane waves, 229
planoconcave, 187
planoconvex, 187
  lens, 93
plastic, 50
  coating cycle time, 50
  disc, 35
  layer, 48
  replica discs, 38, 42
plasticizer, 65, 66
platform, 109
playable, 51
playback, 35, 40, 43, 51, 61, 67, 68, 71, 72, 73,
    127, 128, 172, 229, 296, 297, 299
  only disc, 294
  optical system, 93
  side, 54
  signal, 46, 311
  system, 40, 128
  testing, 35
played back, 46, 53
player servo loop, 112
player's optics, 51, 52
player, 35, 43, 109
playing modes, 109
playing time, 165, 166
play mode, 123
plexiglass, 39
PLL, 113
PMMA, 53
  plastic, 50
Pockels:
  cell, 41
  effect, 234
point, 189
point image, 192

point of incidence, 186
point radiators, 204
point source, 192, 206
point-spread function, 212, 213, 214, 215, 216
Poisson distribution, 254, 274
  PDF, 275
polarization, 43, 94, 185, 226, 233, 234, 235,
    304, 307
polarized, 227, 234
polarizer, 228, 229
polarizing beam splitter, 229, 299, 305
polarizing prism, 227
polar Kerr effect, 305
polymer, 249
  material, 51
polymethyl methacrylate pmma, 39
polyolefins, 65
polyvinyl:
  carbonate, 305
  chloride, 57
  chloride pvc resin, 33
positive, 42
  lens, 194
  meniscus, 187
  and negative-working photoresists, 38, 39
  photoresist, 31, 53
  resist, 39
power, 42, 62
  function, 278
  intensity, 91
  profile, 73
  spectral density, 242, 258, 263, 266, 281, 283
preamplified noise, 124
preamplifier, (preamp), 87, 98, 99, 180
predistortion, 62, 73
pregrooved, 160
  capacitive disc, 34
premastered, 52
pressing molding, 33
pressure, 67
principal plane, 188, 189
principal points, 188
principle of superposition, 201
principle plan, 189
principles, 113
prisms, 185, 187
PRM, 267
process control, 36
processing, 65, 66
PROD, 294, 297, 317, 318
product, 242
product quality, 35
profile, 73
programmable optical disk (recording), 301
programs, 31
propagation of light, 185
proportional and fixed-length asymmetry, 38
protective cassette, 53
protective cover, 34
protective overcoating, 34
protective plastic coating, 51
prow angle, 175

psd, 280
pulse function, 240
pulse-repetition frequency, 143
pulse signals, 55
pulse train, 141, 143, 144, 146
pulse-width modulated, 46
pupil, 223
  function, 213, 216, 218, 226
PVC, 33, 48, 52, 65, 66, 67
  base, 33
  disc, 177
  homopolymers, 67
  substrate, 159

quadrants, 114
quadratic Fresnel approximation, 207
quadratic-phase signal, 196
quadri-layer configuration, 314
quality control, 46, 52
quality inspection, 33
quantum efficiency, 253
quantum optics, 185
quarter-wave, 313
  plate, 76, 94, 227, 299

radial:
  difference signal, 92
  duty cycle, 46
  mirror, 45, 122
  servo control circuit, 119
  system, 121, 123
  tracking, 45, 126
radially diffracted orders, 46
radiance, 227
radiant energy, 199
radiant flux, 190
radiation, 184, 227
radio-frequency, (RF), 114
radiometry, 199
radius, 61
  independent, 107
Raman-Nath regime, 236
random access, 177
  graphics, 293
random process, 269, 270
ran-length-coded NRZ, 324
raster, 160
ray, 185, 186, 192
  aberrations, 192
  of light, 186
Rayleigh:
  criterion, 224
  density function, 289
  PDF, 273, 278
  -Sommerfield region, 204
read beam, 43, 299
read lens, 263
read objective, 116
readout, 35
read spot, 43, 94
read-while-write, 43, 46

reading, 38, 43, 48, 109, 133, 307
  beams, 43, 91, 297
  head, 177
  laser, 36
    beam intensity, 231
  lens, 116
  light beam, 87
  out, 40
  point, 109
  process, 144, 307
  processes, 184
  pulse, 256
  spot diameter, 89, 90
  spot profile, 89, 91
  spot, 43, 88, 91, 255
  stylus, 124
  velocity, 136, 142
  and writing beams, 45
readout, 38, 43, 53, 126
  beam, 45, 86, 302
    energy, 45
  device, 62
  frequency, 94
  head, 177, 179
  laser, 31
  lens, 104
  objective, 125
  radius, 102, 105
  signal, 95
  signals, 96
  system, 184
real-time recording, 36
receiver, 108, 229
  noise, 260
  openings, 229
receiving aperture, 229
recorded master, 33
recorded master disc quality, 46
recorded pits, 256
recorded signal, 46
recorded surface, 33
recorded video signal, 167
recorder, 31
recording, 30, 31, 33, 35, 36, 38, 41, 47, 48, 54,
    60, 184, 299, 301
recording bit rate, 299
  codes, 322
  lens, 36
  master, 33
  material, 31, 35, 38, 59
    and processing, 37
  medium, 37
  proceeds, 39
  process, 38, 55, 74, 184, 315
  sensitivity, 303, 317
  side, 54
  signal, 36, 41
  spot, 41
  systems, 31
records, 33, 297
rectangular function, 240
rectangular or square source, 208

reduction, 109, 119
reference codes, 329
reference quartz oscillator, 111
reference signal, 182
reflecting surfaces, 186
reflected (0 order) rays, 89
reflected beam, 43
reflected light, 119
reflected ray, 186
reflected signal, 94
reflected writing beam, 46
reflected, 185
reflecting boundary, 186
reflection, 45, 109, 226
reflection gratings, 210
reflective, 52, 54
    aluminum layer, 51
    disc, 35, 52, 124
    index, 303
    layer, 38
    optical videodisc, 244
    phase gratings, 88, 89
    player concept, 51
    surface, 34
    system, 51, 83
    and transmissive videodiscs, 54
    videodisc, 87
reflectivity, 39, 50, 88, 316, 317
refracting, 185, 186
    material, 87
    surfaces, 187
refraction, 188, 189, 194, 235
    index, 84, 187, 194, 234, 235, 246, 308
        modulation, 308
        variations, 246
refreshed, 307
registration, 121
relativistic optics, 185
replica, 34, 40, 41, 48
    disc, 50
replicas, 31, 34, 40, 42, 47
    per recording, 48
replicated by molding, 42
replicated discs, 50, 159
replication, 34, 47, 48, 49, 53, 67, 71, 245, 246,
        303
    medium, 72
    process, 51, 52
    rate, 53
    techniques, 34, 41
reproducing transducer, 180
reproduction, 87, 244
    quality, 50
rereads, 307
residual random film structure, 127
residual stress, 67
resin, 33, 34, 65
resist, 39, 40, 42, 46, 57, 63
    coating, 64
    film, 39, 40
        thickness, 39
    master, 64

sensitivity, 40
    substrate interface, 39
    thickness, 40
resistivity, 251
resolution, 31, 42, 94, 129, 130, 162, 190, 295,
        303, 307, 308
    test charts, 225
resolving power, 96
resonance, 165
resonant circuit, 159, 164, 180
resonant curve, 164
resonant frequency, 123, 164, 165
resonant peak, 123, 164
response curve, 96, 164
response function, 215
resultant intensity, 203
retardation, 228, 234
retrieval, 38
rewriting, 109
RF carrier, 165
RF envelope voltage transfer function, 238
rheology, 64, 67
Rice-Nagakami pdf, 288
Rician pdf, 288
rigid mold, 34
ring-electrode structures, 234
ringing, 206, 207
rings, 296
ROD (PROD), 294
RODs, 294
rotary motor, 299
rotation, 108, 241, 242
rotational frequency, 119
rotational speed, 108
roughness, 124
RZ, 325

sagittal, 193
    fan, 192
    image, 193
    plane, 192
sampling, 244
    property, 239
SC, 257
scale change, 241
scan mode, 123
scanning filter, 260
scattered light, 129
scattering, 42
    effect, 88
science and engineering, 214, 215
scratches, 87, 249, 267
screen, 51
SECAM, 296
second law of reflection, 186
sector, 297
selective electroplating, 64
selenium, 42
semiconductor, 316
    detectors, 252
sensitivity, 37, 39, 165, 301, 303, 308, 309
sensor, 111

servo, 116
  control system, 113, 116
  loop, 123
  systems, 73, 111, 297
    in the player, 109
servomechanism, 128
servosystem, 87
set-down, 176
shape or bending of the lens, 192
sharpness, 194
shear, 65, 67, 70
  effects, 65
  rates, 67, 68
  stress, 66, 67
sheet glass, 245
shielding, 66
shift of position, 241
shift theorem, 242
shock, 67, 68
shoe, 175
short noise, 252, 254, 258, 275
  of the photons, 124
short pulse, 42
signal, 31, 36, 40, 61, 74, 83, 109, 111, 134,
    159, 163, 245, 247
  amplitude (SA), 57, 73
    ratio (R), 57
  channel (SC), 257
  degradation, 52
  depth, 59
  editing, 31
  -element, 67
  elements, 33, 36, 38, 160, 162, 163, 174
  frequency, 136
  information, 31, 52, 84
    track, 52
  loss, 166
  path, 158
  periods, 129
  pickup, 175
  processing, 169, 213
    circuitry, 170
  quality, 43
  -quality control, 40
  -to-noise, 128
    -to-noise ratio (SNR), 12, 38, 46, 57, 61, 71,
      94, 95, 123, 124, 127, 128, 153, 154,
      163, 210, 225, 259, 288, 297
  track, 180
  waveform, 88
silicon photodiode, 128
silicon video detectors, 128
silver halide, 55, 302, 317
  emulsion, 249
  materials, 38
silver particles, 56
similarity theorem, 242
simple lens, 187, 192
simulation, 99
sine-wave response, 220
single-centered detector, 129
single-element lenses, 187

single-mode diode lasers, 42
single-pole resonance, 165
sinusoidal disturbance, 168
sinusoidal modulation, 234
skip, 173, 174
sled, 116, 121
slewing, 299
slide motor, 121
sliding rate, 69
slot, 159
slow forward, 177
slow motion, 123, 182
small-angle approximations, 187
Snell's law, 186
SNR, 157, 225, 301, 303, 309, 311, 312, 315,
    317, 329
sound, 63
sound master, 33
space domain, 148
  signal, 144
Sparrow's criterion, 224
spatial density, 267
spatial dependence, 206
spatial frequency, 36, 95, 162, 211, 212, 213,
    214, 215, 217, 219, 220, 221, 225, 229,
    231, 247
  lines, 223
  response, 210
spatial function, 242
spatial modulation sidebands, 229
spatial period, 88, 211, 246
spatial phase variations, 88
spatial spectrum, 213
spatial wavelength, 94
special effects, 121, 177
specific mass, 126
spectra, 98, 140, 153
spectral components, 143, 169, 203
spectral density, 154, 279, 285, 287
spectral density function, 279
spectral domains, 74
spectral flux, 199
spectroscopy, 184
spectrum, 97, 98, 143
spectrum analyzer, 71, 260
speed, 63
speed control, 37
speed of rotation, 116
spherical aberration, 87, 191, 192, 194, 224
spherical mirror, 191
spherical wave, 187, 206
spherical wavefront, 203, 207
spin coating, 40
spin-coating process, 40
spin rate, 40
spindle, 112
spindle motor, 112
spindle servo, 77, 111, 112
spindle speed, 112
spiral channel, 36
spiral pitch, 55
spiral pit pattern, 50

spiral track, 35, 39, 55
spiral track of pits, 31
spiral tracks, 33
spot diagram, 223
spot lens, 43, 45
spot power, 41
sputtering, 70, 304
square aperture, 216
square-law detector, 283
stability, 64, 107, 108, 123, 127, 295
stability problem, 107
stabilization, 113, 125
stabilizer, 126
stabilizers, 124
stage, 112
stains, 72
stalling, 123
stamper, 33, 47, 48, 49, 53, 57, 64, 65, 67, 264,
        267
   damage, 72
   surface, 49
   wear, 50, 52
standard deviation, 276
standards, 294
standing capacitance, 165
standing waves, 39, 80, 202
static pressure, 126
static sag warp, 67
stilbene, 307
still pictures, 109, 182
stochastic deviations, 247
stochastic process, 269
stop, 192, 194, 213
stresses, 65
stress levels, 67
stylus, 34, 35, 56, 59, 60, 61, 67, 68, 69, 70, 72,
        158, 159, 161, 163, 167, 170, 172, 173,
        174, 175, 176, 179, 180, 182, 250
   arm, 172, 173, 175
   arm flylead, 176
   body, 161, 162, 163
   capacitance, 173
   cartridge, 172, 176
      housing, 173
   -disc capacitance, 164
   -disc capacity, 170
   electrode, 69, 163
   flylead, 172, 173
   foot, 161
   kicker, 173
      coils, 176
   lifter, 173
   "prow", 69
   resonant circuit, 173
   shoe, 69
   substrate, 175
   tip, 161, 174, 177
   wear, 70, 176
styrene, 57
styrene dielectric, 159
subcarrier, 112
   oscillator, 112

submaster stamper, 47
submatter, 64
submother, 48, 57
   production, 48
suboxide thin films, 302
substrate, 34, 39, 41, 42, 64, 174, 245, 249, 309
   compound, 52
   disc, 39
   material, 50, 51, 52
   material pmma, 39
superposition, 132, 227, 242, 281
   of light waves, 201
   of N waves, 201
   principle, 185
   theorem, 203
support arm, 176
surface conductivity, 34
surface contamination, 35, 50, 51
surface curvature, 193
surface defects, 39
surface noise, 129
surface quality, 251
switching pulse, 182
symbol interference, 233
synchronization ("sync"), 108
   circuits, 108
   pulse, 108, 109, 112, 264
   problem, 329
synchronous motor, 79
sync signals, 166
sync tip, 73, 169
systems, 34, 53, 127, 187

tachometer, 63, 111
tachometric, 112
tangential, 193
   duty cycle, 47
   image, 192
      surface, 193
   mirror, 45, 123
   plane, 192
   (time-base correction) servo, 122
   velocity, 137, 138, 139, 143, 144
      of the write head, 73
Taylor series, 186
TBC, 112
TBE, 169
techniques, 55
tellurium, 42, 302
   alloys, 302
   based media, 42
   carbon alloy, 302
   copper alloy, 302
temperature birefringence, 235
temperature changes, 50
temporal frequency, 36, 212, 214, 215, 230
   response, 210
tensor, 234
terbium moment, 304
test electronic signal, 46
testing, 70, 184
test masters, 40

test player, 43, 46
test signal, 31
theory of diffraction, 185
thermal and impulse noise, 245
thermal conductivity, 42
thermal cycling, 67
thermal degradation, 65
thermal diffusion length, 311
thermal diffusivity, 309, 311
thermal efficiency, 302, 311, 314
thermal noise, 254, 257, 258, 259, 270
    Johnson-Nyquist, 252
thermal process, 50
thermal replication process, 50
thermal stability, 64
thermodynamic, 257
thermoforming, 34
    process, 33
thermoplastic, 303, 305
    material, 53
thick disc, 51
    optical-disc molding, 49
    videodiscs, 33
thick plastic surface, 51
thin coating of lubricant, 34
thin discs, 49, 50
thin film, 34
thin layers, 42
thin lens, 187
thin metal, 33
thin metal film, 33, 42, 47
thin metal films, 38
thin-metal-films, 42
thin metallic film, 34
thin photoresist film, 39
thin reflective layer, 50
thin videodiscs, 33, 34
thioindigo, 307
third-order harmonic, 62
Thomson stabilizer, 124
three-dimensional optical memory, 319
three-dimensional phase grating, 235
three laws of geometrical optics, 185
three-position modulation (3PM), 329
time base, 167
    correction, 45, 112, 168, 169
    error, 30, 107, 108, 158, 167, 168, 251
        correction, 45, 109, 167
        recorded, 251
        transfer characteristics, 168
time dispersion, 259
time-lag errors, 34
timing error, 108, 109, 122
tip geometry, 174
titanium, 175
tolerance, 113, 128
topographical signal elements, 31
total internal reflection, 187
total power, 89
track, 84, 89, 116, 233, 296, 311, 318
    -and-hold circuit, 170
    density, 55

kissing, 47, 317
    pitch regularity, 46
    spacing, 52
    -to-track crosstrack, 317
    -to-track errors, 47
    velocity, 61
tracking, 45, 94, 109, 111, 130, 174, 176,
        182
    and reading beams, 45
    control circuit, 180
    control mechanism, 182
    deviation, 180, 182
    elements, 70
    error, 179
    error signal, 182
    mirror transducer, 120
    pits, 177
    (radial) servo, 116
    servo, 120, 123, 180
    spots, 126
transducer, 94, 120, 168, 182, 237, 257
transfer:
    characteristic, 62, 96, 107, 162, 165
    characteristics, 168
    efficiency, 128
    function, 102, 104, 120, 150, 162, 207, 215,
        217, 220, 225, 229, 230
transform, 257
translation, 187
translation rate, 78
transmission, 226, 227
    coefficients, 310
    efficiency, 36, 128
    gratings, 210
    line, 97
    -line techniques, 164
    point, 234
    properties, 214, 215
transmissive, 87, 88, 124
    disc, 52, 53, 88, 89, 124, 126
    matter, 53
    videodisc system, 54
transmitivity, 88
transmittance, 227, 228
transmitted intensity, 234
transparent dielectric, 313
transparent polymer, 34
transversal motion, 177
transverse (or lateral) aberration, 192
transverse deviation, 191
transverse modulators, 234
transverse waves, 113
transverse width, 159
trilayer, 70
    antireflection structure, 313
    structure, 313, 314
trimming, 67
tropospheric scattering, 272
true focus, 116
truncated pdf, 274
turntable, 170, 180
turntable speed, 314

TV:
  frame, 46
  picture quality, 46
  program length, 52
  receiver, 108, 167, 168
  screen, 46
  synchronizing pulses, 173
  transmitters, 108
twin-ground plate glass, 39
two-dimensional, 211
  Fourier transform pair, 240
  space domain, 214, 215
two-sided disc, 33

ultraviolet-light source, 42
umbrella, 113
uncombed signal, 172
undercutting, 64
uniform pdf, 278
uniform (rectangular) pdf, 273
unpolarized, 227
  beam, 228
  light, 228
unpolymerized photoresist materials, 43
upper sideband, 56

vacuum coating machine, 50
vacuum deposition, 34, 42
Van Vliet, 253
vapors, 48
varactor diodes, 172, 173
variable oscillator, 168
variable-phase plate, 305
variable-transmission filter, 226
variance, 276
VDC, 252
velocity:
  of light, 227
  of propagation of light waves in vacuum, 199
vertical blanking interval, 31, 173
vertical relief pattern, 61
vertical servoing, 124
vertical tracking (focusing), 124
VHF circuitry, 176
vibration, 67, 68, 244
video and audio demodulators, 170
video and audio signal-to-noise ratio, 35
  player, 128
  program, 43
  quality, 51
  record/playback, 55
  signal, 31, 83, 109, 152, 170
  signal element geometry, 49
  track, 128
video carrier, 72
  frequency, 163
video detector, 128
videodisc, 33, 35, 36, 50, 52, 61, 83, 87, 94,
      109, 111, 129, 176, 177, 185, 229, 230,
      244, 267, 269, 293, 294, 301

channel (VDC), 31, 245
grooves, 94
player, 45, 229
recording, 128
stampers, 33, 52
system, 151, 158, 184, 185, 212, 257, 270,
      275
video FM, 109
  carrier, 172
  signal, 170
vidicon-like, 160
vinyl disc, 40
viscosity, 66
visible light, 211
visible wavelengths, 42
visual inspection, 46
visual-search, 173
voids, 72
volatiles, 67
volatility, 65
voltage gain, 132
voltage-time-history, 269
VTR, 52

warp, 167
warpage, 50
wave, 184, 195, 202, 233, 234
  nature of light, 185
  number, 80, 195
  optics, 185, 195
  property, 233
  trains, 202
  velocity, 227
wavefront, 204, 218
  aberrations, 192
  errors, 218
wavelength, 42, 60, 61, 62, 83, 129, 162, 166
  of light, 36, 224
wear, 69, 83, 174
web press technique, 33
web process, 53
web-roll-fed process, 53
white input, 169
white noise, 154, 252, 283, 284, 285
  process, 250
wideband noise, 250
Wiener-Khintchin theorem, 279
writing, 43
  beam, 43, 45, 73
  head, 35
  laser, 36
  rates, 307
  recording, 42
  velocity, 136, 142, 144, 148

zero crossing, 41
zero modulation, 327
zero position, 121
zero-order beam, 237